PATHOGENS OF MEDICALLY IMPORTANT ARTHROPODS

Edited by

Donald W. Roberts

and

Mary Ann Strand

Boyce Thompson Institute
Yonkers, New York 10701, USA

WORLD HEALTH ORGANIZATION
ORGANISATION MONDIALE DE LA SANTÉ
GENÈVE
1977

ISBN 92 4 068551 0

© World Health Organization 1977

Publications of the World Health Organization enjoy copyright protection in accordance with the provisions of Protocol 2 of the Universal Copyright Convention. Nevertheless, abstracting and other journals may reprint the abstract at the head of articles in the *Bulletin of the World Health Organization* without requesting authorization.

For rights of reproduction or translation of WHO publications, in part or *in toto*, application should be made to the Office of Publications, World Health Organization, Geneva, Switzerland. The World Health Organization welcomes such applications.

Authors alone are responsible for views expressed in signed articles.

The designations employed and the presentation of the material in this publication do not imply the expression of any opinion whatsoever on the part of the Secretariat of the World Health Organization concerning the legal status of any country, territory, city, or area or of its authorities, or concerning the delimitation of its frontiers or boundaries.

The mention of specific companies or of certain manufacturers' products does not imply that they are endorsed or recommended by the World Health Organization in preference to others of a similar nature which are not mentioned. Errors and omissions excepted, the names of proprietary products are distinguished by initial capital letters.

© Organisation mondiale de la Santé 1977

Les publications de l'Organisation mondiale de la Santé bénéficient de la protection prévue par les dispositions du Protocole N° 2 de la Convention universelle pour la Protection du Droit d'Auteur. Les revues de comptes rendus analytiques ou d'autres journaux peuvent toutefois reproduire les textes de présentation figurant en tête des articles du *Bulletin de l'Organisation mondiale de la Santé* sans autorisation préalable.

Pour toute reproduction ou traduction partielle ou intégrale, une autorisation doit être demandée au Bureau des Publications, Organisation mondiale de la Santé, Genève, Suisse. L'Organisation mondiale de la Santé sera toujours très heureuse de recevoir des demandes à cet effet.

Les articles signés n'engagent que leurs auteurs.

Les appellations employées dans cette publication et la présentation des données qui y figurent n'impliquent de la part du Secrétariat de l'Organisation mondiale de la Santé aucune prise de position quant au statut juridique des pays, territoires, villes ou zones, ou de ses autorités, ni quant au tracé de leurs frontières ou limites.

La mention de firmes et de produits commerciaux n'implique pas que ces firmes et produits commerciaux sont agréés ou recommandés par l'Organisation mondiale de la Santé de préférence à d'autres. Sauf erreur ou omission, une majuscule initiale indique qu'il s'agit d'un nom déposé.

PRINTED IN SWITZERLAND

Contents   Table des matières

INTRODUCTION

As evidenced by the approximately 1100 citations in the following reviews of the literature, there has been long-term, widespread interest in the viruses, microorganisms, and nematodes which cause disease, or are suspected to cause disease, in medically important arthropods. These pathogens currently are of great interest to the scientific community because of the need to develop supplements or alternatives to the synthetic chemical pesticides which presently dominate programmes to control medically important arthropods. The literature on the pathogens of these arthropods, unfortunately, is widely scattered over both space and time. This was apparent in 1973 when one of us, Dr D. W. Roberts, organized a "Workshop on Diseases of Medically Important Insects" for the Vth International Colloquium on Insect Pathology. A very useful compilation of literature up to the early 1960s was available (Jenkins, D. W., 1964. Pathogens, parasites, and predators of medically important arthropods. Supplement to Vol. 30 of the B̲ulletin of the World Health Organization, 150 pp.). In the absence of a more recent annotated bibliography, several participants in the Workshop were asked to bring to the meeting updated bibliographies on their fields of specialty. Dr Marshall Laird of the Research Unit on Vector Pathology, St John's, Newfoundland, Canada, who had published a list of references without annotations which supplemented that of Jenkins up to 1967 (In: Burges, H. D. & Hussey, N. W. ed., Microbial control of insects and mites, London, Academic Press, 1971, pp. 751-790) recommended a complete update of Jenkins' list and at his urging discussions were commenced with the Division of Vector Biology and Control (VBC) of the World Health Organization (WHO). It was agreed that the effort should be expanded and an annotated update of the pathogens portion of the Jenkins' list be prepared and published under the sponsorship of WHO. It is our hope that the appearance of this volume will encourage research on pathogens of medically important arthropods.

The following is the combined effort of 25 collaborators. Some differences in style and outlook are apparent, but in general the contributions are very similar. With one exception, each of the following sections can be classified in one of three categories: Host-Pathogen List, References and Abstracts, or Pathogen-Host List. The exception is Section IV (b), a synonomy of the Microsporida which infect mosquitos. Because of the uncertainties associated with the taxonomy of Microsporida, an update of the synonomy of the numerous species which infect mosquitos was considered necessary.

Host-Pathogen List (H-PL). Each author or team of authors was requested to submit a table on a specific topic which included for each host-parasite combination (a) the host, (b) the host stage infected, (c) the pathogen, (d) percentage incidence, (e) locality, (f) whether laboratory or field observations, and (g) references. In most cases the author(s) elected to include the literature cited by Jenkins, but in a few instances the literature coverage is restricted to post-Jenkins publications. The latter should still include all major pathogens. In some cases the references cited include studies on biology and/or ecology as well as new host-pathogen records. Scientific names in a work of this size present numerous problems. An attempt was made to use the newer name of reclassified organisms, but with the older literature it is frequently impossible to know the correct modern identification. Accordingly, many older and, in some cases, dubious names appear. It will be noted that each H-PL has been credited with separate authorship. In quoting from individual H-PLs, the H-PL should be quoted by author in the same style as a book chapter.

References and Abstracts (R). References to all papers cited in the Host-Pathogen Lists are presented in an Abstract section following each list. Also included are other papers not cited in the H-PL such as ultrastructure, biology and/or ecology, inconsequential references,

6

and review articles. Abstracts of post-Jenkins publications and, in some cases, the earlier literature are presented here. Literature not seen by the Abstract author and most literature included in Jenkins' list are presented by title only. The abstracts, the great majority of which were prepared by Mary Ann Strand, are all original (not copies of the summary appearing in the original publications) and were prepared with emphasis on those points presumed important to persons interested in pathogens of medically important arthropods. In general, value judgements were not made. The abstracts are brief statements of information, not complete summaries, and usually do not duplicate information in the title or in the H-PL. Therefore, they are not intended to preclude the reading of the article; however, the abstracts should assist the user in deciding whether access to the full article is needed.

Pathogen-Host List (P-HL). The last portion of the volume consists of nine sections compiled from the H-PLs. The sections are viruses, rickettsiae, bacteria, protozoa other than Microsporida, Microsporida, fungi other than Coelomomyces, Coelomomyces, nematodes, and others. For each pathogen (a) the name, (b) the host group, (c) the host, and (d) reference(s) are given. The reader can find the full citation and abstract to a specific reference by turning to the Abstract section given under "host group."

We wish to offer our appreciation to the many expert collaborators who devoted their time to the completion of this task, and to our librarian, Suzanne Broque, and secretarial staff, particularly Lois Phillipps, who so willingly and capably contributed their skills to the work. Also, the project could not have proceeded without the mental and financial encouragement of VBC/WHO, specifically A. A. Arata, J. W. Wright, M. King and J. Hamon. Their unflagging enthusiasm has been very much appreciated. And, finally, the assistance of the Boyce Thompson Institute administration, particularly R. H. Wellman and D. C. Torgeson, in making available supporting staff is recognized with thanks.

Updating of this volume is to begin immediately. The editors will appreciate comments or information which will improve the quality of succeeding editions or additions to this edition.

Donald W. Roberts
Mary Ann Strand

INTRODUCTION

Comme l'indique le nombre (près de 1100) des titres mentionnés dans les pages qui suivent, beaucoup d'auteurs se sont intéressés depuis longtemps aux virus, micro-organismes et nématodes pathogènes, ou soupçonnés de l'être, pour les arthropodes importants du point de vue médical. Ces dernières années, l'attention de la communauté scientifique a été tout particulièrement attirée sur ces agents pathogènes en raison de la nécessité de trouver de nouvelles méthodes pour compléter ou remplacer l'emploi des pesticides chimiques qui joue aujourd'hui un rôle prépondérant dans les programmes de lutte contre les arthropodes vecteurs de maladies. Malheureusement, la littérature pertinente est très éparpillée dans le temps comme dans l'espace. On l'a remarqué en 1973 quand l'un de nous, le Dr D. W. Roberts, a organisé, à l'occasion du Vème Colloque international sur la Pathologie des Insectes, une conférence-atelier sur les maladies des insectes médicalement importants. Il existait alors un très utile inventaire de la littérature jusqu'au début des années 1960 (Jenkins, D. W., 1964. Pathogens, parasites and predators of medically important arthropods. Supplément au volume 30 du Bulletin de l'Organisation mondiale de la Santé, 150 pages). En l'absence d'une bibliographie annotée plus récente, plusieurs participants à la conférence-atelier ont été invités à communiquer une bibliographie à jour concernant leur discipline. Le Dr Marshall Laird, de l'Unité de Recherches sur la Pathologie des Vecteurs à St-John's (Terre-Neuve, Canada), qui avait publié une liste non annotée de références complétant celle de Jenkins jusqu'en 1967 (Dans : H. D. Burges & N. W. Hussey, ed. Microbial control of insects and mites, Londres, Academic Press, 1971, pp. 751-790) a recommandé une mise à jour complète de la liste de Jenkins. Sur ses instances, on a procédé à des échanges de vues avec la Division de la Biologie des Vecteurs et de la Lutte antivectorielle de l'Organisation mondiale de la Santé (OMS). Il a été décidé qu'une mise à jour annotée de la partie de la liste de Jenkins relative aux agents pathogènes serait préparée et publiée sous les auspices de l'OMS. C'est ainsi qu'est né le présent ouvrage dont la parution, nous l'espérons, encouragera les recherches sur les organismes pathogènes pour les arthropodes importants du point de vue médical.

Vingt-cinq personnes ont collaboré à cet ouvrage. On y décèlera des différences de style et de présentation, mais, dans l'ensemble, les contributions sont très uniformes. A une exception près, chacune des sections peut être classée dans l'une des trois catégories suivantes : liste hôtes-agents pathogènes; références et analyses; liste agents pathogènes-hôtes. L'exception concerne la section IV (b) consacrée aux synonymies entre appellations employées pour les microsporidies qui infectent les moustiques. Etant donné les incertitudes relatives à la taxonomie des microsporidies, une mise à jour de la nomenclature des nombreuses espèces qui infectent les moustiques a été jugée indispensable.

Liste hôtes-agents pathogènes (H-PL). Chaque auteur, ou équipe d'auteurs, devait présenter un tableau sur un sujet particulier, en indiquant pour chaque association hôte-parasite : a) l'hôte, b) le stade infecté de l'hôte, c) l'agent pathogène, d) le taux d'incidence, e) la localité, f) le type d'observations : en laboratoire ou sur le terrain, et g) les références. Dans la plupart des cas, les auteurs ont inclus la littérature citée par Jenkins, mais quelquefois la liste se limite aux publications postérieures. Elle doit alors de toute manière inclure tous les principaux agents pathogènes. Parfois, les références mentionnées comprennent des études sur la biologie ou l'écologie. Dans un ouvrage de cette dimension, les noms scientifiques posent de nombreux problèmes. On s'est efforcé d'employer les noms les plus récents des organismes ayant fait l'objet de reclassifications mais, dans le cas de publications déjà anciennes, une identification moderne correcte se révèle souvent impossible. Il y a donc beaucoup de noms anciens, et parfois douteux. Les lecteurs noteront que, pour chaque H-PL, on indique l'auteur ou les auteurs. Dans toute citation d'une H-PL, il conviendra de mentionner l'auteur ou les auteurs de la même manière que pour un chapitre de livre.

Références et analyses (R). Les références à tous les textes auxquels renvoie une liste sont précisées dans une section spéciale faisant suite à cette liste. On y mentionne aussi d'autres documents non cités dans les H-PL (communications concernant les ultra-structures, la biologie, ou l'écologie, articles se rapportant indirectement au sujet, notes critiques). C'est là que sont présentées des analyses de publications postérieures à la liste Jenkins et de certains textes plus anciens. Pour tout document qui n'a pas été lu par l'auteur d'une analyse et pour la plus grande partie de la littérature figurant sur la liste Jenkins, le titre seul est indiqué. Les analyses, dont la plupart sont dues à Mary Ann Strand, ont toutes été rédigées spécialement pour le présent ouvrage et ne constituent pas des reproductions des résumés que contiennent les publications originales. L'accent y est mis sur les points jugés susceptibles d'intéresser particulièrement les personnes qui s'occupent des organismes pathogènes pour les arthropodes médicalement importants. D'une manière générale, on n'a pas porté de jugements de valeur. Les analyses présentent une information succincte et non des résumés complets; habituellement, elles ne répètent pas l'information donnée dans le titre ou dans la H-PL. Ces analyses n'excluent donc en aucune manière la lecture de l'article mais devraient aider les personnes intéressées à décider si elles doivent ou non chercher à se procurer le texte intégral.

Liste agents pathogènes-hôtes (P-HL). La dernière partie de l'ouvrage renferme neuf sections élaborées à partir des H-PL : virus, rickettsies, bactéries, protozoaires autres que les microsporidies, microsporidies, champignons autres que Coelomomyces, Coelomomyces, nématodes, divers. On y indique, pour chaque agent pathogène : a) le nom, b) le groupe d'hôtes, c) l'hôte, et d) la ou les références. Le lecteur peut trouver la citation complète et l'analyse se rapportant à une référence donnée dans une sous-section spéciale suivant celle qui concerne le groupe d'hôtes.

Nous voudrions exprimer ici notre gratitude aux nombreux spécialistes scientifiques qui ont collaboré à cet ouvrage ainsi qu'à notre bibliothécaire, Suzanne Broque, et au personnel de secrétariat, en particulier Lois Phillipps, qui nous ont aidés avec tant de dévouement et de compétence. Nous tenons aussi à souligner que ce projet n'aurait pu être mené à bien sans les encouragements moraux et financiers de l'OMS et de sa Division de la Biologie des Vecteurs et de la Lutte antivectorielle représentée par A. A. Arata, J. W. Wright, M. King et J. Hamon. Leur inlassable enthousiasme nous a été d'un grand appui. Enfin, qu'il nous soit permis de remercier l'administration de l'Institut Boyce Thompson, en particulier R. H. Wellman et D. C. Torgeson, à qui nous devons d'avoir bénéficié du concours d'un personnel de soutien.

La mise à jour de cet ouvrage doit commencer immédiatement. Nous accueillerons avec reconnaissance toute observation ou information permettant de préparer une nouvelle édition améliorée ou des additifs à la présente édition.

Donald W. Roberts
Mary Ann Strand

I. PATHOGENS OF PSYCHODIDAE (PHLEBOTOMINE SAND FLIES)

D. G. Young [a]

Department of Entomology and Nematology
University of Florida
Gainesville, FL 32611, USA

and

D. J. Lewis [b]

c/o Department of Entomology
British Museum (Natural History)
Cromwell Road, London SW7 5BD, England

[a] Research sponsored by US Army Medical Research and Development Command under contract No. DADA 17-72-C-2139.
[b] Supported by a grant from the Medical Research Council, London.

PATHOGENS OF PSYCHODIDAE (PHLEBOTOMINE SAND FLIES)[a]

Host	Host stage infected	Pathogen	% Incidence	Locality	Lab. or field study	Reference
Lutzomyia camposi (Rodriquez)	Adults	Acephaline gregarines	<1%	Panama	Field	McConnell & Correa (1964)
Lutzomyia cayennensis braci Lewis	"	Nematode	Unknown	Cayman Brac Island	"	Lewis (1967b)
Lutzomyia cruciata (Coq.)	"	Acephaline gregarines	<1%	Panama	"	McConnell & Correa (1964)
"	"	Nematode	<1%	Belize, formerly British Honduras	"	Lewis (1965a)
"	"	Unidentified small bodies	1.6%	"	"	"
"	"	Unidentified parasites	<1%	"	"	"
Lutzomyia flaviscutellata (Mang.)	"	Monocystis chagasi; Adler & Mayrink[b]	11-26%	Brazil (Pará)	"	Lewis et al. (1970)
Lutzomyia gomezi (Nitz.)	"	Fungus	3.7% in resting collections; 0% in biting collections	Panama	"	McConnell & Correa (1964)
Lutzomyia hartmanni (Fchld. & Hertig)	"	Acephaline gregarines	<1%	Panama	"	McConnell & Correa (1964)

a The classification of Phlebotominae follows that of Theodor (1965).

b The generic status of this gregarine and Monocystis mackiei Shortt & Swaminath is unsettled. Tuzet and Rioux (1966) place them in the genus Ascocystis. Other workers, eg. Vávra (1969), put them in Lankesteria. The life cycle of M. mackiei is summarized by Vávra.

PATHOGENS OF PSYCHODIDAE (PHLEBOTOMINE SAND FLIES) (continued)

Host	Host stage infected	Pathogen	% Incidence	Locality	Lab. or field study	Reference
Lutzomyia lainsoni (Fraiha & Ward)	Adults	Microsporida	< 1%	Brazil (Pará)	Field	Ward & Killick-Kendrick (1974)
"	Larvae	Fungus	Not stated	"	Lab.	"
Lutzomyia lichyi (Floch & Abonnenc), as P. vexillarius Fchld. & Hertig	Adults	"	10% in resting collections, 0% in biting collections	Panama	Field	McConnell & Correa (1964)
Lutzomyia longipalpis (Lutz & Neiva)	Larvae	"	Not given	Brazil (Bahia)	Lab.	Sherlock & Sherlock (1959)
"	Adults	Monocystis chagasi	10% field, 20-80% lab.	Brazil (Minas Gerais)	Lab. & field	Adler & Mayrink (1961)
"	Larvae Pupae Adults	"	Not given	"	Field & lab.	Coelho & Falcão (1964)
"	Adults	Fungus	Not given	"	Field?	Adler & Mayrink (1961)
Lutzomyia ovallesi (Ortiz)	"	Minute, unidentified parasites	3.1%	Belize	Field	Lewis (1965a)
"	"	Fungus	43.7% in resting collections; 0% in biting collections	Panama	"	McConnell & Correa (1964)
Lutzomyia panamensis (Shannon)	"	Acephaline gregarines	< 1%	Panama	"	"
Lutzomyia panamensis (Shannon)	"	Fungus	0.5% in biting collections; 0% in resting collections	Panama	"	"
"	"	Nematode	< 1%	"	"	"

PATHOGENS OF PSYCHODIDAE (PHLEBOTOMINE SAND FLIES) (continued)

Host	Host stage infected	Pathogen	% Incidence	Locality	Lab. or field study	Reference
Lutzomyia sallesi (Galvão & Coutinho)	Adults	Monocystis chagasi	Not given	Brazil (Minas Gerais)	Field	Coelho & Falcão (1964)
Lutzomyia sanguinaria (Fchld. & Hertig)	"	Acephaline gregarine	< 1%	Panama	"	McConnell & Correa (1964)
"	"	Fungus	0% in resting collections; 0.2% in biting collections	"	"	"
"	"	Nematode	< 1%	"	"	"
Lutzomyia shannoni (Dyar)	"	Gregarine	Unknown	Belize	"	Garnham & Lewis (1959)
"	"	Gregarine	< 1%	Panama	"	McConnell & Correa (1964)
"	"	Fungus	10% in resting collections	"	"	"
"	"	Fungus	"	"	"	Lewis (1965a)
"	"	Ciliates	"	"	"	"
"	"	Nematode	11%	Costa Rica	Field	Rosabal & Miller (1970)
"	"	Nematode	Unknown	Belize	"	Lewis (1965a)
Lutzomyia steatopyga (Fchld. & Hertig), as Brumptomyia beltrani	"	Acephaline gregarine	< 1%	Panama	"	McConnell & Correa (1964)
Lutzomyia trapidoi (Fchld. & Hertig)	"	Acephaline gregarine	< 1%	"	"	"
Lutzomyia trinidadensis (Newst.)	"	Fungus	3.7%	"	"	"
Lutzomyia vespertilionis (Fchld. & Hertig)	"	Fungus	20%	"	"	"
"	"	Nematode	< 1%	"	"	"

PATHOGENS OF PSYCHODIDAE (PHLEBOTOMINE SAND FLIES) (continued)

Host	Host stage infected	Pathogen	% Incidence	Locality	Lab. or field study	Reference
Lutzomyia vespertilionis (Fchld. & Hertig)	Adults	Nematode, apparently close to genus Tylenchinema	<1%	Panama	Field	McConnell & Correa (1964)
Phlebotomus argentipes	"	Plerocercoids	Unknown	India (Deccan, Hyderabad)	Field?	Subramaniam & Naidu (1944)
Phlebotomus ariasi Tonn.	"	Nematode, Mastophorus muris (Gmelin)	8 positive, 370 examined	France	"	Golvan et al. (1963)
"	"	"	Unknown	France (Herault)	"	Rioux & Golvan (1969)
"	"	Entomophthora sp.	"	"	"	Rioux & Golvan (1969)
"	"	Nematode, Rictularia proni Seurat	"	"	"	Rioux & Golvan (1969)
"	"	"	"	"	"	Rioux et al. (1969)
"	"	Entomophthora sp.	1 positive, 317 examined	"	"	Rioux et al. (1966)
Phlebotomus caucasicus Marzinowsky	"	Gregarine, Diplocystis sp.	0.6%	USSR	"	Lisova (1962)
Phlebotomus chinensis Newst.	"	"	1.3%	"	"	"
Phlebotomus longipes Parrot & Martin	"	Gregarine, Lankesteria sp.	4.5%	Ethiopia	"	Ashford (1974)
Phlebotomus mascittii Grassi	"	Cestode, hymenolepid cysticercoid	<1%	Corsica	"	Quentin et al. (1971)

PATHOGENS OF PSYCHODIDAE (PHLEBOTOMINE SAND FLIES) (continued)

Host	Host stage infected	Pathogen	% Incidence	Locality	Lab. or field study	Reference
Lutzomyia vexator vexator (Coq.)	Probably adults	Rickettsia	Unknown	USA (probably) Washington, DC	Not stated	Hertig (1936)
Lutzomyia vexator occidentis (Fchld. & Hertig)	Adults	Lankesteria sp.	Common	USA (California)	Field	Ayala (1971,1973)
"	"	Fungus	Not stated	"	Lab. & field	Chaniotis & Anderson (1968)
"	Larvae, pupae	Pseudomonas sp.	"	"	Lab.	"
Lutzomyia ylephiletor (Fchld. & Hertig)	Adults	Acephaline gregarine	< 1%	Panama	"	McConnell & Correa (1964)
"	"	Fungus	4.2% in resting collections; 0% in biting collections	"	"	"
Lutzomyia sp. or spp.	Eggs	Coelomomyces citerrii Leão & Pedroso	Not stated	Brazil (Minas Gerais)	Probably field	Leão & Pedroso (1965)
Lutzomyia spp.	Larvae	Fungus, Aspergillus sp.	Not stated	Panama	Lab.	Hertig & Johnson (1961) Hertig (1964)
Phlebotomus argentipes Ann. & Brun.	Adults	Monocystis mackiei	High	India (Assam)	Field	Shortt & Swaminath (1927)
"	"	Unidentified abdominal parasite	Unknown	Malaysia	"	Lewis & Killick-Kendrick (1973)
Phlebotomus orientalis Parrot	"	Nematode	0.8%	Ethiopia	"	Ashford (1974)
"	"	Cestode, hymenolepid cysticercoid	0.1%	"	"	"

PATHOGENS OF PSYCHODIDAE (PHLEBOTOMINE SAND FLIES) (continued)

Host	Host stage infected	Pathogen	% Incidence	Locality	Lab. or field study	Reference
Phlebotomus papatasii (Scopoli)	Adults	Fungus	Unknown	Palestine	?	Adler & Theodor (1927)
"	"	"	"	Palestine & Syria	Field	Adler & Theodor (1929)
"	"	Entomophthora papatasi (= Empusa papatasi)	"	Malta	Lab.	Marett (1915); Larrousse (1921)
"	"	Gregarine, Diplocystis sp.	<1%	USSR	Field	Lisova (1962)
"	"	Monocystis mackiei	1 positive, 450 examined	Italy	"	Missiroli (1929,1932)
"	"	Nematode	—	Iraq (Baghdad)	Field, lab.?	Adler & Theodor (1929)
"	"	"	Unknown	India	Field	Mitra (1956)
"	"	"	"	Pakistan	"	Lewis (1967a)
"	"	Hepatozoon	50% 1 positive, 2 examined	Palestine (Jericho)	"	Adler & Theodor (1929)
Phlebotomus papatasii	Larvae	Fungus, Penicillium glaucum	Not given	USSR (Sebastopol)	Lab.	Zotov (1930)
Phlebotomus perniciosus Newst.	Adults	Microsporida, Adelina sp.	"	Corsica	Field	Rioux et al. (1972)
"	"	Mastophorus muris (Gmelin)	"	France	Field	Golvan et al. (1963)
"	"	"	8 positive, 370 examined	France (Herault)	"	Rioux & Golvan (1969)

PATHOGENS OF PSYCHODIDAE (PHLEBOTOMINE SAND FLIES) (continued)

Host	Host stage infected	Pathogen	% Incidence	Locality	Lab. or field study	Reference
Phlebotomus perniciosus Newst.	Adults	Spirochaeta phlebotomi (=Treponema phlebotomi)	8 positive, 370 examined	France (Marseilles)	Field?	Pringault (1921)
Phlebotomus perniciosus	"	Cestode, hymenolepid cysticercoid	<1%	Corsica	Field	Quentin et al. (1971)
Phlebotomus rodhaini Parrot	"	Gregarine, monocystis?	5 positive, 5 examined	Uganda	"	Barnley (1968)
Phlebotomus sergenti Parrot	"	Fungus	Unknown	Palestine & Syria	"	Adler & Theodor (1929)
"	"	Nematode	"	Pakistan	"	Lewis (1967a)
"	"	Nematode	"	Iraq (Baghdad)	Field, lab.?	Adler & Theodor (1929)
"	"	Gregarine, Diplocystis sp.	5%	USSR (Taškent)	Field	Lisova (1962)
Phlebotomus sp.	"	Fungus	Unknown	Sudan	"	Kirk & Lewis (1947)
Sergentomyia adleri (Theodor)	"	Unidentified parasite	"	"	"	"
"	"	Helminth	"	Uganda	"	Barnley (1968)
"	"	"	16 positive, about 2000 examined	"	"	"
Sergentomyia affinis (Theodor)	"	Tylenchid nematode	Not given	Ethiopia	"	Ashford (1974)
Sergentomyia africana (Newst.)	"	Fungus	"	Sudan	"	Kirk & Lewis (1940)
Sergentomyia bedfordi (Newst.)	"	Helminth	No positive, about 2000 examined	Uganda	"	Barnley (1968)

PATHOGENS OF PSYCHODIDAE (PHLEBOTOMINE SAND FLIES) (continued)

Host	Host stage infected	Pathogen	% Incidence	Locality	Lab. or field study	Reference
Sergentomyia clydei (Sinton)	Adults	Nematode	Unknown	Pakistan	Field	Lewis (1967a)
"	"	"	"	Sudan (Sennar)	"	Lewis (1975b)
"	"	"	"	Kenya	"	Lewis & Minter (1960)
Sergentomyia garnhami (Heisch, Guggisberg & Teesdale)	"	Acephaline, gregarine	"	Kenya	"	"
Sergentomyia graingeri Heisch, (Guggisberg & Teesdale)	"	Nematode	"	"	"	"
Sergentomyia hirta (Parrot & Jolinière)	"	Undetermined[a]	"	Algeria (Ahaggar)	"	Theodor (1948); Kirk & Lewis (1951)
Sergentomyia ingrami (Newst.)	"	Fungus	"	Sudan	"	Kirk & Lewis (1947)
Sergentomyia minuta minuta (Rond.)	"	Adelina sp.	"	Corsica	"	Rioux et al. (1972)
"	"	Mastophorus muris	8 positive, 370 examined	France (Herault)	"	Rioux & Golvan (1969)
Sergentomyia schwetzi (Adler, Theodor & Parrot)	"	Nematode	1 positive, 33 examined	Kenya	"	Lewis (1974c)
"	"	Helminth[b]	Unknown	Uganda	"	Barnley (1968)
Sergentomyia sp., probably S. schwetzi	"	Tylenchid nematode	"	Kenya	"	Lewis & Minter (1960)

[a] Abnormality possibly due to parasitism.

[b] Parasite apparently caused abnormality of male external genitalia.

ABSTRACTS

Mary Ann Strand

Adler, S. & Mayrink, W. (1961).   A gregarine, Monocystis chagasi n. sp., of Phlebotomus
longipalpis.   Remarks on the accessory glands of P. longipalpis.   Rev. Inst. Med. Trop.,
São Paulo, 3: 230-238.

   A new species of gregarine is described from sandflies collected in Brazil.   About 10%
   of the wild specimens and 20-80% of the laboratory specimens were infected.   Oocysts
   were observed in the accessory glands in 20% of infected females.   During oviposition,
   the surface of the eggs becomes contaminated with oocysts from the infected glands.

Adler, S. & Theodor, O. (1927).   The transmission of Leishmania tropica from artificially
infected sandflies to man.   Ann. Trop. Med. Parasitol., 21: 89-110.

   An unidentified fungus was observed infecting Phlebotomus papatasi in Israel.

Adler, S. & Theodor, O. (1929).   The distribution of sandflies and leishmaniasis in
Palestine, Syria and Mesopotamia.   Ann.Trop. Med. Parasitol., 23: 269-306.

   During the survey, several parasites of sandflies were observed: a fungus on eggs,
   nematodes in the haemocoel and ovaries, an oocyst (probably Hepatozoon sp.), and
   Crithidia in the midgut, cardia and stomach.

Ashford, R. W. (1974).   Sandflies (Diptera:Phlebotomidae) from Ethiopia: taxonomic and
biological notes.   J. Med. Entomol., 11: 605-616.

   Gregarines similar to those described by Ayala (1971) were found in the haemocoel of
   Phlebotomus longipes.   Of 669 males examined 30 were infected.   Nematodes were
   found infecting 0.8% of the female P. orientalis dissected.   One nematode type nearly
   destroyed the reproductive organs of its host.   Parasites were rarely observed in other
   species.

Ayala, S. C. (1971).   Gregarine infections in the California sandfly, Lutzomyia vexatrix
occidentis.   J. Invert. Pathol., 17: 440-441.

   Lankesteria gammonts were found in the haemocoel of adult sandflies.   In male and
   nulliparous female flies, this stage persists; however, the onset of the females'
   ovarian cycle apparently stimulates the gregarines' sexual cycle.   The Californian
   gregarines are larger than those from Brazil.

Ayala, S. C. (1973).   The phlebotomine sandfly-protozoan parasite community of central
California grasslands.   Amer. Midland Nat., 89: 266-280.

   The role of sandflies in transmission of protozoan parasites to vertebrates is examined.
   Female sandflies were occasionally killed by feeding on malaria-infected lizards with
   excessively high gametocyte levels.   Lizard trypanosomes caused an upset of the normal
   blood meal digestion, although most parasites cause no apparent effect on the sandflies.

Barnley, G. R. (1968).   Unpublished data.   Attributed malformation of male genitalia of
S. schwetzi to be caused by a helminth parasite.

Chaniotis, B. N. & Anderson, J. R. (1968). Age structure, population dynamics and vector potential of Phlebotomus in northern California. Part II. Field population dynamics and natural flagellate infections in parous females. J. Med. Entomol., 5: 273-292.

Flagellates (appeared to be leptomonads) were observed in 20% of the female Lutzomyia vexator occidentis which had taken a blood meal. Nulliparous females and males were not infected. The infections were located in the intestines of the females, the leptomonad forms in the hindgut and the crithidial forms in the midgut. The flagellates are endosymbiotic. A few wild-caught flies and some laboratory reared ones were also infected with fungus mycelia which either grew over the entire body or only in the thoracic muscles. In other cases, Pseudomonas sp. was isolated from apparently diseased larvae and pupae.

Coelho, M. de V. & Falcão, A. L. (1964). Aspects of the life cycle of Monocystis chagasi, Adler & Mayrink, 1961, in Phlebotomus longipalpis. Rev. Brasil. Biol., 24: 417-421.

Sandflies collected in Minas Gerais, Brazil, were found to harbour the gregarine M. chagasi. Larvae become infected by ingesting oocytes. All developmental stages of the parasite are found in the 4th instar and in adult males. Oocytes are eliminated in the faeces. Adult gregarines migrate to the body cavity, when the host pupates, and they eventually degenerate and are eliminated. M. chagasi was also found in Lutzomyia sallesi.

Feng, L. C. (1951). The role of the peritrophic membrane in Leishmania and trypanosome infections of sandflies. Peking Nat. Hist. Bull., 19: 327-334.

The peritrophic membrane found in some phlebotomines may influence the distribution of ingested parasites.

Forattini, O. P. (1973). Entomologia Médicia, vol. 4. Psychodidae. Phlebotominae. Leishmanioses. Bartonelose. Edgard Blucher, Ltda-, São Paulo, 658 p. (in Portuguese).

General review.

Golvan, Y. J. et al. (1963). Infestation spontanée de Phlébotomes par le Spiruridie Mastophorus muris (Gmelin). Ann. Parasitol. Hum. Comp., 38: 934.

The collection of the nematode M. muris infesting larvae of P. ariasi and P. perniciosus near Roquebrun, France, is reported.

Hertig, M. (1936). The rickettsia, Wolbachia pipientis (gen. et sp. n.) and associated inclusions of the mosquito, Culex pipiens. Parasitology, 28: 453-486.

Hertig mentions observing a rickettsia in the gonads of Lutzomyia vexator.

Hertig, M. (1964). Laboratory colonization of Central American Phlebotomus sandflies. Bull. W.H.O., 31: 569-570.

Although the whole genus Phlebotomus appears to be relatively free of parasites, bacteria and fungi, particularly Aspergillus, can be serious mortality factors in laboratory colonies.

Hertig, M. & Johnson, P. T. (1961).   The rearing of Phlebotomus sandflies (Diptera: Psychodidae) I.   Technique.   Ann. Entomol. Soc. Am., 54: 753-764.

Aspergillus is pathogenic to larvae in culture.   The larvae ingest the spores from mouldy food.   The spores germinate in the intestine and the resulting mycelia invade the muscles of the thoracic region, killing the larvae.

Jenkins, D. W. (1964).   Pathogens, parasites and predators of medically important arthropods. Bull. W.H.O., 30(Suppl.), 150 p.

A review of the literature through 1963 is given.

Kirk, R. & Lewis, D. J. (1940).   Studies in leishmaniasis in the Anglo-Egyptian Sudan. III.   The sandflies (Phlebotomus) of the Sudan.   Trans. R. Soc. Trop. Med. Hyg., 33: 623-634.

An unknown fungus was observed infecting adult Sergentomyia africana.

Kirk, R. & Lewis, D. J. (1947).   Studies in leishmaniasis in the Anglo-Egyptian Sudan.   IX. Further observations on the sandflies (Phlebotomus) of the Sudan.   Trans. R. Soc. Trop. Med. Hyg., 40: 869-888.

During the survey, a fungus was observed occasionally infecting sandflies.

Kirk, R. & Lewis, D. J. (1951).   The Phlebotominae of the Ethiopian Region.   Trans. R. Entomol. Soc. Lond., 102: 383-510.   Abnormality in S. hirta may be due to parasitism.

Larrousse, F. (1921).   Etude systématique et médicale des phlébotomes.   Vigot, Paris, 106 p.

Leão, A. E. A. & Pedroso, M. C. (1965).   Nova espécie do genero Coelomomyces parasito de ovos de Phlebotomus.   Mycopathol. Mycol. Appl., 26: 305-307.

A new species, C. ciferrii, is described.   The fungus was found in Phlebotomus eggs collected in Minas Gerais, Brazil.

Lewis, D. J. (1965a).   Internal structural features of some Central American phlebotomine sandflies.   Ann. Trop. Med. Parasitol., 59: 375-385.

Parasites were observed in a few sandflies taken on human bait.   Fungal, nematode, ciliate, and flagellate endoparasites were found usually in the abdomen or midgut.   None was identified.

Lewis, D. J. (1965b).   Sand-flies and black-flies (Dipt. Phlebotominae and Simuliidae). Proc. South Lond. Entomol. Nat. Hist. Soc., 1964: 14-17.

The biology and taxonomy of phlebotomines are discussed in this general article. Sandfly larvae, not being aquatic, are not exposed to aquatic parasites and phlebotomines perhaps for this reason, appear to have relatively few parasites.

Lewis, D. J. (1967a).   The phlebotomine sand-flies of West Pakistan (Diptera:Psychodidae). Bull. Br. Museum (Nat. Hist.) Entomol., 19: 1-57.

Nematodes were found in 3 sandflies during an extensive survey.   They were large, about 3.7 mm long, and were packed with oval brown eggs.   This may be the same worm seen in the same species by Adler & Theodor (1929).

Lewis, D. J. (1967b).   Phlebotomine sand-flies from Cayman Brac Island (Diptera:Psychodidae). J. Nat. Hist., 2: 73-83.

A nematode was found in the abdomen of a male Lutzomyia cayennensis braci.   This specimen did not resemble nematodes previously reported from sandflies.   It does not appear to be a filarial worm or mermithid.

Lewis, D. J. (1971).   Phlebotomid sandflies.   Bull. W.H.O., 44: 535-551.   The relationship between sandflies and leishmaniasis is reviewed.   Ninety references are given.

Lewis, D. J. (1973).   Phlebotomidae and Psychodidae, p. 155-179.   In Smith, K. G. V., Insects and other arthropods of medical importance, British Museum (Nat. Hist.) Pub. No. 720, London.

General review.

Lewis, D. J. (1974a).   The biology of Phlebotomidae in relation to leishmaniasis.   Ann. Rev. Entomol., 19: 363-384.

Taxonomy and ecology of the flies are reviewed.

Lewis, D. J. (1974b).   The phlebotomid sandflies of the Yemen Arab Republic.   Z. Tropenmed. Parasitol., 25: 187-197.

Notes on distribution, habits, and taxonomy of Phlebotomus and Sergentomyia species found in Yemen are given.   Cibarial filtration in S. antennata, and its possible relation to parasitism, are discussed.

Lewis, D. J. (1974c).   Unpublished data.

Lewis, D. J. (1975a).   Functional morphology of the mouth parts in New World phlebotomine sandflies (Diptera:Psychodidae).   Trans. R. Entomol. Soc. Lond., 126: 497-532.

Lewis discusses the differences in the stylets of the biting fascicle in relation to probable function, blood-sucking habits, and taxonomic grouping.   The cibarial teeth may act as a comb.

Lewis, D. J. (1975b).   Unpublished data.

Lewis, D. J. & Killick-Kendrick, R. (1973).   Some phlebotomid sandflies and other Diptera of Malaysia and Sri Lanka.   Trans. R. Soc. Trop. Med. Hyg., 67: 1-5.

A female Phlebotomus argentipes with a large abdominal parasite was found in Ulu Gombak, Malaya.

Lewis, D. J. et al. (1970).   Determination of parous rates in phlebotomine sandflies with special reference to amazonian species.   Bull. Entomol. Res., 60: 209-219.

Of 524 Lutzomyia flaviscutellata dissected, 24% were infected by Monocystis.   The parous rate of infected and uninfected flies is nearly the same.   This suggested that either the parasites are non-pathogenic or that they have little effect on longevity.

Lewis, D. J. & Minter, D. M. (1960).   Internal structural changes in some African phlebotominae.   Ann. Trop. Med. Parasitol., 54: 351-365.

Lisova, A. I. (1962). ⎣Gregarina (Genus Diplocystis) in the body cavity of sandflies.⎦ Zool. Zh., 41: 1095-1099. (R)

A survey of sandflies in Tashkent revealed a gregarine parasite, Diplocystis sp. Four species of sandflies were infected: Phlebotomus papatasi, P. caucasicus, P. chinensis, and P. sergenti. Gregarines were found in the body cavity glands and ovaries. They caused marked changes in the organs affected. Frequency of infection is greatest in P. sergenti, about 5%.

Marett, P. J. (1915). The bionomics of the Maltese Phlebotomi. Br. Med. J., 2: 172-173.

Phlebotomus papatasi was found infected by Entomophthora (Empusa) papatasi. Ingested blood appeared to remain in thorax of infected flies. Motile spores were found in the salivary glands and mycelium was seen in the intestines.

McConnell, E. & Correa, M. (1964). Trypanosomes and other microorganisms from Panamanian Phlebotomus sandflies. J. Parasitol., 50: 523-528.

Acephaline gregarines of one or more species were found in the hemocoel of 10 species of sandflies. Only 18 of more than 6000 female flies dissected contained this parasite. No pathogenic conditions were noticed in the flies. Also, mycelia of fungi occurred in individuals of 10 sandfly species. The fungus was almost always associated with the gonads. The infection may disturb the normal physiology so as to prevent the female from seeking a blood meal. Nematodes were observed in hemocoels of three species of sandfly.

Missiroli, A. (1929). I protozoi parassiti del Phlebotomus papatasii. Ann. Ig., 39: 481-487.

In Italy no parasites were found in sandflies collected in areas where human leishmaniasi occurs. In other areas, parasites resembling a developmental stage of Trypanosoma lewis and Monocystis mackiei were observed.

Missiroli, A. (1932). Sullo sviluppo di una gregarina del Phlebotomus. Ann. Ig., 42: 373-377.

The developmental stages of Monocystis mackiei in a sandfly (probably P. papatasi) are described.

Mitra, R. D. (1956). Notes on sandflies. Sandflies of the Poona District. Z. Tropenmed. Parasitol., 7: 229-240.

Pringault, E. (1921). Présence de spirochètes chez Phlebotomus perniciosus Newstead. C. R. Séances Soc. Biol. Ses Fil., 84: 209-210.

Spirochaeta phlebotomi was found in P. perniciosus collected in the Marseilles region.

Quentin, J. C. et al. (1971). Présence du cysticercoïde d'Hymenolepsis brusatae Vaucher, 1971, chez Phlebotomus perniciosus Newstead, 1911 et Phlebotomus mascittii Grassi, 1908. Ann. Parasitol. Hum. Comp., 46: 589-594.

Of 15 436 sandflies collected in Corsica, 2 adult females were infected with H. brusatae. This parasite in its adult stage infects the shrew mouse. The cystercercoids were found in the abdomen of the sandflies.

Rioux, J. A. et al. (1969).   Infestation spontanée de Phlebotomus ariasi par Rictularia proni, Spiruride parasite du mulot.   Les terriers de mulots sont-ils des gîtes larvaires à phlébotomes?   Ann. Parasitol. Hum. Comp., 44: 757-760.

A description is given of the nematode R. proni which was found infecting P. ariasi larvae from France.   This nematode also infects the field mouse, Apodemus sylvaticus.

Rioux, J. A. & Golvan, Y. J. (1969).   Epidemiologie des leishmanioses dans le sud de la France.   Mon. Inst. National Santé Rech. Méd., No. 37, 220 p.

Parasitism of P. ariasi and S. minuta by Mastophorus muris is reported.   Study of parasites can help to determine the breeding places of sandfly hosts.

Rioux, J. A. et al. (1972).   Adelina sp. parasite de phlébotomes.   Ann. Parasitol. Hum. Comp., 47: 347-350.

The development of Adelina sp. in Phlebotomus perniciosus is described.

Rioux, J. A. et al. (1966).   Infestation à Entomophthora sp. chez Phlebotomus ariasi tonnoir, 1921.   Ann. Parasit. Hum. Comp., 41: 251-253.

Of 317 female sandflies dissected, one was found to be infected with Entomophthora.   The body cavity of the fly was filled with resting spores.

Rosabel, R. & Miller, A. (1970).   Phlebotomine sandflies in Louisiana (Diptera:Psychodidae).   Mosq. News, 30: 180-187.

A key to the species of Lutzomyia found in the US is given.   No parasites were found in any of 29 female L. shannoni collected in Louisiana.   However, nematodes were found in the hemocoels of 2 females of 18 examined in Costa Rica.

Sherlock, I. A. & Sherlock, V. A. (1959).   Criação e biologia, em laboratorio do Phlebotomus longipalpis Lutz and Neiva, 1912 (Diptera:Psychodidae).   Rev. Brasil. Biol., 19: 229-250.

Fungal growths fatal to larvae occurred in cultures of Lutzomyia longipalpis.

Shortt, H. E. & Swaminath, C. S. (1927).   Monocystis mackiei n. sp., parasitic in Phlebotomus argentipes Ann and Brun.   Indian J. Med. Res., 15: 539-552.

The parasite was found in about 25% of the adult sandflies caught in Assam.   Oocysts are deposited with eggs and are ingested by the newly emerged larvae.   The infection does not apparently affect feeding or longevity.

Subramaniam, M. K. & Naidu, M. B. (1944).   On a new plerocercoid from a sand-fly.   Curr. Sci., 13: 260-261.

Plerocercoids were found in the adipose tissue of a sandfly collected near Hyderabad, India.   It resembles Sparaganum proliferum; however this specimen possesses distinct segmentation.

Theodor, O. (1948).   Classification of the Old World species of the subfamily Phlebotominae (Diptera, Psychodidae).   Bull. Entomol. Res., 39: 85-115.

In his discussion of the genus Sergentomyia, Theodor mentions a species, S. hirta, which is probably an abnormality.

Theodor, O. (1965).   On the classification of American Phlebotominae.   *J. Med. Entomol.*, **2**: 171-197.

A general classification of the subfamily is given.

Tuzet, O. & Rioux, J. A. (1966).   Les gregarines des Culicidae, Ceratopogonidae, Simuliidae, et Psychodidae.   Unpublished document WHO/EBL/66.50, 18 pp.

A taxonomic review of the gregarines is given.

Vávra, J. (1969).   *Lankesteria barretti*, n. sp. (Eugregarinida, Diplocystidae), a parasite of the mosquito *Aedes triseriatus* (Say) and a review of the genus *Lankesteria* Mingazzini.   *J. Protozool.*, **16**: 546-570.

Ward, R. D. & Killick-Kendrick, R. (1974).   Field and laboratory observations on *Psychodopygus lainsoni* Fraiha and Ward and other sandflies (Diptera, Phlebotomidae) from the Transamazonica highway, Pará state, Brazil.   *Bull. Entomol. Res.*, **64**: 213-221.

A microsporidian parasite was found rarely infecting *Lutzomyia lainsoni* adults.

Zotov, M. P. (1930).   /Experiments in breeding sandflies in the laboratory./   *Vestn. Mikrobiol.*, **9**: 236-243.   (R)

Penicillium glaucum was found to be pathogenic to larvae in culture.

II. VIRUS PATHOGENS OF CULICIDAE (MOSQUITOS) [a]

Brian A. Federici

Department of Entomology
Division of Biological Control
University of California
Riverside, CA 92502, USA

---

[a] The following table and bibliography are mainly based on literature published up to and including 1973, although a few references to work published in 1974 are included.

VIRUS PATHOGENS OF CULICIDAE (MOSQUITOS)

Host	Host stage infected[a]	Virus[b]	% Incidence[c]	Locality	Lab. or field study	Reference
Aedes aegypti	48-h-old larvae	Baculovirus from Aedes sollicitans	0.26	USA (Louisiana)	Lab.	Clark & Fukuda (1971)
"	Larvae and pupae	Entomopoxvirus	-	USSR (Kiev)	Lab.	Lebedeva & Zelenko (1972)
"	Larvae	Parvovirus (densonucleosis)	-	USSR (Kiev)	Lab.	Lebedeva et al. (1973)
"	24-h-old larvae	Chilo iridescent virus (type 6)	3.08	USA (Louisiana)	Lab.	Fukuda (1971)
"	Second-instar larvae	Sericesthis iridescent virus (type 2)	-	Australia (Canberra)	Lab.	Day & Mercer (1964)
Aedes annulipes	Third- and fourth-instar larvae	Iridescent virus (type 5)	≤1	Czechoslovakia (Bohemia)	Field	Weiser (1965)
Aedes cantans	Third- and fourth-instar larvae	Iridescent virus (type 4)	≤1	Czechoslovakia (Bohemia)	Field	Weiser (1965)
"	Third- and fourth-instar larvae	Iridescent virus (type 12)	<1	England (Kent)	Field	Tinsley et al. (1971)

[a] The stage exposed to the virus is indicated in this column, if it was known. However, in most field studies this was unknown, in which case the stage indicated is the stage in which the disease was recognized. In both laboratory and field studies, infections were not usually patent until the late third- or fourth-instar.

[b] See Wildy (1971) for currently recommended nomenclature for viruses. In the case of the iridoviruses the interim nomenclature system type number (Tinsley & Kelly, 1970; Kelly & Robertson, 1973) is in parentheses.

[c] In studies where the number and type of observations varied greatly, a range for the incidence of infection is indicated. The original paper should be consulted for more meaningful information.

VIRUS PATHOGENS OF CULICIDAE (MOSQUITOS) (continued)

Host	Host stage infected[a]	Virus[b]	% Incidence[c]	Locality	Lab. or field study	Reference
Aedes detritus	Larvae	Iridescent virus (type 15)	-	Tunisia	Field	Vago et al. (1969)
"	Third- and fourth-instar larvae	Iridescent virus (type 14)	-	France (Camargue)	Field	Hasan et al. (1960)
"	Larvae	Iridescent virus	-	England (Brownsea Island)	Field	Service (1968)
"	Third-instar larvae	Iridescent virus (type 14)	34.2	France (St Christol)	Lab.	Hasan et al. (1971)
"	Third-instar larvae	Iridescent virus (type 14)	8.0	France (St Christol)	Lab.	Hasan et al. (1971)
Aedes dorsalis	Fourth-instar larvae	Iridescent virus (unconfirmed)	-	USA (Nevada)	Field	Chapman et al. (1966)
Aedes fulvus pallens	Larvae	Iridescent virus	-	USA (Louisiana)	Field	Chapman et al. (1966)
Aedes nigromaculis	Early-instar larvae	Baculovirus from Aedes sollicitans	90	USA (California)	Lab.	Clark (1972)
Aedes sierrensis	Fourth-instar larvae	Tetragonal virus (Parvovirus ?)	-	USA (California)	Field	Kellen et al. (1963)
"	24-h-old larva	Chilo iridescent virus (type 6)	0.58	USA (Louisiana)	Lab.	Fukunda (1971)
Aedes sollicitans	Mixed larval instars	Cytoplasmic polyhedrosis virus	0.2-15.9	USA (Hackberry, Louisiana)	Field	Clark & Fukuda (1971)
"	Mixed larval instars	Cytoplasmic polyhedrosis virus	0.7-3.8	USA (Grand Lake, Louisiana)	Field	Clark & Fukuda (1971)
"	Mixed larval instars	Cytoplasmic polyhedrosis virus and baculovirus	0.2-56.7	USA (Hackberry, Louisiana)	Field	Clark & Fukuda (1971)

VIRUS PATHOGENS OF CULICIDAE (MOSQUITOS) (continued)

Host	Host stage infected[a]	Virus[b]	% Incidence[c]	Locality	Lab. or field study	Reference
Aedes sollicitans	Mixed larval instars	Cytoplasmic polydrosis virus and baculovirus	0.7-12.5	USA (Grand Lake, Louisiana)	Field	Clark & Fukuda (1971)
"	Mixed larval instars	Cytoplasmic polydrosis virus and baculovirus	~70	USA (Louisiana)	Field	Chapman (1974)
"	Larvae	Baculovirus	5	USA (Grand Lake, Louisiana)	Lab. and field	Clark et al. (1969)
"	24-h-old larvae	Baculovirus	29	USA (Louisiana)	Lab.	Clark & Fukuda (1971)
"	48-h-old larvae	Baculovirus	14.25	USA (Louisiana)	Lab.	Clark & Fukuda (1971)
"	24-h-old larvae	Chilo iridescent virus (type 6)	0.54	USA (Louisiana)	Lab.	Fukuda (1971)
"	Early-instar larvae	Iridescent viruses from Aedes taeniorhynchus	8	USA (Louisiana)	Lab.	Woodard & Chapman (1968)
Aedes sticticus	Larvae	Cytoplasmic polyhedrosis virus	-	USA (Louisiana)	Field	Chapman (1972)
"	Larvae	Iridescent virus	-	USA (Louisiana)	Field	Chapman et al. (1969)
Aedes stimulans	Third-instar larvae	Iridescent virus (type 11)	<1	USA (Connecticut)	Field	Anderson (1970)
Aedes taeniorhynchus	Fourth-instar larvae	Cytoplasmic polyhedrosis virus	-	USA (Louisiana)	Field	Chapman (1972)
"	First-instar larvae	Cytoplasmic polyhedrosis virus	1-12 (average 4.3)	USA (New York)	Lab.	Federici (1973)

VIRUS PATHOGENS OF CULICIDAE (MOSQUITOS) (continued)

Host	Host stage infected[a]	Virus[b]	% Incidence[c]	Locality	Lab. or field study	Reference
Aedes taeniorhynchus	Fourth-instar	Cytoplasmic polyhedrosis virus	—	USA (Louisiana)	Field	Federici (1973)
"	Adults	Cytoplasmic polyhedrosis virus	<1	USA (New York)	Lab.	Federici (1973)
"	48-h-old larvae	Cytoplasmic polyhedrosis virus of Aedes sollicitans	1.81	USA (Louisiana)	Lab.	Clark & Fukuda (1971)
"	Larvae	Baculovirus	—	USA (Louisiana)	Field	Chapman (1974)
"	24-h-old larvae	Chilo iridescent virus (type 6)	2.70	USA (Louisiana)	Lab.	Fukuda (1971)
"	Larvae	Iridescent virus from Aedes vexans	<1	USA (Louisiana)	Lab.	Woodard & Chapman (1968)
"	Larvae	Iridescent virus from Psorophora ferox	1	USA (Louisiana)	Lab.	Woodard & Chapman (1968)
"	First-instar larvae	Regular mosquito iridescent virus (type 3) and turquoise mosquito iridescent virus	—	USA (Florida)	Lab.	Anthony & Hall (1970)
"	Fourth-instar larvae	Regular mosquito iridescent virus (type 3)	—	USA (Florida)	Field	Clark et al. (1965)
"	24-h-old larvae	Regular mosquito iridescent virus (type 3)	—	USA (Florida)	Lab.	Hall & Anthony (1971)
"	3-day-old larvae	Regular mosquito iridescent virus (type 3)	—	USA (Florida)	Lab.	Hall & Anthony (1971)
"	Adults	Regular mosquito iridescent virus (type 3)	—	USA (Florida)	Lab.	Hall & Anthony (1971)

VIRUS PATHOGENS OF CULICIDAE (MOSQUITOS) (continued)

Host	Host stage infected[a]	Virus[b]	% Incidence[c]	Locality	Lab. or field study	Reference
Aedes taeniorhynchus	Larvae	Regular mosquito iridescent virus (type 3)	<1	USA (North Key, Florida) USA (Atsena Otie Key, Florida)	Field	Hall & Lowe (1971)
"	Larvae	Regular mosquito iridescent virus (type 3) and turquoise mosquito iridescent virus	-	USA (Florida)	Lab.	Hall & Lowe (1972)
"	First-instar larvae	Regular mosquito iridescent virus (type 3)	0-33.3	USA (Florida)	Lab.	Linley & Nielsen (1968a)
"	First-instar larvae	Regular mosquito iridescent virus (type 3)	0-15.6	USA (Florida)	Lab. and field	Linley & Nielsen (1968b)
"	Larvae (transovarial transmission)	Regular mosquito iridescent virus (type 3)	16.6	USA (Florida)	Lab.	Lowe et al. (1970)
"	18-24-h-old larvae	Regular mosquito iridescent virus (type 3)	53.8 39.3	USA (Florida)	Lab.	Matta & Lowe (1970)
"	First-instar larvae	Regular mosquito iridescent virus (type 3)	-	USA (Texas)	Lab.	Stoltz (1971)
"	Larvae	Regular mosquito iridescent virus (type 3) and turquoise MIV	-	USA (Texas)	Lab.	Stoltz (1973)
"	First-instar larvae	Regular mosquito iridescent virus (type 3)	5.2-19.7 (average 8.1)	USA (Texas)	Lab.	Stoltz & Summers (1974)

VIRUS PATHOGENS OF CULICIDAE (MOSQUITOS) (continued)

Host	Host stage infected[a]	Virus[b]	% Incidence[c]	Locality	Lab. or field study	Reference
Aedes taeniorhynchus	24-h-old larvae	Regular mosquito iridescent virus (type 3) and turquoise MIV	5-25	USA (Indiana)	Lab.	Wagner et al. (1973)
"	First-instar and early L$_2$	RMIV TMIV	16 (0-60) 21	USA (Louisiana)	Lab.	Woodard & Chapman (1968)
"	First-instar larvae	RMIV	10	USA (Louisiana)	Lab.	Woodard & Chapman (1968)
"	Second-instar larvae	RMIV	32	USA (Louisiana)	Lab.	Woodard & Chapman (1968)
"	Third-instar larvae	RMIV	3	USA (Louisiana)	Lab.	Woodard & Chapman (1968)
"	Fourth-instar larvae	RMIV	0	USA (Louisiana)	Lab.	Woodard & Chapman (1968)
Aedes thibaulti	Larvae	Cytoplasmic polyhedrosis virus	-	USA (Louisiana)	Field	Chapman (1972)
Aedes tormentor	48-h-old larvae	Baculovirus from Aedes sollicitans	14.09	USA (Louisiana)	Lab.	Clark & Fukuda (1971)
Aedes triseriatus	Larvae	Cytoplasmic polyhedrosis virus	-	USA (Louisiana)	Field	Clark et al. (1969)
"	48-h-old larvae	Baculovirus from Aedes sollicitans	6.62	USA (Louisiana)	Lab.	Clark & Fukuda (1971)
"	Larvae	Baculovirus	-	USA (Louisiana)	Field	Chapman (1974)
"	Larvae	Baculovirus from Aedes sollicitans	-	USA (Florida)	Lab.	Federici & Anthony (1972)
"	24-h-old larvae	Baculovirus from Aedes sollicitans	36	USA (Florida)	Lab.	Federici & Lowe (1972)
"	48-h-old larvae	Baculovirus from Aedes sollicitans	34	USA (Florida)	Lab.	Federici & Lowe (1972)

VIRUS PATHOGENS OF CULICIDAE (MOSQUITOS) (continued)

Host	Host stage infected[a]	Virus[b]	% Incidence[c]	Locality	Lab. or field study	Reference
Aedes triseriatus	72-h-old larvae	Baculovirus from Aedes sollicitans	11	USA (Florida)	Lab.	Federici & Lowe (1972)
"	96-h-old larvae	Baculovirus from Aedes sollicitans	8	USA (Florida)	Lab.	Federici & Lowe (1972)
Aedes vexans	Larvae	Iridescent virus	<1	USA (Louisiana)	Field	Chapman et al. (1966)
"	Larvae	Iridescent virus	<1	USA (Louisiana)	Lab.	Woodard & Chapman (1968)
"	Larvae	Iridescent virus from Psorophora ferox	1	USA (Louisiana)	Lab.	Woodard & Chapman (1968)
Anopheles albimanus	24-h-old larvae	Chilo iridescent virus (type 6)	0.75	USA (Louisiana)	Lab.	Fukuda (1971)
Anopheles bradleyi	Larvae	Cytoplasmic polyhedrosis virus	-	USA (Louisiana)	Field	Chapman (1972)
Anopheles crucians	Larvae	Cytoplasmic polyhedrosis virus	-	USA (Louisiana)	Field	Chapman (1972)
"	Larvae	Baculovirus	-	USA (Louisiana)	Field	Chapman (1974)
"	Larvae	Tetragonal virus	-	USA (Louisiana)	Field	Chapman et al. (1970)
Anopheles freeborni	Fourth-instar larvae	Tetragonal virus	-	USA (California)	Field	Kellen et al. (1963)
Anopheles quadrimaculatus	Second-instar larvae	Cytoplasmic polyhedrosis virus	-	USA (Florida)	Lab.	Anthony et al. (1973)
"	24-h-old first instar	Chilo iridescent virus (type 3)	0.12	USA (Louisiana)	Lab.	Fukuda (1971)
Anopheles subpictus	Larvae	Baculovirus (?) unconfirmed or possibly a parvovirus	-	India (Dhakuria)	Field	Dasgupta & Ray (1957)

VIRUS PATHOGENS OF CULICIDAE (MOSQUITOS) (continued)

Host	Host stage infected[a]	Virus[b]	% Incidence[c]	Locality	Lab. or field study	Reference
Anopheles stephensi	Adults	Cytoplasmic poly- hedrosis virus	–	England (Epsom)	Lab.	Bird et al. (1972)
"	Adults	Cytoplasmic poly- hedrosis virus	–	England (Liverpool)	Lab.	Davies et al. (1971)
"	Adults	Non-occluded virus	–	England (Liverpool)	Lab.	Davies et al. (1971)
Culex erraticus	Larvae	Cytoplasmic poly- hedrosis virus	–	USA (Louisiana)	Field	Chapman (1972)
Culex peccator	Larvae	Cytoplasmic poly- hedrosis virus	–	USA (Louisiana)	Field	Chapman (1972)
"	24-h-old larvae	Chilo iridescent virus (type 6)	2.17	USA (Louisiana)	Lab.	Fukuda (1971)
Culex pipiens quinquefasciatus	Larvae	Baculovirus	–	USA (Louisiana)	Field	Chapman (1974)
Culex restuans	Larvae	Cytoplasmic poly- hedrosis virus	–	USA (Louisiana)	Field	Clark et al. (1969)
Culex salinarius	Larvae	Cytoplasmic poly- hedrosis virus	–	USA (Louisiana)	Field	Clark et al. (1969)
"	48-h-old larvae	CPVs of _Culiseta melanura_ _Culex salinarius_ _Culex territans_	–	USA (Louisiana)	Lab.	Clark & Fukuda (1971)
"	Larva	Baculovirus	1 specimen	USA (Louisiana)	Field	Clark & Fukuda (1971)
"	Larvae	Tetragonal virus (Parvovirus ?)	–	USA (Louisiana)	Field and lab.	Clark & Chapman (1969)
"	Larvae	Tetragonal virus (Parvovirus ?)		USA (Louisiana)	Field	Chapman et al. (1970)

VIRUS PATHOGENS OF CULICIDAE (MOSQUITOS) (continued)

Host	Host stage infected[a]	Virus[b]	% Incidence[c]	Locality	Lab. or field study	Reference
Culex salinarius	24-h-old larvae	Chilo iridescent virus (type 3)	0.30	USA (Louisiana)	Lab.	Fukuda (1971)
"	Larvae	Tetragonal virus	-	USA (Texas)	Lab.	Stoltz et al. (1974)
Culex tarsalis	Fourth-instar larvae	Tetragonal virus (reported as a CPV) (Parvovirus ?)	~26	USA (California)	Field and lab.	Kellen et al. (1963)
"	Fourth-instar larvae	Tetragonal virus (referred to as a CPV)	-	USA (California)	Lab.	Kellen et al. (1966)
"	Larvae	Tetragonal virus of Culex salinarius	-	USA (Louisiana)	Lab.	Clark & Chapman (1969)
"	Adults	Non-occluded virus	24	USA (California)	Lab.	Richardson et al. (1974)
"	Adults	Non-occluded virus	5	USA (California)	Lab.	Richardson et al. (1974)
Culex territans	Larvae	Cytoplasmic polyhedrosis virus	-	USA (Louisiana)	Field	Clark et al. (1969)
"	Larvae	CPV of Culiseta melanura Culex salinarius Culex territans	-	USA (Louisiana)	Lab.	Clark & Fukuda (1971)
Culiseta inornata	Larvae	Cytoplasmic polyhedrosis virus	-	USA (Louisiana)	Field	Chapman (1972)
"	Fourth-instar larvae	Chilo iridescent virus (type 6)	0.57	USA (Louisiana)	Lab.	Fukuda (1971)
"	Fourth-instar larvae	Non-occluded virus	-	USA (New York)	Lab.	Federici (1974)
Culiseta melanura	Larvae	Cytoplasmic polyhedrosis virus	-	USA (Louisiana)	Field	Chapman (1972)

VIRUS PATHOGENS OF CULICIDAE (MOSQUITOS) (continued)

Host	Host stage infected[a]	Virus[b]	% Incidence[c]	Locality	Lab. or field study	Reference
Culiseta melanura	Larvae	Cytoplasmic polyhedrosis virus	-	USA (Louisiana)	Lab.	Clark & Fukuda (1971)
"	24-h-old larvae	Chilo iridescent virus (type 6)	0.95	USA (Louisiana)	Lab.	Fukuda (1971)
Orthopodomyia signifera	Larvae	Cytoplasmic polyhedrosis virus	-	USA (Louisiana)	Field	Clark et al. (1969)
Psorophora continnis	Larvae	Cytoplasmic polyhedrosis virus	-	USA (Louisiana)	Field	Clark et al. (1969)
"	Larva	Baculovirus	1 specimen	USA (Louisiana)	Field	Clark & Fukuda (1971)
"	Larvae	Iridescent virus	-	USA (Louisiana)	Field	Chapman (1974)
Psorophora ferox	Larvae	Cytoplasmic polyhedrosis virus	-	USA (Louisiana)	Field	Chapman (1972)
"	48-h-old larvae	Cytoplasmic polyhedrosis virus from Aedes sollicitans	1.07	USA (Louisiana)	Lab.	Clark & Fukuda (1971)
"	Larvae	Baculovirus	-	USA (Louisiana)	Field	Chapman (1974)
"	48-h-old larvae	Baculovirus of Aedes sollicitans	0.42	USA (Louisiana)	Lab.	Clark & Fukuda (1971)
"	Larvae	Iridescent virus	-	USA (Louisiana)	Field	Chapman et al. (1966)
"	Fourth-instar larvae	Iridescent virus	<1	USA (Florida)	Field	Federici (1970)
"	Larvae	Iridescent virus	15 0-50	USA (Louisiana)	Lab.	Woodard & Chapman (1968)
"	24-h-old larvae	Chilo iridescent virus (type 6)	.99	USA (Louisiana)	Lab.	Fukuda (1971)
Psorophora horrida	Larvae	Iridescent virus	-	USA (Louisiana)	Field	Chapman et al. (1969)

VIRUS PATHOGENS OF CULICIDAE (MOSQUITOS) (continued)

Host	Host stage infected[a]	Virus[b]	% Incidence[c]	Locality	Lab. or field study	Reference
Psorophora varipes	48-h-old larvae	Baculovirus from Aedes sollicitans	9.81	USA (Louisiana)	Lab.	Clark & Fukuda (1971)
"	Larvae	Iridescent virus	-	USA (Louisiana)	Field	Chapman et al. (1969)
"	24-h-old larvae	Chilo iridescent virus (type 6)	0.40	USA (Louisiana)	Lab.	Fukuda (1971)
Uranotaenia sapphirina	Larvae	Cytoplasmic polyhedrosis virus	-	USA (Louisiana)	Field	Clark et al. (1969)
"	Fourth-instar larvae	Cytoplasmic polyhedrosis virus	-	USA (Florida)	Field	Hazard (1972)
"	Fourth-instar larvae	Baculovirus	-	USA (Florida)	Field	Hazard (1972)
Wyeomyia smithii	First-instar larvae	Baculovirus	~70	USA (Massachusetts)	Lab.	Hall & Fish (1974)

ABSTRACTS

Mary Ann Strand

Anderson, J. F. (1970).   An iridescent virus infecting the mosquito Aedes stimulans.
J. Invertebr. Pathol., 15: 219-224.

An iridescent virus with a particle size of 135-140 nm was isolated from larvae of
Aedes stimulans collected from woodland pools in Connecticut.   The virus appears to be
a different strain than those isolated from other insect species and the name Aedes
stimulans iridescent virus is proposed.   Infected larvae were recognized by their
opalescent turquoise colour.   The incidence in the field was considerably less than 1%.
The presence of the virus in ovarian tissue suggests transovarial passage.

Anthony, D. W. & Hall, D. W. (1970).   Electron microscope studies of the "R" and "T" strains
of mosquito iridescent virus in Aedes taeniorhynchus (Wied.) larvae.   Proc. 4th Int. Colloq.
Insect Pathol., pp. 386-395.

Electron microscope studies of the "R" (regular brown-orange iridescence) and "T"
(turquoise iridescence) strains of mosquito iridescent virus in 4th-instar
A. taeniorhynchus larvae showed the fat body to be the primary site of replication for
both strains.   In tissue sections, both strains appear hexagonal or pentagonal in shape,
which suggests that each particle is an icosahedron.   "T" strain is approximately
30-40 nm smaller than the "R" strain.

Anthony, D. W. et al. (1973).   A virus disease in Anopheles quadrimaculatus.   J. Invertebr.
Pathol., 22: 1-5.

Free and occluded virus particles were found in the cytoplasm of the midgut epithelial
cells of 2nd instar larvae of A. quadrimaculatus infected with Thelohania legeri.
Spherical crystals, which contained these particles, showed a macromolecular para-
crystalline lattice typical of polyhedral protein.   In a few instances, the cuboidal
crystals appeared to have coalesced to form larger crystals.   The observations suggest
that the free particles and their occluded forms may represent stages of a cytoplasmic
polyhedrosis virus.

Bertram, D. S. (1965).   Double infection of mosquitoes with a virus and a malarial parasite.
Proc. 12th Int. Cong. Entomol. (London), pp. 766-767.

Malarial transmission by Aedes aegypti can be suppressed by concurrent infection of the
vector with an arbovirus.   The interaction may be fatal to the mosquito.

Bird, R. G. et al. (1970).   Evidence of insect viruses in colonies of Anopheles stephensi.
Trans. Roy. Soc. Trop. Med. Hyg., 64(1): 28-29 (abstract).

Three different virus-like particles were found in gut lesions of adult A. stephensi and
a cell line derived from them.   One was identified as a cytoplasmic polyhedrosis virus
from the midgut and was associated with difficulty in malarial transmission.   Another
was found in the gut of individuals of the same strain but different colonies.   The
third type was observed in cell cultures and appeared to be TMV type rods.

Chapman, H. C. (1972).   Personal communication.

Chapman, H. C. (1974).   Biological control of mosquito larvae.   Ann. Rev. Entomol., 19:
33-59.

    Review article.

Chapman, H. C. et al. (1970).   Protozoans, nematodes, and viruses of anophelines.   Misc.
Publ. Entomol. Soc. Amer., 7: 134-139.

    Viruses such as the mosquito iridescent virus, cytoplasmic polyhedrosis viruses of the
    hypodermal and gut cells, and a nuclear polyhedrosis virus are reported only from
    culicines, but 2 probable cytoplasmic polyhedrosis viruses, one in the hypodermal cells *
    and the other in the gut cells are reported from larvae of Anopheles crucians.

Chapman, H. C. et al. (1969).   A two-year survey of pathogens and parasites of Culicidae,
Chaoboridae, and Ceratopogonidae in Louisiana.   New Jersey Mosquito Extermin. Assoc. Proc.,
56: 203-212.

    Adult and larval populations of mosquitos were sampled from 1967 to 1969.   Forty-four
    species of mosquitos were captured and 35 were hosts to one or more pathogens.
    Inclusion and noninclusion viruses were recorded.

Chapman, H. C. et al. (1966).   Additional hosts of the mosquito iridescent virus.
J. Invertebr. Pathol., 8: 545-546.

    Mosquito iridescent virus has been verified in Aedes taeniorhynchus, A. fulvus pallens,
    A. vexans, and Psorophora ferox from Louisiana.   Infected larvae had an external
    iridescent hue that was usually first visible in the thorax of late 3rd- or early 4th-
    instar larvae.   Most patently infected larvae die before pupation.

Chapman, H. C. et al. (1972).   Predators and pathogens for mosquito control.   Amer. J. Trop.
Med. Hyg., 21(5): 777-781.

    The pathogens that can be or show promise of being manipulated and mass-reared in the
    laboratory are presented.

Chapman, H. C. et al. (1967).   Pathogens and parasites in Louisiana Culicidae and
Chaoboridae.   New Jersey Mosquito Exterm. Assoc. Proc., 54: 54-60.

    Twenty-seven species of mosquitos were found infected with internal parasites or
    pathogens in southwestern Louisiana.   Mosquito iridescent virus was reported in some
    samples.

Clark, T. B. (1972).   Personal communication.

Clark, T. B. & Chapman, H. C. (1969).   A polyhedrosis in Culex salinarius of Louisiana.
J. Invertebr. Pathol., 13: 312.

    A viral disease characterized by the presence of teragonal inclusion bodies in limb anlage
    and hypodermal cells was found in the larvae of Culex tarsalis in California.   A similar
    virus has been found in C. salinarius in Louisiana.

---

    * Not a true cytoplasmic polyhedrosis virus in that the inclusions studied in this paper
are made up entirely of virus particles.   This virus is most commonly referred to as a
"tetragonal virus" because of the tetragonal nature of the inclusions.   It is possibly a
parvovirus.

Clark, T. B. et al. (1969). Nuclear polyhedrosis and cytoplasmic polyhedrosis virus infections in Louisiana mosquitoes. J. Invertebr. Pathol., 14: 284-286.

A nuclear polyhedrosis virus infecting the gastric caeca and midgut of Aedes sollicitans and a cytoplasmic polyhedrosis virus infecting the gastric caeca and midgut of Culex salinarius were found in larvae collected in Louisiana. The most obvious sign of the disease was the whiteness of the gut wall due to the large number of inclusion bodies. Attempts at laboratory transmission of the nuclear polyhedrosis virus have yielded infection in only about 5% of the exposed larvae, but those infected died before pupation. The cytoplasmic polyhedrosis virus was easily transmitted in the laboratory but even heavily infected individuals pupated and emerged as apparently healthy adults.

Clark, T. B. & Fukuda, T. (1971). Field and laboratory observations of two viral diseases in Aedes sollicitans (Walker) in southwestern Louisiana. Mosquito News, 31: 193-199.

A viral epizootic involving both a cytoplasmic polyhedrosis and a nuclear polyhedrosis occurred in populations of Aedes sollicitans in Louisiana. Experimental results suggest that a series of overlapping broods of Aedes sollicitans may lead to a buildup of infective viral material in the habitat, but a period of drying between broods appears to reduce it very significantly. Transovum transmission and lateral transmission could explain the levels of infection found after the dry period. The introduction of infective material into an area previously almost free of disease resulted in a significant rise in the rate of infection.

Clark, T. B. et al. (1965). A mosquito iridescent virus (MIV) from Aedes taeniorhynchus (Wiedemann). J. Invertebr. Pathol., 7: 519-521.

Diseased 4th-instar larvae of A. taeniorhynchus were collected in Florida. Examination revealed a noninclusion virus with a dense central core surrounded by a 6-sided capsule which is composed of at least 2 layers. The principal site of infection is the cytoplasm of fat cells. Infected larvae became iridescent orange during the 4th-instar and sometimes appeared milky white late in the disease. Per os transmission of MIV was achieved in only a small percentage of trials.

Cunningham, J. C. & Tinsley, T. W. (1968). A serological comparison of some iridescent nonoccluded insect viruses. J. Gen. Virol., 3: 1-8.

Tipula iridescent virus and Sericesthis iridescent virus were found to be serologically related by complement-fixation, tube-precipitation and agar-gel diffusion tests. They were unrelated to mosquito iridescent virus when compared by complement fixation.

Dasgupta, B. (1968). A possible virus disease of the malaria parasite. Trans. Roy. Soc. Med. Hyg., 62: 730.

Inclusion bodies have been observed in oocysts of Plasmodium. The abnormal oocysts showed partial to total loss of nuclei.

Dasgupta, B. & Ray, H. N. (1957). The intranuclear inclusions in the mid-gut of the larva of Anopheles subpictus. Parasitology, 47: 194-195.

Feulgen-positive intranuclear inclusions were observed in the secretory cells of the midgut of larvae collected near Calcutta. A mature inclusion body appeared as a lump of DNA formed of globular bodies. Staining reactions suggested that the amount of DNA in the nucleus of other cells in the neighbourhood of the affected nucleus was reduced considerably.

Davies, E. E. et al. (1971).   Microbial infections associated with plasmodial development in
Anopheles stephensi.   Ann. Trop. Med. Parasitol., 63: 403-408.

Two forms of virus-like particles and a rickettsia-like organism have been observed in
the midgut epithelial cells of a laboratory colony of A. stephensi.   One of the viral
particles does not appear to affect either the mosquito or the plasmodial infection.
The other occurs in oocysts which show evidence of degenerative changes.

Day, M. F. & Mercer, E. H. (1964).   Properties of an iridescent virus from the beetle,
Sericesthis purinosa.   Aust. J. Biol. Sci., 17: 892-902.

Sericesthis iridescent virus (SIV) was added to the medium in which second instar larvae
of Aedes aegypti were developing.   After two weeks the fat body in two living larvae
became iridescent blue indicating infection by SIV.

Faust, R. M. et al. (1968).   Nucleic acids in the blue-green and orange mosquito iridescent
viruses (MIV) isolated from larvae of Aedes taeniorhynchus.   J. Invertebr. Pathol., 10: 160.

Analysis of the nucleic acids revealed that DNA was present in both blue-green and orange
MIV and that RNA was absent.   DNA constituted 10.5% of the blue-green form and 11.7% of
the orange.

Federici, B. A. (1970).   Unpublished observations.

Federici, B. A. (1973).   Preliminary studies on a cytoplasmic polyhedrosis virus of Aedes
taeniorhynchus.   Abstracts of papers, 5th Int. Colloq. Insect Pathol. Microbial Cont.
(Oxford), p. 34.

The 4th-instar larva of A. taeniorhynchus from Louisiana was found infected with a cytoplasmic polyhedrosis virus (CPV) in the posterior portion of the stomach.   Infection
trials gave an average rate of patent infection of 4.3% and mortality rate less than 1%.
Occlusion bodies were present in the posterior stomach and occasionally in the gastric
caeca.   The coalescing process during occlusion formation is similar to that observed
in a baculovirus in larvae of A. triseriatus.   This similarity suggests that the
mosquito host plays a role in the formation of occlusions.

Federici, B. A. (1974).   Virus pathogens of mosquitoes and their potential use in mosquito
control.   In: Aubin, A. et al., ed. Le contrôle des moustiques/Mosquito control.   Quebec,
Univ. Quebec Press, pp. 93-135.

Review article with 56 references.

Federici, B. A. & Anthony, D. W. (1972).   Formation of virion-occluding proteinic spindles in
a baculovirus of Aedes triseriatus.   J. Invertebr. Pathol., 20: 129-138.

An unusual process of inclusion formation was studied in A. triseriatus larvae infected
with a Baculovirus (BV) similar to the nuclear polyhedrosis virus (NPV) type.   In this
disease virion-occluding proteinic inclusions initially developed individually.
However, as the disease progressed the proteinic inclusions gradually coalesced
eventually forming large rugose ellipsoids and finally, large smooth-surfaced spindles.
Nuclei in late stages of infection usually contained two to five rugose ellipsoidal
inclusions, frequently measuring 5 μm to 7 μm in diameter by 10 μm to 15 μm in length.
The ellipsoidal forms exhibited different chemical behaviour from the spindles.

Federici, B. A. & Lowe, R. E. (1972).   Studies on the pathology of a baculovirus in
Aedes triseriatus.   J. Invertebr. Pathol., 20: 14-21.

   The pathology of a Baculovirus (BV) in A. triseriatus was studied.   The virus infected
   the cardia, gastric caeca, and the entire stomach of larval midgut epithelium.   The
   progress of the disease was similar to that of other Baculoviruses of the nuclear polyhedrosis virus (NPV) type.   The disease differed from other BVs of the NPV type in that
   small proteinic inclusions gradually coalesced as they grew, forming large fusiform
   inclusions.

Fukuda, T. (1971).   Per os transmission of Chilo iridescent virus to mosquitoes.
J. Invertebr. Pathol., 18: 152-153.

   Chilo iridescent virus (CIV) was successfully transmitted to 13 species of mosquitos,
   however the percentage contracting the disease was low.   The CIV in mosquito larvae was
   first detected at the time of 3rd-instar when the blue-violet iridescence began to show.
   Death usually occurred in the 4th-instar.   The virus had a wide host range in mosquitos
   and can be more easily mass-produced than MIV.

Hall, D. W. & Anthony, D. W. (1971).   Pathology of a mosquito iridescent virus (MIV) infecting
Aedes taeniorhynchus.   J. Invertebr. Pathol., 18: 61-69.

   RMIV was capable of infecting a variety of tissues within its host.   Cells of the fat
   body, tracheal epithelium, imaginal discs, and epidermis were the primary sites of viral
   replication.   Extensive destruction of the fat body by this virus resulted in the death
   of most infected mosquitos before they reached the adult stage.   The transovarial transmission of RMIV was confirmed, and when transovarial transmission occurred, either all
   or none of the progeny of a given female were infected.

Hall, D. W. & Fish, D. D. (1974).   A Baculovirus from the mosquito Wyeomyia smithii.
J. Invertebr. Pathol., 23: 383-388.

   A Baculovirus was found in larvae of W. smithii collected from a sphagnum bog in
   Massachusetts.   This virus is similar in size and appearance to Baculoviruses from
   other mosquitos.   A unique feature of the virus is the formation of polymorphic
   inclusions.   The prepatent period for this virus is 3-5 days.

Hall, D. W. & Lowe, R. E. (1971).   A new distribution record for the mosquito iridescent
virus (MIV).   Mosquito News, 31: 448-449.

   Aedes taeniorhynchus larvae collected off the west coast of Florida were found to be
   infected with RMIV.   About 0.12% of the specimens examined were infected.   The location
   of these collections, in addition to those previously reported, suggests that RMIV is
   present throughout the range of A. taeniorhynchus.   TMIV has not been found in Florida.

Hall, D. W. & Lowe, R. E. (1972).   Physical and serological comparisons of "R" and "T"
strains of mosquito iridescent virus from Aedes taeniorhynchus.   J. Invertebr. Pathol.,
19: 317-324.

   Gel diffusion studies with alkaline degraded virus preparations exhibited four antigens
   common to both viruses but no unique antigens were detected for either virus.   Electron
   micrographs of infected tissue sections showed tubular structures associated with both
   RMIV and TMIV.   RMIV and TMIV sedimented at different rates in sucrose density gradients
   and RMIV was found to be slightly more dense than TMIV by equilibrium ultracentrifugation
   in cesium chloride.   Mixtures of the two viruses can be separated by sucrose density
   gradient centrifugation.

Hasan, S. et al. (1970).   Infection a virus irisant dans une population naturelle d'_Aedes_
detritus Haliday en France.   _Ann. Zool. Ecol. anim._, _2_: 295-299 (English summary).

   An iridescent virus has been isolated from a naturally infected population of _A. detritus_
   in southern France.   Morphologically and serologically the virus is identical to the
   iridescent virus described from this mosquito in Tunisia.   (See Vago, Rioux, Duthoit
   & Dedet, 1969.)

Hasan, S. et al. (1971).   Infection of _Aedes detritus_ Hal. with mosquito iridescent virus.
Bull. World Health Organ., _45_(2): 268-269.

   Attempts were made to infect larvae of _A. detritus_ collected in France with MIV.   Some
   larvae were kept for 24 hrs in water in which infected larvae had been macerated;
   others were infected with a virus suspension.   In both cases, only 4th-instars showed
   disease symptoms, which appeared after 10 days for those injected and 20 days for those
   in the suspension.   Infected larvae died within 2 weeks of appearance of symptoms.

Hazard, E. I. (1972).   Personal communication.

Kellen, W. R. et al. (1963).   A possible polyhedrosis in _Culex tarsalis_ Coquillet (Diptera:
Culicidae).   _J. Insect Pathol._, _5_: 98-103.

   Tetragonal intranuclear inclusion bodies have been observed in larvae collected from
   stagnant ponds in California.   Infected larvae succumb in the 4th-instar, and successful transmissions of the disease agent have been obtained in the laboratory.   It was
   concluded that an infectious agent is definitely involved, and that it might be a virus.

Kellen, W. R. et al. (1966).   A cytoplasmic-polyhedrosis virus of _Culex tarsalis_ (Diptera:
Culicidae).   _J. Invertebr. Pathol._, _8_: 390-394.

   The tetragonal crystals described by Kellen, Clark & Lindegren (1963) have been confirmed
   as inclusion bodies of a cytoplasmic-polyhedrosis virus* upon further examination.
   Polyhedra were present in hypodermal cells, in the developing leg, wing, and antennal
   buds.   This is the first cytoplasmic polyhedrosis reported which is not restricted to
   gut tissue.

Kelly, D. C. & Robertson, J. S. (1973).   Icosahedral cytoplasmic deoxyriboviruses.
J. Gen. Virol., _20_(Suppl.): 17-41.

   A comprehensive catalogue of icosahedral cytoplasmic deoxyriboviruses (ICDV) isolated
   from animals and plants is given.   The list includes 7 from mosquitos.

Lebedeva, O. P. et al. (1973).   Investigation of a virus disease of the densonucleosis type
in a laboratory culture of _Aedes aegypti_.   _Acta Virol._, _17_: 253-256.

   Histological changes in infected larvae were observed by light and electron microscopy.
   The most obvious pathological changes were in the cells of the fat body.   The virus
   particles were about 200 Å in size, having a paraspherical shape and polygonal outlines.
   Observations revealed a similarity between the viral infection found in _Aedes aegypti_
   larvae and densonucleosis of _Galleria mellonella_.

---

*   See footnote to Chapman et al., 1970.

Lebedeva, O. P. & Zelenko, A. P. (1972). Virus-like formations in larvae of <u>Aedes</u> and <u>Culex</u> mosquitoes. <u>Med. Parazitol. Parazit. Bolenzi.</u>, <u>41</u>: 490-492 (Russian, with English summary).

Inclusion bodies in the cytoplasm of fat body cells were found in 4th-instar larvae and pupae of <u>Aedes aegypti</u> and <u>Culex pipiens molestus</u> from laboratory colonies and in some <u>Culex</u> sp. from natural water bodies. The inclusions resembled in size, shape, and staining properties those of Entomopoxvirus described from other insects. Also, the diseased mosquitos exhibited the hypertrophied nuclei structurally similar to those described in densonucleosis of <u>Galleria mellonella</u>. (See Vago, 1963, and Vago et al., 1964.)

Linley, J. R. & Nielsen, H. T. (1968a). Transmission of a mosquito iridescent virus in <u>Aedes taeniorhynchus</u>. I. Laboratory experiments. <u>J. Invertebr. Pathol.</u>, <u>12</u>: 7-16.

The virus can be effectively transmitted to the larvae <u>per os</u>; transovarial transmission occurs from an infected female to her eggs, and evidence was obtained that all, or a very high proportion, of the eggs are infected. Larvae infected <u>per os</u> in their early instars may develop signs and symptoms of disease, and die before pupation. Exposure to infection progressively later in their development results in later development of signs and symptoms, and a reduction in the total number of larvae that show disease before pupation. Rearing larvae under different conditions of temperature, diet, and crowding produced no detectable differences in infection rates among either groups of larvae exposed from the time of hatching to the same dosage of virus, or groups of larvae hatched from a large batch of eggs in which a proportion carrying infection had been homogeneously distributed.

Linley, J. R. & Nielsen, H. T. (1968b). Transmission of a mosquito iridescent virus in <u>Aedes taeniorhynchus</u>. II. Experiments related to transmission in nature. <u>J. Invertebr. Pathol.</u>, <u>12</u>: 17-24.

A field experiment showed that infection could be acquired <u>per os</u> by larvae exposed under natural conditions, and that when healthy 4th-instar larvae were given access to intact diseased cadavers for a short time before pupation, they became infected and produced adults that laid infected eggs. Transovarial transmission of virus gives rise to diseased larvae, which die in the 4th-instar. The cadavers so formed provide the source of new infection, which is acquired <u>per os</u> by healthy larvae just before pupation. Infected adults from these larvae complete the cycle by depositing infected eggs.

Lowe, R. E. et al. (1970). Comparison of the mosquito iridescent virus (MIV) with other iridescent viruses. <u>Proc. 4th Int. Colloq. Insect Pathol.</u>, pp. 163-170.

The R isolate of the mosquito iridescent virus, although closely related to other insect iridescent viruses, possesses distinct physico-chemical characteristics. The diameter of RMIV (195 nm), molecular weight (2.486×10^9 daltons), sedimentation coefficient (4.458), and DNA content (15.97%) are greater than the respective values for similar viruses. Although biologically similar, and showing no unique antigens serologically, the two colour isolates of MIV differ in size, sedimentation rate, and density.

Matta, J. F. (1970). The characterization of a mosquito iridescent virus. II. Physico-chemical characterization. <u>J. Invertebr. Pathol.</u>, <u>16</u>: 157-164.

A procedure for the purification of the R type of MIV is presented. The RMIV was distinctly different from the other iridescent viruses in terms of several physical parameters, and it would be difficult to justify considering it as a strain of these viruses. There was, however, a similarity in amino acid composition, indicating that they have a common phylogeny.

Matta, J. F. & Lowe, R. E. (1969). A differential staining technique for a mosquito iridescent virus. J. Invertebr. Pathol., 13: 457-458.

A staining technique, that allows routine screening for MIV infections using a light microscope, was devised.

Matta, J. F. & Lowe, R. E. (1970). The characterization of a mosquito iridescent virus (MIV). I. Biological characteristics, infectivity, and pathology. J. Invertebr. Pathol., 16: 38-41.

Infected Aedes taeniorhynchus were mass-produced by rearing them in dishes containing homogenized infected 4th-instar larvae. The average mortality was dependent on crowding of the larvae during infection. The increase in virus over the inoculum varied between 70 and 310%. The RMIV infects only the fat body and imaginal discs in A. taeniorhynchus and destruction of these tissues is usually complete.

Richardson, J. et al. (1974). Evidence of two inapparent nonoccluded viral infections of Culex tarsalis. J. Invertebr. Pathol., 23: 213-224.

C. tarsalis from a laboratory colony in California were found to be infected with 2 noninclusion cytoplasmic viruses. Both viruses were seen in thin sectioned material from salivary glands, fat body, and nervous tissue of infected mosquitos. C. tarsalis virus 1 (CTV 1) was retained through 6 serial passages by needle inoculations of infected mosquito suspensions into virus-free adults. Both CTV1 and CTV2 appeared to multiply in the hosts. Infection may cause degeneration of salivary glands in adult females.

Service, M. W. (1968). The ecology of the immature stages of Aedes detritus (Diptera: Culicidae). J. Appl. Ecol., 5: 613-630.

Large numbers of larvae collected on Brownsea Island were infected by various pathogens including non-inclusion iridescent viruses.

Stoltz, D. B. (1971). The structure of icosahedral cytoplasmic deoxyriboviruses. J. Ultrastruct. Res., 37: 219-239.

The presence of two unit membranes in mosquito iridescent virus particles is clearly illustrated. The unit membranes in this study are defined on the basis of morphology alone. (See Stoltz, 1973.)

Stoltz, D. B. (1973). The structure of icosahedral cytoplasmic deoxyriboviruses. II. An alternative model. J. Ultrastruct. Res., 43: 58-74.

The available evidence seems to indicate that ICDV particles contain only a single structural unit membrane, associated with the viral nucleoid.

Stoltz, D. B. et al. (1974). Virus-like particles in the mosquito Culex salinarius. J. Microscopie, 19: 109-112.

Larvae of C. salinarius collected in Louisiana were found to have identical symptoms as those described in Culex tarsalis by Kellen, Clark & Lindegren (1966). The causative agent is readily transmissible to C. tarsalis. Although electron microscopic examination revealed virus-like particles, the symptoms are not those expected from a typical cytoplasmic polyhedrosis virus.

Stoltz, D. B. & Summer, M. D. (1971). Pathway of infection of mosquito iridescent virus.
I. Preliminary observations on the fate of ingested virus. J. Virology, 8: 900-909.

MIV is ingested in large amounts by 1st- and 2nd-instar Aedes taeniorhynchus larvae
without causing a high rate of infection. Preliminary observations suggest that most,
if not all, ingested particles are degraded shortly after entering the midgut. MIV and
other virus particles were apparently unable to penetrate the peritrophic membrane; consequently none was observed inside, or in contact with, midgut epithelial cells.

Tinsley, T. W. & Kelly, D. C. (1970). An interim nomenclature system for the iridescent
group of insect viruses. J. Invertebr. Pathol., 16: 470-472.

In the new system each iridescent virus is given a type number so additional host citing
would pose no problem. This system is suggested to be employed until a system of more
permanent names can be assigned based on comparisons of physical, chemical, and biological
properties of the various isolates.

Tinsley, T. W. et al. (1971). An iridescent virus of Aedes cantans in Great Britain.
J. Invertebr. Pathol., 18: 427-428.

Larvae of A. cantans were collected from a pond in Kent, England, and found to be
infected with a virus of the iridescent group. They developed a lime-green colour at an
advanced stage of infection.

Vago, C. et al. (1969). Infection spontanee a virus irisant dans une population d'Aedes
detritus (Hal., 1883) des environs de Tunis. Ann. Parasit. Hum. Comp., 44: 667-676
(English summary).

A virus disease was observed in natural populations of A. detritus in Tunisia. Diseased
larvae could be identified by their slower movements and milky white coloration.
Extensive damage to the fat bodies was observed. The purified virus has a cubical
symmetry and measured 180 nm in diameter. It appears to be related to the iridescent
viruses by its morphology, its location, and the iridescence of affected tissue.

Wagner, G. W. et al. (1973). Biochemical and biophysical properties of two strains of
mosquito iridescent virus. Virology, 52: 72-80.

Several differing biophysical properties of the two strains (RMIV and TMIV) were observed
due to size differences. The two strains appeared to be antigenically identical but
the percent protein, DNA, and lipid of the two were dissimilar. Other than size
variation, no morphological differences between RMIV and TMIV could be detected.

Weiser, J. (1965). A new virus infection of mosquito larvae. Bull. World Health Organ.,
33(2): 586-588.

Diseased Aedes annulipes and A. cantans were found in South-western Bohemia. Infected
individuals are opaque and opalescent, and are usually less motile. In their normal
habitat, they appear to die before pupation. Nearly every tissue seems to be affected.
Infection is not common, not more than 1% of the larvae are infected. In general
appearance, the virus is similar to the Tipula iridescent virus.

Wildy, P. (1971). Classification and nomenclature of viruses. Monogr. Virol., 5: 1-81.

Woodard, D. B. & Chapman, H. C. (1968).   Laboratory studies with mosquito iridescent virus (MIV).   J. Invertebr. Pathol., 11: 296-301.

RMIV and a blue MIV of Aedes taeniorhynchus were serially passed through 68 and 30 generation, respectively, of larvae of A. taeniorhynchus.   The average rate of infection for RMIV was 16% and 21% for the blue MIV.   The maximum rate of infection was usually reached when large numbers of early-instar larvae were exposed for 48 hours to substantial amounts of MIV material and the larvae were reared to the 4th stadium at 25°C. Attempts to transmit MIV to larvae of 9 species of mosquitos not known to be hosts were successful only with A. sollicitans.   About 20% patent infection occurred in the larval progeny of adult A. taeniorhynchus derived from early 4th-instar larvae exposed to MIV. Such transovarial transmission was observed in the MIV of Psorophora ferox.

III. BACTERIAL PATHOGENS OF CULICIDAE (MOSQUITOS) [a]

Samuel Singer

Department of Biological Sciences
Western Illinois University
Macomb, IL 61455, USA

[a] The literature for this table and bibliography covers the period 1962–1975.

BACTERIAL PATHOGENS OF CULICIDAE (MOSQUITOS)

Host	Host stage infected	Pathogen	LD50 or percent incidence	Locality	Lab. or field study	Reference
Aedes aegypti	Larvae	Bacillus alvei-circulans-brevis	10^7 cells/ml	USA	Lab.	Singer (1974)
"	Larvae	B. sphaericus	-	India	Field	Singer (1975)
"	"	"	10^6 cells/ml	USA	Lab.	Singer (1974, 1975)
"	"	"	-	USA	Lab.	Kellen et al. (1965)
"	"	"	550 ppm dried spores	USA	Lab.	Kellen & Myers (1964)
"	Larvae	B. thuringiensis	10^7 cells/ml	USA	Lab.	Singer (1974)
"	"	" (toxin)	-	USA	Lab.	Davidson & Singer (1973)
"	"	" (bakthane-L-69)	≤1.6 mg/ml	Canada	Lab.	Shaikh & Morrison (1966)
"	"	" (BA 068)	10^7 spores/ml	USA	Lab.	Reeves (1970)
"	Adults	Corynebacterium sp.	-	USA (lab. colony)	Lab.	Micks & Ferguson (1973)
"	Larvae	Mycobacterium sp.	-	USSR	Lab.	Mikhnovka et al. (1972)
"	Larvae	Pseudomonas sp.	-	USSR	Lab.	Mikhnovka et al. (1972)
"	Larvae	Spore-former	-	USSR	Lab.	Mikhnovka et al. (1972)
"	Larvae	Vibrio sp.	-	USA (Louisiana)	Field	Chapman et al. (1967)
"	Larvae	Spore forming bacillus	-	India	Field	Briggs (1972)
"	Larvae	Long, rod shaped bacterium	-	"	"	Briggs (1973b)
Aedes atropalpus	Larvae	Bacillus sphaericus	-	USA	Lab.	Kellen et al. (1965)

48

BACTERIAL PATHOGENS OF CULICIDAE (MOSQUITOS) (continued)

Host	Host stage infected	Pathogen	LD$_{50}$ or percent, incidence	Locality	Lab. or field study	Reference
Aedes dorsalis	Larvae	Bacillus thuringiensis	-	USA	Lab.	Singer (1975)
"	"	"	-	USA	Field	Briggs (1973b)
Aedes freeborni	Adults	Flavobacterium near lutescens	-	USA (Maryland)	Field	Steinhaus & Marsh (1962)
Aedes fulvus pallens	Larvae	Vibrio sp.	Rare	USA (Louisiana)	Field	Chapman et al. (1967)
Aedes nigromaculus	"	Bacillus thuringiensis (BA 068)	10^7 spores/ml	USA	Lab.	Reeves (1970)
Aedes sierrensis	Larvae	Bacillus sphaericus	-	USA	Lab.	Kellen et al. (1965)
"	"	"	500 ppm dried spores	USA	Lab.	Kellen & Myers (1964)
"	"	Bacillus thuringiensis (BA 068)	10^7 spores/ml	USA	Lab.	Reeves (1970)
Aedes sollicitans	Larvae	Vibrio sp.	Rare	USA (Louisiana)	Field	Chapman et al. (1969)
Aedes squaniger	Larvae	Bacillus sphaericus	-	USA	Lab.	Kellen et al. (1965)
Aedes stimulans	Larvae	Bacillus thuringiensis (Bakthane L-69)	⌐1.66 mg/ml	Canada	Lab.	Skaikh & Morrison (1966)
Aedes taeniorhynchus	"	Vibrio sp.	Rare	USA (Louisiana)	Field	Chapman et al. (1967)
Aedes triseriatus	"	Bacillus sphaericus	-	USA	Lab.	Kellen et al. (1965)
"	"	Bacillus thuringiensis (BA 068)	10^7 spores/ml	USA	Lab.	Reeves (1970)

BACTERIAL PATHOGENS OF CULICIDAE (MOSQUITOS) (continued)

Host	Host stage infected	Pathogen	LD50 or percent. incidence	Locality	Lab. or field study	Reference
Aedes triseriatus	Larvae	Vibrio sp.	Rare	USA (Louisiana)	Field	Chapman et al. (1967)
"	"	Aerobic spore former	-	USA	Field	Briggs (1966)
Aedes vexans	-	Pseudomonas aeruginosa	-	USSR	Field	Povazhna (1974)
Aedes spp.	Larvae	Bacillus thuringiensis (BA 068)	10^4–10^7 spores/ml	USA	Lab.	Reeves & Garcia (1971a, b)
Anopheles annulipes	Larvae	Aeromonas punctata	<1	Australia	Lab.	Kalucy & Daniel (1972)
Anopheles barberi	Larvae	Vibrio sp.	Rare	USA (Louisiana)	Field	Chapman et al. (1969)
Anopheles bradleyi	Larvae	Vibrio sp.	Rare	USA (Louisiana)	Field	Chapman et al. (1969)
Anopheles gambiae	Larvae	Spore-former	-	Tanzania	Field	Briggs (1972)
"	Larvae, pupae	Bacteria (?)	-	Uganda	Field	Briggs (1967)
"	Adults	Non-spore forming short rods	-	England	Field	Briggs (1973b)
Anopheles punctipennis	Larvae	Serratia near marcescens	-	USA	Field	Briggs (1974)
Anopheles spp.	Larvae	Bacillus sphaericus (SSII-1)	-	Nigeria	Lab.	Singer (1975)
"	"	Bacillus thuringiensis	10^6 cells/ml	USSR	Field & lab.	Lavrent'yev et al. (1965)
Culiseta incidens	Larvae	Bacillus sphaericus	-	USA (California)	Field & lab.	Reeves (1970)

BACTERIAL PATHOGENS OF CULICIDAE (MOSQUITOS) (continued)

Host	Host stage infected	Pathogen	LD$_{50}$ or percent. incidence	Locality	Lab. or field study	Reference
Culiseta incidens	Larvae	Bacillus sphaericus	500 ppm dried spores	USA (California)	Lab.	Kellen & Myers (1964)
Culiseta inornata	Larvae	Vibrio sp.	Rare	USA (Louisiana)	Field	Chapman et al. (1967)
Culex annuterostria	Larvae	Bacillus sp.	-	Australia	Field	Singer (1975)
"	"	Spore former	-	"	"	Briggs (1974)
Culex erythrothorax	"	Bacillus sphaericus	0.5-3.3 X 10^7 spores/ml	USA	Lab.	Kellen et al. (1965)
"	"	"	500 ppm dried spores	USA	Lab.	Kellen & Myers (1964)
Culex p. fatigans	Larvae	Bacillus sphaericus var. fusiformis	10^6 spores/ml	USA	Lab.	Rogoff et al. (1969)
"	"	Bacillus thuringiensis var. thuringiensis	-	USA	Lab.	Rogoff et al. (1969)
"	"	Bacillus sp.	-	Burma	Field	Singer (1975)
"	"	"	-	India	"	"
"	"	"	-	Korea	"	"
"	Larvae, pupae	Spore-former	-	Tanzania	"	Briggs (1973b)
Culex modestus	Larvae	Bacillus thuringiensis (entobacterin, exotoxin)	-	USSR	Lab.	Saubenova et al. (1973)
Culex pipiens molestus	Larvae	Bacillus thuringiensis (exotoxin)	100 mortality 1:1 dilution	USSR	Lab.	Gurgenidze (1970)

BACTERIAL PATHOGENS OF CULICIDAE (MOSQUITOS) (continued)

Host	Host stage infected	Pathogen	LD50 or percent. incidence	Locality	Lab. or field study	Reference
Culex pipiens molestus	Larvae	Bacillus thuringiensis (exotoxin)	-	USSR	Lab.	Yarnykh & Tonhozhenko (1969)
"	"	Pseudomonas sp.	-	USSR	Lab.	Mikhnovka et al. (1972)
"	"	Mycobacterium sp.	-	"	"	"
"	"	Spore-former	-	"	"	"
Culex partecips	Larvae	Bacillus sphaericus	500 ppm dried spores	USA	Lab.	Kellen & Myers (1964)
Culex peus	Larvae	Bacillus sphaericus	"	"	"	"
Culex pipiens	Larvae	Bacillus cereus	-	USSR	Lab.	Zharov (1969)
"	"	Bacillus thuringiensis dendrolimus	-	"	"	"
"	"	Bacillus sphaericus var. fusiformis	10^2-10^4 cells/ml	USA	Lab.	Goldberg et al. (1974)
"	"	Bacillus sphaericus	0.5-3.3 X 10^7 spores/ml	USA	Lab.	Kellen et al. (1965)
"	"	Bacillus thuringiensis	-	USSR	Lab.	Zharov (1969)
"	"	Bacillus sp.	-	India	Field	Briggs (1964)
"	"	Pseudomonas aeruginosa	-	USSR	Field	Povazhna (1974)
"	"	Entero-bacteriaceae	<5	Burma	Field	Muspratt (1964)
Culex p. quinquefasciatus	Larvae	Bacillus alvei-circulans-brevis	10^6 cells/ml	USA	Lab.	Singer (1973, 1974, 1975)

BACTERIAL PATHOGENS OF CULICIDAE (MOSQUITOS) (continued)

Host	Host stage infected	Pathogen	LD$_{50}$ or percent, incidence	Locality	Lab. or field study	Reference
Culex p. quinquefasciatus	Larvae	Bacillus cereus var. mycoides	-	USA (Hawaii)	Field	Thomas & Poinar (1973)
"	"	Bacillus sphaericus (SS-1)	10^2 cells/ml	USA	Lab.	Singer (1975)
"	"	Bacillus sphaericus (SSII-1)	10^2-10^3	USA	Lab.	Davidson & Singer (1973), Singer (1973, 1974a)
"	"	Bacillus thuringiensis	10^7 cells/ml	"	"	Singer (1974)
"	"	Bacillus spp.	-	Philippines, Indonesia, Thailand	Field	Singer (1975)
"	"	Bacillus spp.	-	USA (Mississippi)	Field	Fulton et al. (1974)
"	"	Pseudomonas aeruginosa	-	USA (Hawaii)	Field	Thomas & Poinar (1973)
"	"	Vibrio sp.	-	USA (Mississippi)	Field	Fulton et al. (1974)
"	"	"	Rare (1-2)	USA (Louisiana)	"	Chapman et al. (1967, 1969)
Culex restuans	Larvae	Bacillus sp.	-	USA (Mississippi)	Field	Fulton et al. (1974)
"	"	Vibrio sp.	Rare	USA (Louisiana)	"	Chapman et al. (1967)
Culex salinarius	Larvae	Bacillus sp.	-	USA (Mississippi)	"	Fulton et al. (1974)
"	"	Vibrio sp.	-	USA (Louisiana)	"	Chapman et al. (1967, 1969)
Culex tarsalis	"	Bacillus sphaericus	10^2-10^4 cells/ml	USA	Lab.	Goldberg et al. (1974)

BACTERIAL PATHOGENS OF CULICIDAE (MOSQUITOS) (continued)

Host	Host stage infected	Pathogen	LD50 or percent. incidence	Locality	Lab. or field study	Reference
Culex tarsalis	Larvae	Bacillus sphaericus	500 ppm dried spores	USA	Lab.	Kellen & Myers (1964)
"	"	Bacillus thuringiensis (toxin)	-	USA	Lab.	Ahmad et al. (1971)
"	"	Bacillus thuringiensis (HD-1)	-	"	"	Goldberg et al. (1974)
Culex territans	Larvae	Vibrio sp.	Rare	USA (Louisiana)	Field	Chapman et al. (1967)
Culex sp.	"	Bacillus sphaericus (SSII-1)	-	Nigeria	Lab.	Singer (1975)
Culex spp.	"	Bacillus thuringiensis (BA 068)	10^4-10^7 spores/ml	USA	Lab.	Reeves & Garcia (1971a, b)
"	"	Bacillus thuringiensis	10^6 cells/ml	USSR	Field & lab.	Lavrent'yev et al. (1965)
Mosquito?	Larvae	Clostridium sp.	-	USA (California)	Field	Thomas & Poinar (1973)
"	"	Pseudomonas aeruginosa	-	"	"	"

BACTERIAL GUT FLORA OF CULICIDAE (MOSQUITOS)

Host	Stage	Pathogen	Percent. incidence	Locality	Lab. or field study	Reference
Aedes aegypti	Larvae	Alcaligenes viscolactis[a]	-	Pakistan	Field	Shafti et al. (1966)
"	"	Proteus mirabilis	-	"	"	"
"	"	Flavobacterium arborescens[b]	-	"	"	Zuberi et al. (1969)
"	Larvae, pupae	Bacillus cereus	-	"	"	"
"	Pupae	Bacillus megaterium	-	"	"	"
"	"	Brevibacterium tegumenticola[b]	-	"	"	"
"	"	Proteus inconstans	-	"	"	Zuberi et al. (1969)
"	Adults	Alcaligenes (Achromobacter) superficiales[a]	-	"	"	Shafti et al. (1966)
"	"	Alcaligenes (Achromobacter) sp.	-	"	"	"
"	"	Klebsiella pneumoniae (=Aerobacter aerogens)	-	"	"	Zuberi et al. (1969)
"	"	Enterobacter (Aerobacter) cloacae	-	"	"	"
"	"	Alcaligenes bookeri[a]	-	"	"	Shafti et al. (1966)
"	"	Alcaligenes faecalis	-	"	"	Zuberi et al. (1969)
"	"	Escherichia coli	-	"	"	"
"	"	Flavobacterium dormitator[b]	-	"	"	Shafti et al. (1966)
"	"	Proteus rettgeri	-	"	"	"
"	"	Salmonella paratyphi	-	"	"	Zuberi et al. (1969)
Anopheles annularius	"	Alcaligenes bookeri[a]	-	"	Field	Shafti et al. (1966)

BACTERIAL GUT FLORA OF CULICIDAE (MOSQUITOS) (continued)

Host	Stage	Pathogen	Percent. incidence	Locality	Lab. or field study	Reference
Anopheles annularius	Adults	Corynebacterium sp.	-	Pakistan	Field	Shafti et al. (1966)
"	"	Citrobacter intermedius (=Escherichia intermedia)	-	"	"	"
"	"	Paracolobactrum sp.[a]	-	"	"	"
"	"	Staphylococcus epidermidis	-	"	"	"
Anopheles quadrimaculatus	"	Aerobacter sp.[a]	-	USA (lab. colony)	Lab.	Micks & Ferguson (1963)
"	"	Pseudomonas sp.	-	"	"	"
"	"	Streptococcus sp.	-	"	"	"
Culex fatigans	Larvae	Achromobacter liquefaciens[a]	-	Pakistan	Field	Shafti et al. (1966)
"	"	Pseudomonas putida	-	"	"	"
"	"	Pseudomonas reptilivora[b]	-	"	"	"
"	Adults	Alcaligenes bookeri[a]	-	"	"	"
"	"	Alcaligenes faecalis	-	"	"	"
"	"	Alcaligenes sp.	-	USA	Lab.	Ferguson & Micks (1961)
"	"	Paracolobactrum aerogenoides[a]	-	Pakistan	Field	Shafti et al. (1966)
"	"	Pseudomonas myxogenes[b]	-	"	"	"
"	"	Pseudomonas sp.	-	USA	Lab.	Ferguson & Micks (1961)
"	"	Salmonella typhimurium	-	Pakistan	Field	Shafti et al. (1966)
"	"	2 unidentified gram-negative rods	-	USA	Lab.	Ferguson & Micks (1961)

[a] Not a recognized taxon.
[b] Species incertae sedis.

ABSTRACTS

Mary Ann Strand

Ahmad, R. et al. (1971).   Toxin production by Bacillus thuringiensis in defined media.
Bacteriol. Proc., 71: 2.

    B. thuringiensis grown on basal salts media with casamino acids and thiamine or with
valine and citrate produced toxins which caused 100% mortality in Culex tarsalis and
Aedes aegypti larvae in 6 days at dosages of 0.016 mg/ml.   Other media tested resulted
in less potent toxin production.

Briggs, J. D. (1966).   1965 activities of the WHO International Reference Centre for
diagnosis of diseases of vectors.   Unpublished document WHO/EBL/66.73, 8 pp.

Briggs, J. D. (1967).   1966 activities of the WHO International Reference Centre for
diagnosis of diseases of vectors.   Unpublished document WHO/VBC/67.8, 8 pp.

Briggs, J. D. (1969).   1968 activities of the WHO International Reference Centre for
diagnosis of diseases of vectors.   Unpublished document WHO/VBC/69.171, 6 pp.

Briggs, J. D. (1974).   1973 activities of the WHO International Reference Centre for
diagnosis of diseases of vectors.   WHO mimeographed document, 19 pp.

    Tentative determinations of pathogens including bacterial diseases of mosquitos are
given.

Chao, J. et al. (1963).   Failure to isolate microorganisms from within mosquito eggs.
Ann. Entomol. Soc. Am., 56: 559-561.

    No microorganisms were isolated from mosquito eggs when they were completely surface
sterilized.

Chapman, H. C. et al. (1969).   A two-year survey of pathogens and parasites of Culicidae,
Chaoboridae, and Ceratopogonidae in Louisiana.   Proc. N. J. Mosq. Exterm. Assoc., 56:
203-212.

    Vibrio sp. was found infecting 6 species of mosquitos.   The highest incidence occurred
in Culex pipiens quinquefasciatus.

Chapman, H. C. et al. (1967).   Pathogens and parasites in Louisiana Culicidae and Chaoboridae.
Proc. N. J. Mosq. Exterm. Assoc., 54: 54-60.

    The haemocoels of 13 species of mosquito larvae were invaded by Vibrio sp.   Patently
infected larvae become opaquish white and die before pupating.   Infection incidence was
low, at most 1-2%.

Davidson, E. W. & Singer, S. (1973).   Pathogenesis of Bacillus sphaericus infections in
mosquito larvae.   (Abstr.) Pap. 5th Int. Colloq. Insect Pathol. Microb. Control (Oxford),
p. 5.

    Insecticidal activity of the bacterium is apparently due to toxic material closely
associated with the bacterial cell at the outset of sporulation.   Nonpathogenic bacteria
were digested rapidly by larvae, while viable B. sphaericus cells multiplied in the gut.
After treatment which killed the bacterial cells but not the spores, the insecticidal
activity was reduced.

Davidson, E. W. et al. (1975). Pathogenesis of Bacillus sphaericus strain SSII-1 infection in Culex pipiens quinquefasciatus (= C. pipiens fatigans) larvae. J. Invertebr. Pathol., 25: 179-184.

> Single doses of B. sphaericus were fed to starved larvae and the fate of the ingested cells was observed. Cells of a non-pathogenic strain were consumed, but their numbers rapidly decreased due mainly to digestion. The number of pathogenic SSII-1 cells also declined after ingestion when fed at low dosages (LD_{20}). At higher dosages (LD_{70}) the bacilli multiplied in the larval gut. Multiplication was not necessary for pathogenicity, however.

Dyl'ko, M. I. (1971). /Testing suitability of entobacterin and boverin for biological control of mosquito larvae./ Vesti Akad. Navuk Belarusk. Ser. Biyalag. Navuk, 4: 85-89. (Br).

Ferguson, M. J. & Micks, D. W. (1961). Microorganisms associated with mosquitoes: I. Bacteria isolated from the mid-gut of adult Culex fatigans Wiedemann. J. Insect Pathol., 3: 112-119.

> Lactobacillus sp., Alcaligenes sp., Pseudomonas sp., and 2 unidentified species of gram-negative rods were isolated from the mid-gut of adult female Culex fatigans. No attempts were made to cultivate obligate anaerobes.

Fulton, H. R. et al. (1974). A survey of north Mississippi mosquitoes for pathogenic micro-organisms. Mosq. News, 34: 86-90.

> Bacillus sp., Vibrio sp., and Spirillum sp. were isolated from mosquito larvae during this survey.

Goldberg, L. J. & Ford, I. (1973). Aquatic control of Culex tarsalis mosquito larvae using combinations of Bacillus thuringiensis (HD-1) with two selected growth regulators, Altosid SR-10 (Zoecon) and Monsanto 585. (Abstr.) Pap. 5th Int. Colloq. Insect Pathol. Microb. Control (Oxford), p. 46.

> The bacteria showed effective larvicidal activity when applied to early larval instars (10^6 spores/ml). Combinations of the bacillus and the growth regulators demonstrated independent (simple additive) mode of action.

Goldberg, L. J. et al. (1974). Bacillus sphaericus var. fusiformis as a potential pathogen against Culex tarsalis and Culex pipiens. Proc. Calif. Mosq. Control Assoc., 42: 81-82.

> Dosages required for 90% mortality of 2nd and 4th instar larvae were found for B. sphaericus cultures grown under different conditions. These ED_{90} values ranged from 5×10^3 ml to 1.9×10^4 ml.

Gurgenidze, T. V. (1970). /The effect of thermostable exotoxin of Bacillus thuringiensis on mosquitos Culex pipiens molestus./ Tr. Vses. Nauchno-Issled. Inst. Vet. Sanit., 37: 50-55. (R, e).

> The exotoxin had no effect on adult mosquitos, but some mortality in larval stages was recorded. In most cases, 100% mortality was found at 1:1 dilution of the exotoxin and none at 1:100 dilution.

Kalucy, E. C. & Daniel, A. (1972).   The reaction of Anopheles annulipes larvae to infection of Aeromonas punctata.   J. Invertebr. Pathol., 19(2): 189-197.

   The bacterium was isolated from the haemolymph of mosquito larvae with internal black lesions.   Less than 1% of the colony was affected.   The lesions were formed by deposition of black pigment around invading bacteria.   The bacteria were not seriously pathogenic to healthy mosquitos.

Kellen, W. R. et al. (1965).   Bacillus sphaericus Neide as a pathogen of mosquitoes. J. Invertebr. Pathol., 7: 442-448.

   The bacterium was isolated from Culiseta incidens larvae found near Fresno, California. It invades the larvae via the alimentary canal and the infection is sometimes fatal. Infected larvae lose their normal motility and eventually turn dark brown.   Field tests to determine whether epizootics could be started in natural larval populations showed, in most cases, no apparent population reduction.   This was probably due to the settling of the inoculum to the bottom of the pool, so it was not available to surface feeding larvae.

Kellen, W. R. & Myers, C. H. (1964).   Bacillus sphaericus Neide as a pathogen of mosquitoes. Proc. Calif. Mosq. Control Assoc., 32: 37.

Krieg, A. (1967).   Neues über Bacillus thuringiensis und seine Anwendung.   Mitt. Biol. Bundesanst. Land-Forstwirsch. Berl.-Dahlem., 125: 82-83.

   A brief review of the use of the bacterium and its toxins against mosquitos is given.

Lavrent'ev, P. A. et al. (1965).   /The use of bacteria for mosquito control./   Veterinariya, 42(8): 107-108.   (R).

   Bacillus thuringiensis, B. t. dendrolimus, and B. cereus were tested for pathogenicity to Aedes, Culex, and Anopheles larvae.   Laboratory and field tests were made using various preparations of spores, vegetative cells, and crystals.   Good control of the mosquito population was maintained using spore preparations when the rearing water was above 24°.

Micks, D. W. & Ferguson, M. J. (1963).   Microorganisms associated with mosquitoes. IV. Bacteria isolated from the midgut of adult Culex molestus Forskal, Aedes aegypti (Linnaeus) and Anopheles quadrimaculatus Say.   J. Insect Pathol., 5: 483-488.

   Several species of bacteria were isolated from the mid-guts of female mosquitos.   The cultures were characterized by their biochemical and antibiotic sensitivity reactions.

Mikhnovs'ka, N. D. et al. (1972).   /Bacterial flora of mosquito larvae and its entopatho-genic properties./   Mikrobiol. Zh. (Kiev), 34(1): 121.   (U).

   Pseudomonas, Mycobacterium, and an unidentified spore forming bacterium were isolated from diseased mosquito larvae.   These isolates were pathogenic to Culex pipiens molestus and Aedes aegypti larvae in laboratory tests.

Muspratt, J. (1964).   Parasitology of larval mosquitos, especially Culex pipiens fatigans Wied., at Rangoon, Burma.   Unpublished document WHO/EBL/18, 19 pp.

   In a one month survey of mosquitos, two strains of Enterobacteriaceae were found infecting Culex pipiens.   The bacteria were isolated from larvae in polluted water. Large concentrations of these bacteria were observed in the haemolymph of infected larvae.   Patently infected individuals were distinguished by their whitish opacity.

Povazhna, T. M. (1974). /Bacteria of the genus Pseudomonas in insect pathology./
Mikrobiol. Zh. (Kiev), 35(6): 793-797. (U).

   Literature concerning bacterial infections of various insects including some of medical
   importance is reviewed. Fifty-one references mainly to Russian literature are given.
   P. aeruginosa is reported from Culex pipiens and Aedes vexans.

Rajak, R. L. & Perti, S. L. (1968). Toxicity of bacterial spores to fly and mosquito larvae.
Pesticides (Bombay), 1(9): 45.

   Bioassays for toxicity of spores of Bacillus cereus var. thuringiensis against third
   instar larvae of Culex fatigans showed that even under high spore concentrations
   (10^4 ppm), mortality of the larvae was not significant.

Reeves, E. L. (1970). Pathogens of mosquitoes. Proc. Calif. Mosq. Control Assoc., 38:
20-22.

   A preliminary report on a bicrystalliferous bacillus isolated from moribund Culex
   tarsalis larvae is given. More details are in Reeves & Garcia, 1971a and b.

Reeves, E. L. & Garcia, C. (1971a). Pathogenicity of bicrystalliferous Bacillus isolate for
Aedes aegypti and other aedine mosquito larvae. Proc. 4th Int. Colloq. Insect Pathol.
(College Park, Md.), pp. 219-228.

   B. thuringiensis (BA-068)* isolate proved to be pathogenic to Aedes (aegypti, triseriatus
   and negromaculis) larvae and to Culex (peus and tarsalis) larvae in laboratory tests.
   Aedes larvae were more susceptible to the isolate than the Culex larvae. Three small-
   scale field tests showed considerable mortality within 24 hours. The development of
   surviving larvae was retarded.

Reeves, E. L. & Garcia, C. (1971b). Susceptibility of Aedes mosquito larvae to certain
crystalliferous Bacillus pathogens. Proc. Calif. Mosq. Control Assoc., 39: 118-120.

   The activity of Bacillus thuringiensis var. thuringiensis Serotype H-1 is primarily
   contained in the protein crystals with no activity displayed by the spores alone. The
   proteinaceous crystals are non-toxic to the insects in their native form, but they can
   be hydrolysed and activated by certain factors found in the guts of susceptible insects.
   Although the toxic effect is contained in the crystals, the presence of spores reduces
   the dosage required for complete mortality.

Rogoff, M. H. et al. (1969). Insecticidal activity of thirty-one strains of Bacillus against
five insect species. J. Invertebr. Pathol., 14: 122-129.

   The insecticidal activity of various strains of Bacillus grown on δ-endotoxin-promoting
   medium and on β-exotoxin-promoting medium was tested against Culex pipiens fatigans by
   mixing Bacillus cultures with the insect rearing diet. Seven strains from the first
   medium and nine strains from the second medium produced mortality.

---

   * See Reeves & Garcia 1971b for serotype identification.

Saubenova, O. G. et al. (1973). /Tests on the effect of microbial preparations on larvae of mosquitos of the genus Culex in south-eastern Kazakhstan./ Parazitologiya, 7(3): 227-230. (R, e).

C. modestus and C. pipiens larvae were tested for susceptibility to 4 toxins. Entobacterin from Bacillus thuringiensis was the most effective toxin against C. modestus and caused 79.5% mortality in first and second instars at a 1% dose. Against C. pipiens, exotoxin from B. t. var. galleriae was the most effective and caused 86% mortality at 0.1% dosage. Dendrobacillin and boverin caused little mortality.

Shafti, R. et al. (1966). Studies on the bacterial isolates of the common mosquitoes of Pakistan. Zentralbl. Bakteriol. Parasitenkd. Infektionskr. Hyg. Abt. I Orig., 199: 514-521. (E, f, g, s).

Most of the isolates were gram-negative coccobacilli. All were apparently symbiotes of the mosquito.

Shaikh, M. U. & Morrison, F. O. (1966). Susceptibility of nine insect species to infection by Bacillus thuringiensis var. thuringiensis. J. Invertebr. Pathol., 8: 347-350.

Aedes stimulans and A. aegypti larvae were exposed to a commercial preparation, Bakthane L-69. No bacterial growth was observed on living or recently dead larvae. Although larval mortality was not high, only 20% and 2% of the adults emerged from treated A. stimulans and A. aegypti larvae, respectively, at 1.66 mg powder/ml.

Singer, S. (1973). Insecticidal activity of recent bacterial isolates and their toxins against mosquito larvae. Nature, 244: 110-111.

Bacteria isolated from infected mosquito larvae collected in India were tested for pathogenicity to Culex pipiens quinquefasciatus. Ninety-one isolates killed the mosquitos: 40 of them were identified as Bacillus alvei, B. circulans, or intermediate strains and 51 belonged to the B. sphaericus group. The insecticidal activity of B. alvei-circulans was found in the supernatant, while the activity of B. sphaericus seemed to be related to the cells themselves.

Singer, S. (1974a). Entomogenous bacilli against mosquito larvae. Dev. Ind. Microbiol., 15: 187-194

Three morphological groups in the genus Bacillus were examined for insecticidal activity against larvae of Culex pipiens and Aedes aegypti. All showed activity but to different degrees. The LD_{50} values and the nature of the pathogenic material are discussed.

Singer, S. (1975). Use of bacteria for control of aquatic insect pests, pp. 5-22. In Bourquin, A. W. et al. (ed.), Impact of the use of microorganisms on the aquatic environment. EPA Ecological Research Series 660-3-75-001, Corvallis, Oregon.

Two new groups of bacilli were isolated from mosquito larvae collected in Delhi, India. One group, B. sphaericus/SS11-1 has been found to be 10^3 times more active than previously examined strains. The second group is part of the complex which belongs to B. alvei - B. circulans - B. brevis and has also shown larvicidal activity. The worldwide distribution of these bacilli is discussed.

Singer, S. et al. (1966).   Defined media for the study of bacilli pathogenic to insects.
Ann. N. Y. Acad. Sci., 139: 16-23.

> Bacillus sphaericus could be grown readily on a variety of complex media.   They were
> based on the requirements of the bacillus for biotin and thiamin, inability to utilize
> the usual carbohydrate sources, and a preference for organic nitrogen sources.   Media
> for Bacillus thuringiensis were designed not to interfere with recovery of the toxins.
> A completely soluble medium containing casein hydrolysate as the only complex component
> was devised.

Steinhaus, E. S. & Marsh, G. A. (1962).   Report of diagnoses of diseased insects 1951-1961.
Hilgardia, 33: 349-490.

> The location, host, submission date, and individual requesting the diagnosis are given
> for each tentative diagnosis.   For details see the host-parasite table.

Thomas, G. M. & Poinar, G. O., jr (1973).   Report of diagnoses of diseased insects 1962-1972.
Hilgardia, 42: 261-360.

> The location, host, submission date, and individual requesting the diagnosis are given
> for each tentative diagnosis.   For details see the host-parasite table.

Wistreich, G. A. & Chao, J. (1963).   Microorganisms from the midgut of larval and adult
Aedes aegypti (Linnaeus).   J. Insect Pathol., 5: 56-60.

> All midgut samples from larval mosquitos and 9 out of 30 from adults contained
> microorganisms.   Body wall sections and other tissues were free of bacteria and fungi.
> No pathological conditions were reported.

Yarnykh, V. S. & Tonkonozhenko, A. P. (1969).   /Microbiological methods in veterinary insect
control./   Sel'skokhoz. Biol., 4(1): 98-103.   (R, e).

> Results are given of a 3 year study on the use of bacterial preparations against various
> insects including Culex pipiens molestus.

Zharov, V. G. (1969).   /The effect of some microorganisms pathological for insects on the
larvae and pupae of Culex pipiens./   Tr. Vses. Nauchno-Issled. Inst. Vet. Sanit., 32: 612-615.
(R).

> The susceptibility of C. pipiens to entobacterin-3, biotrol, and cultures of Bacillus
> thuringiensis, B. t. dendrolimus, B. alvei and B. cereus was tested.   None of the
> preparations was strongly pathogenic to the mosquitos and larval mortality in the tests
> ranged from 7 to 32%.

Zuberi, R. I. et al. (1969).   Bacterial and fungal isolates from laboratory-reared Aedes
aegypti (Linnaeus), Musca domestica (Linnaeus), and Periplaneta americana (Linnaeus).   Pak. J.
Sci. Ind. Res., 12: 77-82.

> Aerobic bacteria were isolated from surface sterilized and macerated eggs, larvae, pupae
> and adults of A. aegypti.   Only Pseudomonas aeruginosa was isolated from the eggs and
> with increasing life stage development there was an increase in the aerobic flora.   Many
> of the Bacillus spp. found in the mosquitos may enter with their food as saprophytic
> contaminants.

IV(a). MICROSPORIDAN PATHOGENS OF CULICIDAE (MOSQUITOS)

E. I. Hazard

USDA, Agricultural Research Service
Insects Affecting Man Research Laboratory
Gainesville, FL 32604, USA

H. C. Chapman

USDA, Agricultural Research Service
Gulf Coast Mosquito Research Laboratory
Lake Charles, LA 70601, USA

MICROSPORIDAN PATHOGENS OF CULICIDAE (MOSQUITOS)

Host	Pathogens	Stage	Tissue	Locality	Lab.-field incidence	References
Aedes abserratus	Amblyospora sp.	Larvae	Adipose	USA (Connecticut)	<1% Field	Anderson (1968)
				USA (Massachusetts)	Field	Hazard & Oldacre (1975)
Aedes aegypti	Nosema aedis	Larvae	Adipose	Puerto Rico	Field	Kudo (1930)
"	Nosema algerae	Larvae, Adults	Ventral nerve ganglia, mid-gut epithelium, malpighian tubules	England	Lab.	Canning & Hulls (1970)
				USA (Florida)	Lab.	Hazard & Lofgren (1971)
"	Nosema lutzi (may be an invalid sp.)	Adults	Intestine	Brazil	Field	Lutz & Splendore (1908)
"	Pleistophora culicis	Not given	Not given	England	Lab.	Garnham (1959)
"	Pleistophora stegomyiae	Larvae, Adults	Stomach, oesophagus, air sacs, haemocoel, ovaries, thoracic muscles, nerve ganglia of head, salivary glands, proboscis	Brazil	8% Field	Marchoux et al. (1903)
Aedes australis	Parathelohania barra	Larvae	Adipose	New Zealand	3.5% Field	Pillai (1968)
Aedes canadensis	Amblyospora canadensis	Larvae	Adipose	USA (California)	Field	Kellen & Wills (1962a)
				USA (Connecticut)	<1% Field	Anderson (1968)
				USA (Louisiana)	<1% Field	Chapman et al. (1966)
				USA (Massachusetts)	Field	Hazard & Oldacre (1975)
				USA (Maryland)	Field	Bailey et al. (1967a)
				USA (Pennsylvania)	<1% Field	Wills & Beaudoin (1965)
"	Pleistophora sp.	Larvae	Gastric caeca	USA (Louisiana)	<1% Field	Chapman et al. (1967)
Aedes cantans	Amblyospora sp.	Adult ♀	Oenocytes & ovaries	Czechoslovakia	3% Field	Hazard & Oldacre (1975)
				Germany	Field	Nöller (1920)
"	Thelohania barbata (nomen nudum)	Larvae	Adipose	Czechoslovakia		Weiser (1971)

MICROSPORIDAN PATHOGENS OF CULICIDAE (MOSQUITOS) (continued)

Host	Pathogen	Stage	Tissue	Locality	Lab.-field incidence	References
Aedes cantator	Amblyospora sp.	Larvae	Adipose	USA (Connecticut)	<1% Field	Anderson (1968)
				USA (Delaware)	Field	Chapman et al. (1973)
Aedes caspius	Amblyospora sp.	Larvae	Adipose	France	16% Field	Tour et al. (1971)
Aedes cataphylla	Amblyospora sp.	Larvae	Adipose	USA (California)	Field	Kellen et al. (1965)
				USA (Alaska)	Field	Chapman et al. (1973)
Aedes cinereus	Amblyospora sp.	Larvae	Adipose	USA (Nevada)	Field	Chapman (1966)
				USA (Connecticut)	<1% Field	Anderson (1968)
				USA (Massachusetts)	Field	Hazard & Oldacre (1975)
Aedes communis	Amblyospora khaliulini	Larvae	Adipose	Canada	3-11% Field	Welch (1960)
				Czechoslovakia	Field	Weiser (1946)
				Germany	Field	Nöller (1920)
				USA (Alaska)	Field	Chapman et al. (1973)
				USA (Massachusetts)	Field	Hazard & Oldacre (1975)
				USSR (Mari)	Field	Khaliulin & Ivanov (1971)
Aedes detritus	Amblyospora sp.	Larvae	Adipose	France	6.4% Field	Tour et al. (1971)
"	Stempellia magna (doubtful parasite of this host)	Larvae	Adipose	France	1% Field	Tour et al. (1971)
"	Stempellia tuzetae	Larvae	Adipose	France	Field	Tour et al. (1971)
Aedes dorsalis	Amblyospora sp.	Larvae	Adipose	USA (California)	Field	Kellen et al. (1965)
				USA (Nevada)	<1% Field	Chapman (1966)
				USA (Utah)	3% Field	Tsai et al. (1969)
Aedes excrucians	Amblyospora sp.	Larvae	Adipose	USA (Alaska)	Field	Chapman et al. (1973)
				USA (Connecticut)	<1% Field	Anderson (1968)
				USA (Massachusetts)	Field	Hazard & Oldacre (1975)
Aedes fitchii	Amblyospora sp.	Larvae	Adipose	USA (Alaska)	Field	Chapman et al. (1973)
Aedes grossbecki	Amblyospora sp.	Larvae	Adipose	USA (Louisiana)	<1% Field	Chapman et al. (1966)
Aedes hexodontus	Amblyospora sp.	Larvae	Adipose	USA (Alaska)	Field	Chapman et al. (1973)
				USA (California)	Field	Kellen et al. (1965)

MICROSPORIDAN PATHOGENS OF CULICIDAE (MOSQUITOS) (continued)

Host	Pathogen	Stage	Tissue	Locality	Lab.-field incidence	References
Aedes increpitus	Amblyospora sp.	Larvae	Adipose	USA (California)	Field	Kellen et al. (1965)
Aedes melanimon	Amblyospora unica	Larvae	Adipose	USA (California)	Field	Kellen & Wills (1962a)
				USA (Nevada)	Field	Chapman (1966)
Aedes notoscriptus	Pleistophora milesi	Larvae	Adipose	New Zealand	Field	Pillai (1974)
Aedes pullatus	Amblyospora sp.	Larvae	Adipose	USA (Alaska)	Field	Chapman et al. (1973)
Aedes punctor	Amblyospora sp.	Larvae	Adipose	USA (Alaska)	Field	Chapman et al. (1973)
				USA (Massachusetts)	Field	Hazard & Oldacre (1975)
Aedes riparius	Amblyospora sp.	Larvae	Adipose	USA (Alaska)	Field	Chapman et al. (1973)
Aedes sierrensis	Stempellia magna (doubtful parasite of this host)	Larvae, Adults	Adipose	USA (California)	27% Field	Clark & Fukuda (1967)
Aedes sollicitans	Amblyospora sp.	Larvae, Adult ♀	Adipose Oenocytes & ovaries	USA (Louisiana) USA (Florida)	<1% Field Field	Kellen et al. (1966a) Hazard & Oldacre (1975)
Aedes squamiger	Amblyospora bolinasae	Larvae	Adipose	USA (California)	Field	Kellen & Wills (1962a)
Aedes sticticus	Amblyospora sp.	Larvae	Adipose	USA (Louisiana)	<1% Field	Chapman et al. (1966)
Aedes stimulans	Amblyospora sp.	Larvae	Adipose	USA (New Jersey)	Field	Franz & Hagmann (1962)
				USA (Connecticut)	<1% Field	Anderson (1968)
Aedes taenio-rhynchus	Amblyospora sp.	Larvae	Adipose	USA (Florida) USA (Louisiana)	Field <1% Field	Hazard & Oldacre (1975) Kellen et al. (1966a)
Aedes thibaulti	Amblyospora sp.	Larvae	Adipose	USA (Louisiana)	<1% Field	Initial report
Aedes triseriatus	Pilosporella chapmani	Larvae	Adipose	USA (Louisiana)	Field	Hazard & Oldacre (1975)
"	Pleistophora culicis	Larvae	Gut	USA (Louisiana)	Field	Chapman (1974a)
Aedes ventro-vittis	Amblyospora sp.	Larvae	Adipose	USA (California)	Field	Kellen et al. (1965)

MICROSPORIDAN PATHOGENS OF CULICIDAE (MOSQUITOS) (continued)

Host	Pathogen	Stage	Tissue	Locality	Lab.-field incidence	Reference
Aedes vexans	Thelohania barbata (nomen nudum)	Larvae	Adipose	Czechoslovakia	Field	Weiser (1961)
"	Pleistophora sp.	Larvae	Gastric caeca	USA (Louisiana)	<1% Field	Chapman et al. (1967)
Aedomyia squamipennis	Amblyospora keenani	Larvae	Adipose	Panama	Field	Hazard & Oldacre (1975)
"	Parathelohania chagrasensis	Larvae	Adipose	Panama	3-6% Field	Hazard & Oldacre (1975)
Anopheles albimanus	Nosema algerae	Larvae, Adults	Anterior intestine blood cells, fat body, gastric caeca, hypodermis, malpighian tubules, muscle, nerve ganglia, oesophagus, salivary glands, tracheal epithelium, ventriculus	USA (Maryland) Panama El Salvador	Lab. Lab. Lab.	Hazard (1970) Initial report Initial report
"	Nosema sp.	Larvae, Adults	Undetermined	Panama	Field	Initial report
"	Parathelohania sp.	Larvae, Adult ♀	Adipose Oenocytes & ovaries	Panama El Salvador	3-6% Field Field	Hazard & Oldacre (1975) Hazard & Oldacre (1975)
"	Pleistophora culicis	Larvae	Midgut, fat body, abdominal wall, malpighian tubules	USA (Louisiana) Italy	Lab. Lab.	Chapman et al. (1970a) Weiser & Coluzzi (1972)
Anopheles annularis	Parathelohania sp.	Larvae	Adipose	India	Field	Kudo (1929)
Anopheles balabacensis	Nosema algerae	Larvae	Fat body, hypodermis, malpighian tubules, tracheal epithelium, ventriculus	USA (Maryland)	Lab.	Hazard (1970)
Anopheles barbirostris	Parathelohania sp.	Larvae	Adipose	India	Field	Kudo (1929)

MICROSPORIDAN PATHOGENS OF CULICIDAE (MOSQUITOS) (continued)

Host	Pathogen	Stage	Tissue	Locality	Lab.-field incidence	References
Anopheles bradleyi	Parathelohania sp.	Larvae	Adipose	USA (Louisiana)	<1% Field	Chapman et al. (1966)
Anopheles crucians	Nosema algerae	Larvae	Adipose	USA (Florida)	Lab.	Chapman et al. (1970a)
"	Parathelohania obesa	Larvae, Adult ♀	Adipose & oenocytes; Oenocytes & ovaries	USA (Louisiana); USA (Florida & Georgia)	<1% Field; 3-6% Field	Chapman et al. (1966a); Hazard & Weiser (1968)
Anopheles dureni	Pleistophora culicis	Adult	Not given	Congo	Field	Reynolds (1966)
Anopheles earlei	Parathelohania sp.	Larvae	Adipose	Canada	Field	Initial report
Anopheles eiseni	Amblyospora mojingensis	Larvae	Adipose	Panama	Field	Hazard & Oldacre (1975)
Anopheles franciscanus	Parathelohania periculosa	Larvae, adult ♀	Oenocytes & adipose; Oenocytes	USA (California)	1-5% Field	Kellen & Wills (1962a)
"	Pleistophora culicis	Not given	Not given	Not given	Lab.	Chapman et al. (1970a)
Anopheles funestus	Parathelohania sp.	Larvae, Adult ♀	Adipose; Oenocytes & ovaries	Nigeria	Field	Hazard & Oldacre (1975)
"	Pleistophora sp.	Larvae	Gut epithelium	Nigeria	Field	Hazard (1972)
Anopheles gambiae	Nosema algerae	Larvae, Adult	Epithelium of all regions of gut; malpighian tubules; salivary glands; heart muscles; fat body; muscles of head, thorax, abdomen; ganglia; interstitial tissue of ovaries, nurse cells; immature follicles; tracheal epithelium; blood cells; hypodermis	Liberia; USA (Maryland); Tanzania	Field & Lab.; Lab.; Lab.	Fox & Weiser (1959); Hazard (1970); Canning & Hulls (1970)

MICROSPORIDAN PATHOGENS OF CULICIDAE (MOSQUITOS) (continued)

Host	Pathogen	Stage	Tissue	Locality	Lab.-field incidence	References
Anopheles gambiae	Parathelohania africana	Larvae, Adult ♀	Adipose Oenocytes & ovaries	Nigeria	3-5% Field	Hazard & Anthony (1974)
"	Pleistophora culicis	Larvae, pupae, adults,	Midgut, malpighian tubules, fat body	England Italy	Lab. 50-100% Lab. Lab.	Garnham (1957) Canning (1957b) Weiser & Coluzzi (1972)
"	Pleistophora sp.	Larvae	Gut epithelium	Nigeria	Field	Hazard (1972)
Anopheles hyrcanus	Parathelohania indica	Larvae	Adipose	India	Field	Kudo (1929)
Anopheles labranchiae atroparvus	Parathelohania sp.	Larvae	Adipose	France	Field	Tour et al. (1971)
Anopheles maculipennis	Parathelohania legeri	Larvae, Adult ♀	Adipose & oenocytes Oenocytes & ovaries	France Italy Czechoslovakia	Field Field Field	Hesse (1904a) Missiroli (1929) Weiser (1961)
"	Thelohania grassi (doubtful generic placement)	Eggs		Italy	Field	Missiroli (1929)
Anopheles melas	Nosema algerae	Larvae, Adults	Gut, malpighian tubules, fat body, tracheal epithelium, dorsal blood vessel, nervous system, muscles, ovaries, ova, salivary glands, cuticular epithelium, oenocytes, testes	Liberia Tanzania	Lab. Lab.	Fox & Weiser (1959) Canning & Hulls (1970)
Anopheles nili	Parathelohania sp.	Larvae, Adult ♀	Adipose Oenocytes & ovaries	Nigeria	3-6% Field	Hazard & Oldacre (1975)
Anopheles pharoensis	Parathelohania sp.	Larvae, Adult ♀	Adipose & oenocytes Oenocytes & ovaries	Nigeria	3-6% Field	Hazard & Oldacre (1975)

MICROSPORIDAN PATHOGENS OF CULICIDAE (MOSQUITOS) (continued)

Host	Pathogen	Stage	Tissue	Locality	Lab.-field incidence	References
Anopheles pretoriensis	Parathelohania octolagenella	Larvae, Adult ♀	Adipose & oenocytes Oenocytes & ovaries	Nigeria	3-6% Field	Hazard & Anthony (1974)
Anopheles pseudo-punctipennis pseudopunctipennis	Parathelohania sp.	Larvae	Adipose	Guatemala El Salvador	44% Field 3-6% Field	Camey-Pacheco (1965) Initial report
Anopheles punctipennis	Parathelohania illinoisensis	Larvae, Adult ♀	Adipose & oenocytes Oenocytes & ovaries	USA (Illinois) USA (Lousiana) USA (Connecticut) Canada (Quebec)	Field Field Field Field	Kudo (1921) Hazard & Anthony (1974) Anderson (1968) Fantham et al. (1941)
"	Parathelohania legeri (doubtful parasite of this host)	Larvae	Adipose & haemocoele	USA (Illinois)	Field	Simmers (1974a)
"	Stempellia magna (doubtful parasite of this host)	Larvae	Adipose & haemocoele	USA (Illinois)	Field	Simmers (1974c)
Anopheles quadrimaculatus	Nosema algerae	Larvae, Adults	Fat body, gut, ganglia, hypodermis, muscles, malpighian tubules, salivary glands, tracheal epithelium, ventral diverticula, accessory glands	USA (Washington, DC) USA (Florida)	Lab. Lab.	Hazard (1970) Hazard & Lofgren (1971)
"	Parathelohania anophelis	Larvae, Adult ♀	Adipose & oenocytes Oenocytes & ovaries	USA (Georgia & New York) USA (Lousiana) USA (Florida)	Field Field Field	Kudo (1924c) Chapman (1966) Hazard & Anthony (1974)

MICROSPORIDAN PATHOGENS OF CULICIDAE (MOSQUITOS) (continued)

Host	Pathogen	Stage	Tissue	Locality	Lab.-field incidence	References
Anopheles quadrimaculatus	Parathelohania obesa	Larvae, Adult ♀	Adipose & oenocytes Oenocytes & ovaries	USA (Georgia) USA (Pennsylvania) USA (Lousiana) USA (Florida)	Field Field <1% Field 3-6% Field	Kudo (1924c) Wills & Beaudoin (1965) Chapman et al. (1966) Hazard & Weiser (1968)
Anopheles ramsayi	Parathelohania anomala	Larvae	Oenocytes in area of fat body	India	Field	Sen (1941)
Anopheles sinensis	Parathelohania sp.	Larvae	Adipose	China (Province of Taiwan)	Field	Initial report
Anopheles stephensi	Nosema algerae	Larvae, Adults	Epithelial & muscular layers of gut, malpighian tubules, hypodermal cells, muscles, fat body, tracheal epithelium, pericardial cells, gonads, salivary glands, oenocytes, ganglia, nerves, accessory glands, blood cells	USA (Illinois) England USA (Washington, DC) USA (Maryland)	Lab. Lab. Lab. Lab.	Vavra & Undeen (1970) Canning & Hulls (1970) Hazard (1970) Hazard (1970)
"	Pleistophora culicis	Larvae, Pupae, Adults	Midgut, malpighian tubules, fat body, muscles, salivary glands, neural ganglia, ovaries	England USA (New York) Italy	Lab. Lab. Lab.	Garnham (1956) Reynolds (1966) Weiser & Coluzzi (1972)
Anopheles subpictus	Parathelohania sp.	Larvae	Adipose	India	Field	Kudo (1929)
Anopheles triannualatus	Parathelohania sp.	Larvae	Adipose	Panama	3-6% Field	Hazard & Oldacre (1975)
Anopheles vagus	Parathelohania sp.	Larvae	Adipose	India	Field	Sen (1941)
Anopheles varuna	Parathelohania obscura	Larvae	Not given	India	Field	Kudo (1929)

MICROSPORIDIAN PATHOGENS OF CULICIDAE (MOSQUITOS) (continued)

Host	Pathogen	Stage	Tissue	Locality	Lab.-field incidence	References
Anopheles walkeri	Parathelohania sp.	Larvae	Not given	USA (Minnesota)	Field	Laird (1961)
Anopheles sp.	Thelohania pyriformis (doubtful generic placement)	Larvae	Adipose	USA (Georgia)	Field	Kudo (1924c)
Coquillettidia perturbans	Amblyospora sp.	Adult ♀	Oenocytes & ovaries	USA (Louisiana) USA (Florida)	<1% Field Field	Chapman et al. (1967) Hazard & Oldacre (1975)
Culex apicalis	Amblyospora benigna	Larvae	Adipose	USA (California)	1-3% Field	Kellen & Wills (1962a)
"	Thelohania sp. (doubtful generic placement)	Larvae	Adipose	USA (California)	Field	Kellen et al. (1965)
Culex annulirostris	Amblyospora sp.	Larvae	Haemocoel	New Hebrides	Field	Laird (1956)
Culex erraticus	Amblyospora minuta	Larvae	Adipose	USA (Georgia) USA (Louisiana)	Field <1% Field	Kudo (1924c) Chapman et al. (1967)
Culex erythrothorax	Amblyospora gigantea	Larvae	Adipose	USA (California)	Field	Kellen & Wills (1962a)
Culex gelidus	Pleistophora collessi	Adult ♀	Ovary	Singapore	1% Field	Laird (1959a)
Culex nigripalpus	Pleistophora sp.	Larvae	Gut	USA (Florida)	Field	Initial report
Culex peccator	Amblyospora sp.	Larvae	Adipose	USA (Louisiana)	<1% Field	Chapman et al. (1969)
Culex peus	Amblyospora sp.	Larvae	Adipose	USA (California)	Field	Kellen et al. (1965)
Culex pilosus	Stempellia lunata	Larvae	Adipose	USA (Florida)	24% Field	Hazard & Savage (1970)
Culex pipiens pipiens	Amblyospora opacita (doubtful parasite of this host)	Larvae	Adipose & haemocoel	USA (Illinois)	Field	Simmers (1974b)
"	Amblyospora sp.	Larvae	Adipose	USA (Pennsylvania)	Field	Initial report

MICROSPORIDAN PATHOGENS OF CULICIDAE (MOSQUITOS) (continued)

Host	Pathogen	Stage	Tissue	Locality	Lab.-field incidence	References
Culex pipiens pipiens	Nosema algerae	Larvae, Pupae, Adults	Gut, malpighian tubules, fat body, tracheal epithelium, dorsal blood vessel, nervous system, muscles, ovaries, ova, salivary glands, cuticular epithelium, oenocytes, testes	England	Lab.	Canning & Hulls (1970)
"	Nosema culicis (doubtful generic placement)	Larvae	Haemocoel contents	Germany	Field	Bresslau & Bushkiel (1919)
"	Pleistophora culicis	Larvae	Malpighian tubules	Czechoslovakia	<1% Field	Weiser (1947)
"	Stempellia magna (doubtful parasite of this host)	Larvae	Adipose	USA (Illinois)	Field	Kudo (1921)
"	Stempellia milleri	Larvae	Adipose	USA (Louisiana)	14% Lab.	Hazard & Fukuda (1974)
"	Thelohania sp. (doubtful generic placement)	Larvae	Adipose	Venezuela	Field	Iturbe & Gonzales (1921)
"	Weiseria spinosa	Larvae, Pupae, Adults	Adipose	USSR	93.4% Field 27.5% Field <1% Field	Gol'berg (1971)
Culex pipiens quinquefasciatus	Nosema algerae	Larvae, Adults	Fat body, gut, malpighian tubules, oesophagus, rectum, ganglia, ventriculus	USA (Florida)	Lab.	Hazard & Lofgren (1971)
"	Pleistophora culicis	Larvae, Pupae, Adults	Fat body, gut, malpighian tubules & most other organs	England	Lab.	Reynolds (1966)
"	Pleistophora sp.	Larvae	Gastric caeca & midgut	USA (Louisiana)	<1% Field	Initial report

MICROSPORIDAN PATHOGENS OF CULICIDAE (MOSQUITOS) (continued)

Host	Pathogen	Stage	Tissue	Locality	Lab.-field incidence	References
Culex pipiens quinquefasciatus	Stempellia milleri	Larvae	Adipose	USA (Texas)	Field	Hazard & Fukuda (1974)
				USA (Louisiana)	1-10% Field	
				Thailand	1-10% Field	
Culex restuans	Pleistophora sp.	Larvae	Gastric caeca & midgut	USA (Louisiana)	1% Field	Initial report
"	Stempellia magna	Larvae, Adults	Adipose Ovaries	USA (Pennsylvania)	5-16% Field	Kudo (1922)
				USA (Florida)	3-6% Field	Wills & Beaudoin (1965)
				USA (Louisiana)	Field	Chapman et al. (1967)
				USA (Connecticut)	Field	Anderson (1968)
				USA (Virginia)	Field	Bailey et al. (1967b)
"	Stempellia milleri	Larvae	Adipose	USA (Louisiana)	Field	Initial report
Culex salinarius	Amblyospora sp.	Larvae, Adult ♀	Adipose Oenocytes & ovaries	USA (Louisiana)	Field	Kellen et al. (1966a)
				USA (Florida)	Field	Hazard & Oldacre (1975)
"	Nosema algerae	Larvae, Adults	Fat body, gut, malpighian tubules, accessory glands, ganglion, oesophagus, rectum, ventriculus	USA (Florida)	Lab.	Hazard & Lofgren (1971)
"	Nosema sp. (doubtful generic placement)	Larvae	Adipose	USA (Louisiana)	<1% Field	Chapman et al. (1967)
"	Pleistophora culicis	Larvae	Gut	USA (Louisiana)	<1% Field	Chapman (1974a)
"	Pleistophora sp.	Larvae	Gut	USA (Louisiana)	<1% Field	Chapman et al. (1969)
"	Stempellia milleri	Larvae	Adipose	USA (Louisiana)	4% Lab.	Hazard & Fukuda (1974)
Culex secutor	Undetermined microsporidium	Larvae	Not given	Puerto Rico	Field	Briggs (1972)

MICROSPORIDAN PATHOGENS OF CULICIDAE (MOSQUITOS) (continued)

Host	Pathogen	Stage	Tissue	Locality	Lab.-field incidence	References
Culex tarsalis	Amblyospora californica	Larvae, Adult ♀	Adipose, Oenocytes & ovaries	USA (California)	5-25% Field	Kellen & Lipa (1960)
				USA (Louisiana)	<1% Field	Chapman et al. (1967)
				USA (Nevada)	<1% Field	Chapman (1966)
				USA (Texas)	<1% Field	Initial report
				USA (Utah)	2-12% Field	Tsai et al. (1969)
"	Stempellia milleri	Larvae	Adipose	USA (Louisiana)	3% Lab.	Hazard & Fukuda (1974)
Culex territans	Amblyospora opacita	Larvae, Adult ♀	Adipose, Oenocytes & ovaries	USA (Illinois & New York)	Field	Kudo (1922)
				USA (Louisiana)	<1% Field	Chapman et al. (1969)
				USA (Connecticut)	4-5% Field	Anderson (1968)
				USA (Florida)	Field	Hazard & Oldacre (1975)
"	Pleistophora chapmani	Larvae	Midgut & gastric caeca	USA (Louisiana)	Field	Clark & Fukuda (1971)
"	Pleistophora culicis	Larvae	Gut	USA (Louisiana)	1% Field	Chapman et al. (1969)
"	Stempellia milleri	Larvae	Adipose	USA (Louisiana)	25% Lab.	Hazard & Fukuda (1974)
"	Thelohania sp. (doubtful generic placement)	Larvae	Adipose	USA (Louisiana)	Field	Kellen et al. (1966a)
					<1% Field	Chapman et al. (1967)
Culex thriambus	Amblyospora noxia	Larvae	Adipose	USA (California)	Field	Kellen & Wills (1962a)
Culex tritaeniorhynchus summorosus	Pleistophora collessi	Adults	Ovaries	Singapore	1% Field	Laird (1959a)
Culiseta annulata	Amblyospora sp.	Larvae	Adipose	Federal Republic of Germany	Field	Bresslau & Buschkeilm (1919)
Culiseta impatiens	Amblyospora sp.	Larvae	Adipose	USA (Wyoming)	2% Field	Tsai et al. (1969)

MICROSPORIDAN PATHOGENS OF CULICIDAE (MOSQUITOS) (continued)

Host	Pathogen	Stage	Tissue	Locality	Lab.-field incidence	References
Culiseta incidens	Amblyospora campbelli	Larvae	Adipose	USA (California)	<1-80% Field	Kellen & Wills (1962a)
				USA (Nevada)	Field	Chapman (1966)
Culiseta inornata	Amblyospora inimica	Larvae	Adipose	USA (California)	<1% Field	Kellen & Wills (1962a)
				USA (Florida)	Field	Hazard & Oldacre (1975)
				USA (Louisiana)	20% Field	Chapman et al. (1966)
				USA (Nevada)	Field	Chapman (1966)
				USA (Utah)	<1% Field	Tsai et al. (1969)
"	Amblyospora opacita (doubtful parasite of this host)	Larvae	Adipose & haemocoel	USA (Illinois)	Field	Simmers (1974b)
"	Pleistophora caecorum	Larvae, Pupae, Adults	Gastric caeca & midgut	USA (Louisiana)	2-5% Field	Chapman & Kellen (1967)
"	Pleistophora chapmani	Larvae	Midgut & gastric caeca	USA (Louisiana)	<1% Field	Initial report
"	Stempellia magna (doubtful parasite of this host)	Larvae	Adipose & haemocoel	USA (Illinois)	Field	Simmers (1974c)
Culiseta longiareolata	Pleistophora culicis	Larvae, Pupae, Adults	Fat body, muscles /esp in 4th & 5th abdominal segments/, gut wall, neural ganglia	Italy	Field	Weisser & Coluzzi (1964)
Culiseta melanura	Hyalinocysta chapmani	Larvae	Adipose	USA (Louisiana)	<1% Field	Hazard & Oldacre (1975)
Culiseta particeps	Amblyospora sp.	Larvae	Adipose	USA (California)	Field	Kellen et al. (1965)
Mansonia dyari	Amblyospora sp.	Larvae	Adipose	Panama	3-6% Field	Hazard & Oldacre (1975)
"	Stempellia sp.	Larvae	Adipose	Panama	Field	Initial report
Mansonia leberi	Amblyospora sp.	Larvae	Adipose	Panama	3-6% Field	Hazard & Oldacre (1975)
Maorigoeldia argyropus	Pleistophora milesi	Larvae	Adipose	New Zealand	Field	Pillai (1974)

MICROSPORIDAN PATHOGENS OF CULICIDAE (MOSQUITOS) (continued)

Host	Pathogen	Stage	Tissue	Locality	Lab.-field incidence	References
Orthopodomyia signifera	Pleistophora sp.	Larvae	Adipose	USA (Louisiana)	1% Field	Chapman et al. (1967)
Psorophora ciliata	Parathelohania legeri (doubtful parasite of this host)	Larvae	Adipose	USA (Illinois)	Field	Simmers (1974a)
"	Pleistophora sp.	Larvae	Gastric caeca	USA (Florida)	Field	Initial report
Psorophora columbiae	Amblyospora sp.	Larvae, Adult ♀	Adipose Oenocytes & ovaries	USA (Louisiana) USA (Florida)	1% Field Field	Chapman et al. (1966) Hazard & Oldacre (1975)
Psorophora ferox	Stempellia sp.	Larvae	Adipose	USA (Louisiana)	1% Field	Chapman et al. (1967)
Psorophora horrida	Stempellia sp.	Larvae	Adipose	USA (Louisiana)	4% Field	Chapman et al. (1969)
Toxorhynchites rutilus septentrionalis	Pleistophora sp.	Larvae	Adipose	USA (Louisiana)	1% Field	Chapman et al. (1967)
Uranotaenia sapphirina	Pleistophora sp.	Larvae	Adipose	USA (Louisiana)	1% Field	Initial report
"	Stempellia sp.	Larvae	Adipose	USA (Florida) USA (Louisiana)	Field Field	Initial report Chapman et al. (1969)
Wyeomyia vanduzeei	Pilosporella fishi	Larvae	Adipose	USA (Florida)	Field	Hazard & Oldacre (1975)

IV(b). SYNONYMY AND HOST RECORDS OF MICROSPORIDA AFFECTING CULICIDAE

E. I. Hazard

USDA, Agricultural Research Service
Insects Affecting Man Research Laboratory
Gainesville, FL 32604, USA

SYNONOMY AND HOST RECORDS OF MICROSPORIDA AFFECTING CULICIDAE

NOSEMATIDAE

Nosema aedis Kudo, 1930

      Host: Aedes aegypti Linnaeus

Nosema algerae Vavra & Undeen, 1970
    Nosema algerae: Canning & Hulls, 1970; Ward & Savage, 1972
    Nosema stegomyiae Fox & Weiser, 1959; Reynolds, 1971
    Nosema sp. Savage & Lowe, 1970; Hazard, 1970; Hazard & Lofgren, 1971

      Host: Aedes aegypti Linnaeus, Anopheles albimanus Wiedemann, Anopheles balabacensis
             Baisas, Anopheles crucians Wiedemann, Anopheles gambiae Giles, Anopheles
             melas (Theobald), Anopheles quadrimaculatus Say, Anopheles stephensi Liston,
             Culex pipiens pipiens Linnaeus, Culex pipiens quinquefasciatus Say /= Culex
             pipiens fatigans Wiedemann/, Culex salinarius Coquillett

Nosema culicis Bresslau & Buschkiel, 1919

      Host: Culex pipiens pipiens Linnaeus

Nosema lutzi Kudo, 1929
    Nosema stegomyiae Lutz & Splendore, 1908

      Host: Aedes aegypti Linnaeus (= Stegomyia fasciata of Lutz & Splendore, 1908 and
             Aedes calopus of Kudo, 1929)

[a]Nosema sp. Chapman & Hazard, hoc loco

      Host: Anopheles albimanus Wiedemann

[b]Nosema sp. Chapman et al., 1967

      Host: Culex salinarius Coquillett

Stempellia lunata Hazard & Savage, 1970

      Host: Culex pilosus (Dyar & Knab)

Stempellia magna (Kudo, 1920) Kudo, 1924c
    Thelohania magna Kudo, 1920
    Stempellia magna (Kudo, 1920) Kudo, 1924c; Kudo, 1925b; Wills & Beaudoin, 1965;
                 Chapman et al., 1967; Bailey et al., 1967b; Anderson, 1968
    Pyrotheca magna (Kudo, 1920) Hesse, 1935

      Host: Culex restuans Theobald (= Culex territans of Kudo, 1924c)

---

[a] As initially reported in the host list.

[b] Doubtful generic placement.

<u>Stempellia milleri</u> Hazard & Fukuda, 1974

>   Host: <u>Culex pipiens pipiens</u> Linnaeus, <u>Culex pipiens quinquefasciatus</u> Say, <u>Culex restuans</u> Theobald, <u>Culex salinarius</u> Coquillett, <u>Culex tarsalis</u> Coquillett, <u>Culex territans</u> Walker

<u>Stempellia tuzetae</u> Tour et al., 1971

>   Host: <u>Aedes detritus</u> (Haliday)

[a]<u>Stempellia</u> sp. Chapman & Hazard, <u>hoc loco</u>

>   Host: <u>Mansonia dyari</u> Belkin, Heinemann & Page

<u>Stempellia</u> sp. Chapman et al., 1967; Chapman et al., 1969

>   Host: <u>Psorophora ferox</u> (Humboldt)

<u>Stempellia</u> sp. Chapman et al., 1969

>   Host: <u>Psorophora horrida</u> (Dyar & Knab)

<u>Stempellia</u> sp. Chapman et al., 1969

>   Host: <u>Uranotaenia sapphirina</u> (Osten-Sacken)

<u>Weiseria spinosa</u> Gol'berg, 1971

>   Host: <u>Culex pipiens pipiens</u> Linnaeus

PLEISTOPHORIDAE

<u>Pleistophora caecorum</u> (Chapman & Kellen, 1967) Clark & Fukuda, 1971
    <u>Plistophora caecorum</u> Chapman & Kellen, 1967
    <u>Plistophora</u> sp. Chapman et al., 1967; Chapman et al., 1969

>   Host: <u>Culiseta inornata</u> (Williston)

<u>Pleistophora chapmani</u> Clark & Fukuda, 1971

>   Host: <u>Culex territans</u> Walker, <u>Culiseta inornata</u> (Williston)

<u>Pleistophora collessi</u> (Laird, 1959a) Clark & Fukuda, 1971
    <u>Plistophora collessi</u> Laird, 1959a

>   Host: <u>Culex gelidus</u> Theobald, <u>Culex tritaeniorhynchus summorosus</u> Dyar

<u>Pleistophora culicis</u> (Weiser, 1947) Clark & Fukuda, 1971
    <u>Plistophora culicis</u> Weiser, 1947; Garnham, 1957; Canning, 1957b; Bano, 1958;
                Garnham, 1959; Weiser, 1961; Reynolds, 1966; Chapman et al., 1967;
                Weiser & Coluzzi, 1972
    <u>Plistophora kudoi</u> Weiser, 1946
    <u>Plistophora culisetae</u> Weiser & Coluzzi, 1964

Host: Aedes aegypti Linnaeus, Aedes triseriatus (Say), Anopheles albimanus Wiedemann, Anopheles dureni Edwards, Anopheles franciscanus McCracken, Anopheles gambiae Giles, Anopheles stephensi Liston, Culex pipiens pipiens Linnaeus, Culex pipiens quinquefasciatus Say (= Culex pipiens fatigans Wiedemann), Culex salinarius Coquillett, Culex territans Walker, Culiseta longiareolata (Macquart)

Pleistophora milesi Pillai, 1974

Host: Aedes notoscriptus (Skuse), Maorigoeldia argyropus Walker

Pleistophora stegomyiae (Marchoux et al., 1903) Chatton, 1911
    Nosema stegomyiae Marchoux et al., 1903; Marchoux & Simond, 1906
    Plistophora stegomyiae Kudo, 1924c

Host: Aedes aegypti Linnaeus (= Stegomyia fasciata of Marchoux et al., 1903)

Pleistophora sp. Chapman & Hazard, hoc loco
    Plistophora sp. Chapman et al., 1967

Host: Aedes canadensis (Theobald)

Pleistophora sp. Chapman & Hazard, hoc loco
    Plistophora sp. Chapman et al., 1967

Host: Aedes vexans (Meigen)

Pleistophora sp. Chapman & Hazard, hoc loco
    Plistophora sp. Briggs, 1972

Host: Anopheles funestus Giles

Pleistophora sp. Chapman & Hazard, hoc loco
    Plistophora sp. Briggs, 1972

Host: Anopheles gambiae Giles

[a]Pleistophora sp. Chapman & Hazard, hoc loco

Host: Culex nigripalpus Theobald

[a]Pleistophora sp. Chapman & Hazard, hoc loco

Host: Culex pipiens quinquefasciatus Say

[a]Pleistophora sp. Chapman & Hazard, hoc loco

Host: Culex restuans Theobald

Pleistophora sp. Chapman & Hazard, hoc loco
    Plistophora sp. Chapman et al., 1969

Host: Culex salinarius Coquillett

Pleistophora sp. Chapman & Hazard, hoc loco
    Plistophora sp. Chapman et al., 1967; Chapman et al., 1969

       Host: Orthopodomyia signifera (Coquillett)

[a]Pleistophora sp. Chapman & Hazard, hoc loco

       Host: Psorophora ciliata (Fabricius)

Pleistophora sp. Chapman & Hazard, hoc loco
    Plistophora sp. Chapman et al., 1967; Chapman et al., 1969

       Host: Toxorhynchites rutilus septentrionalis (Dyar & Knab)

[a]Pleistophora sp. Chapman & Hazard, hoc loco

       Host: Uranotaenia sapphirina (Osten-Sacken)

THELOHANIIDAE

Amblyospora benigna (Kellen & Wills, 1962a) Hazard & Oldacre, 1975
    Thelohania benigna Kellen & Wills, 1962a

       Host: Culex apicalis Adams

Amblyospora bolinasae (Kellen & Wills, 1962a) Hazard & Oldacre, 1975
    Thelohania bolinasae Kellen & Wills, 1962a

       Host: Aedes squamiger (Coquillett)

Amblyospora californica (Kellen & Lipa, 1960) Hazard & Oldacre, 1975
    Thelohania californica Kellen & Lipa, 1960
    Nosema lunatum Kellen et al., 1967
    Thelohania sp. Chapman, 1966

       Host: Culex tarsalis Coquillett

Amblyospora campbelli (Kellen & Wills, 1962a) Hazard & Oldacre, 1975
    Thelohania campbelli Kellen & Wills, 1962a
    Thelohania sp. Chapman, 1966

       Host: Culiseta incidens (Thomson)

Amblyospora canadensis (Wills & Beaudoin, 1965) Hazard & Oldacre, 1975
    Thelohania inimica canadensis Wills & Beaudoin, 1965
    Thelohania inimica: Bailey et al., 1967
    Thelohania sp. Chapman et al., 1966; Anderson, 1968

       Host: Aedes canadensis (Theobald)

Amblyospora gigantea (Kellen & Wills, 1962a) Hazard & Oldacre, 1975
    Thelohania gigantea Kellen & Wills, 1962a

       Host: Culex erythrothorax Dyar

Amblyospora inimica (Kellen & Wills, 1962a) Hazard & Oldacre, 1975
    Thelohania inimica Kellen & Wills, 1962a; Chapman et al., 1966
    Thelohania sp. Chapman, 1966; Chapman et al., 1967

      Host: Culiseta inornata (Williston)

Amblyospora keenani Hazard & Oldacre, 1975

      Host: Aedeomyia squamipennis (Lynch Arribalzaga)

Amblyospora khaliulini Hazard & Oldacre, 1975
    Thelohania opacita var. mariensis Khaliulin & Ivanov, 1971
    Thelohania sp. Nöller, 1920; Welch, 1960; Chapman et al., 1973

      Host: Aedes communis DeGeer (= Aedes nemorosus of Nöller, 1920)

Amblyospora minuta (Kudo, 1924c) Hazard & Oldacre, 1975
    Thelohania minuta Kudo, 1924c; Kudo, 1925a
    Thelohania rotunda Kudo, 1924c; Kudo, 1925a

      Host: Culex erraticus (Dyar & Knab)

Amblyospora mojingensis Hazard & Oldacre, 1975

      Host: Anopheles eiseni Coquillett

Amblyospora noxia (Kellen & Wills, 1962a) Hazard & Oldacre, 1975
    Thelohania noxia Kellen & Wills, 1962a

      Host: Culex thriambus Dyar

Amblyospora opacita (Kudo, 1922) Hazard & Oldacre, 1975
    Thelohania opacita Kudo, 1922; Kudo, 1924c; Chapman et al., 1966; Anderson, 1968

      Host: Culex territans Walker (= Culex apicalis of Kudo)

Amblyospora unica (Kellen & Wills, 1962a) Hazard & Oldacre, 1975
    Thelohania unica Kellen & Wills, 1962a
    Thelohania sp. Chapman et al., 1966

      Host: Aedes melanimon Dyar

Amblyospora sp. Hazard & Oldacre, 1975
    Thelohania sp. Anderson, 1968

      Host: Aedes abserratus (Felt & Young)

Amblyospora sp. Hazard & Oldacre, 1975
    Nosema sp. Nöller, 1920

      Host: Aedes cantans (Meigen)

Amblyospora sp. Hazard & Oldacre, 1975
    Thelohania sp. Anderson, 1968; Chapman et al., 1973

    Host: Aedes cantator (Coquillett)

Amblyospora sp. Hazard & Oldacre, 1975
    Thelohania opacita: Tour et al., 1971

        Host: Aedes caspius (Pallas)

Amblyospora sp. Hazard & Oldacre, 1975
    Thelohania sp. Kellen et al., 1965; Chapman et al., 1973

        Host: Aedes cataphylla Dyar

Amblyospora sp. Hazard & Oldacre, 1975
    Thelohania sp. Chapman, 1966; Anderson, 1968

        Host: Aedes cinereus (Meigen)

Amblyospora sp. Hazard & Oldacre, 1975
    Thelohania opacita Tour et al., 1971

        Host: Aedes detritus (Haliday)

Amblyospora sp. Hazard & Oldacre, 1975
    Thelohania sp. Kellen et al., 1965; Chapman, 1966; Tsai et al., 1969

        Host: Aedes dorsalis (Meigen)

Amblyospora sp. Hazard & Oldacre, 1975
    Thelohania sp. Anderson, 1968; Chapman et al., 1973

        Host: Aedes excrucians (Walker)

Amblyospora sp. Hazard & Oldacre, 1975
    Thelohania sp. Chapman et al., 1973

        Host: Aedes fitchii (Felt & Young)

Amblyospora sp. Hazard & Oldacre, 1975
    Thelohania sp. Chapman et al., 1966; Chapman et al., 1967

        Host: Aedes grossbecki Dyar & Knab

Amblyospora sp. Hazard & Oldacre, 1975
    Thelohania sp. Kellen et al., 1965; Chapman et al., 1973

        Host: Aedes hexodontus Dyar

Amblyospora sp. Hazard & Oldacre, 1975
    Thelohania sp. Kellen et al., 1965

        Host: Aedes increpitus Dyar

Amblyospora sp. Hazard & Oldacre, 1975
    Thelohania sp. Chapman et al., 1973

        Host: Aedes pullatus (Coquillett)

<u>Amblyospora</u> sp. Hazard & Oldacre, 1975
    <u>Thelohania</u> sp. Chapman et al., 1973

       Host: <u>Aedes punctor</u> (Kirby)

<u>Amblyospora</u> sp. Hazard & Oldacre, 1975
    <u>Thelohania</u> sp. Chapman et al., 1973

       Host: <u>Aedes riparius</u> Dyar & Knab

<u>Amblyospora</u> sp. Hazard & Oldacre, 1975
    <u>Thelohania</u> sp. Kellen et al., 1966a; Chapman et al., 1966; Chapman et al., 1967;
           Chapman et al., 1969

       Host: <u>Aedes sollicitans</u> (Walker)

<u>Amblyospora</u> sp. Hazard & Oldacre, 1975
    <u>Thelohania</u> sp. Chapman et al., 1966; Chapman et al., 1967; Chapman et al., 1969

       Host: <u>Aedes sticticus</u> (Meigen)

<u>Amblyospora</u> sp. Hazard & Oldacre, 1975
    <u>Thelohania</u> sp. Franz & Hagmann, 1962; Anderson, 1968

       Host: <u>Aedes stimulans</u> (Walker)

<u>Amblyospora</u> sp. Hazard & Oldacre, 1975
    <u>Thelohania</u> sp. Kellen et al., 1966a; Chapman et al., 1966; Chapman et al., 1967;
           Chapman et al., 1969

       Host: <u>Aedes taeniorhynchus</u> (Wiedemann)

[a]<u>Amblyospora</u> sp. Chapman & Hazard, <u>hoc loco</u>

       Host: <u>Aedes thibaulti</u> Dyar & Knab

<u>Amblyospora</u> sp. Hazard & Oldacre, 1975
    <u>Thelohania</u> sp. Kellen et al., 1965

       Host: <u>Aedes ventrovittis</u> Dyar

<u>Amblyospora</u> sp. Hazard & Oldacre, 1975
    <u>Stempellia</u> sp. Chapman et al., 1967

       Host: <u>Coquillettidia perturbans</u> (Walker) /= <u>Mansonia perturbans</u>/

<u>Amblyospora</u> sp. Hazard & Oldacre, 1975
    <u>Thelohania opacita</u> Laird, 1956

       Host: <u>Culex annulirostris</u> Skuse

<u>Amblyospora</u> sp. Hazard & Oldacre, 1975
    <u>Thelohania</u> sp. Chapman et al., 1969

       Host: <u>Culex peccator</u> Dyar & Knab

Amblyospora sp. Hazard & Oldacre, 1975
    Thelohania sp. Kellen et al., 1965

       Host: _Culex peus_ Speiser

[a]_Amblyospora_ sp. Chapman & Hazard, _hoc loco_

       Host: _Culex pipiens pipiens_ Linnaeus

Amblyospora sp. Hazard & Oldacre, 1975
    Thelohania sp. Kellen et al., 1966a; Chapman et al., 1966; Chapman et al., 1967;
           Chapman et al., 1969

       Host: _Culex salinarius_ Coquillett

Amblyospora sp. Hazard & Oldacre, 1975
    Thelohania sp. Bresslau & Buschkiel, 1919
    Thelohania opacita: Weiser, 1961; Weiser, 1966

       Host: _Culiseta annulata_ (Schrank)

Amblyospora sp. Hazard & Oldacre, 1975
    Thelohania sp. Tsai et al., 1969

       Host: _Culiseta impatiens_ (Walker)

Amblyospora sp. Hazard & Oldacre, 1975
    Thelohania sp. Kellen et al., 1965

       Host: _Culiseta particeps_ (Adams)

Amblyospora sp. Hazard & Oldacre, 1975

       Host: _Mansonia dyari_ Belkin, Heinemann & Page

Amblyospora sp. Hazard & Oldacre, 1975

       Host: _Mansonia leberi_ Boreham

Amblyospora sp. Hazard & Oldacre, 1975
    Thelohania sp. Chapman et al., 1966; Chapman et al., 1967; Chapman et al., 1969

       Host: _Psorophora columbiae_ (Dyar & Knab) /= _P. confinnis_/

Hyalinocysta chapmani Hazard & Oldacre, 1975

       Host: _Culiseta melanura_ (Coquillett)

Parathelohania africana Hazard & Anthony, 1974

       Host: _Anopheles gambiae_ Giles

Parathelohania anomala (Sen, 1941) Hazard & Anthony, 1974
    Thelohania anomala Sen, 1941

       Host: _Anopheles ramsayi_ Covell

Parathelohania anophelis (Kudo, 1924c) Hazard & Anthony, 1974
    Nosema anophelis Kudo, 1924c
    Thelohania legeri: Kudo, 1924c; Chapman et al., 1966; Hazard & Weiser, 1968

        Host: Anopheles quadrimaculatus Say

Parathelohania barra (Pillai, 1968) Hazard & Oldacre, 1975
    Thelohania barra Pillai, 1968

        Host: Aedes australis (Erichson)

Parthelohania chagrasensis Hazard & Oldacre, 1975

        Host: Aedeomyia squamipennis (Lynch Arribalzaga)

Parathelohania illinoisensis (Kudo, 1921) Hazard & Anthony, 1974
    Thelohania illinoisensis Kudo, 1921
    Thelohania legeri: Kudo, 1924a; Fantham et al., 1941; Kudo, 1962; Anderson, 1968

        Host: Anopheles punctipennis Say

Parathelohania indica (Kudo, 1929) Hazard & Anthony, 1974
    Thelohania indica Kudo, 1929; Sen, 1941
    Thelohania obesa: Weiser, 1961

        Host: Anopheles hyrcanus Pallas

Parathelohania legeri (Hesse, 1904a) Codreanu, 1966
    Thelohania legeri Hesse, 1904a
    Toxoglugea missiroli Weiser, 1961

        Host: Anopheles maculipennis Meigen

Parathelohania obesa (Kudo, 1924c) Hazard & Anthony, 1974
    Thelohania obesa Kudo, 1924c; Wills & Beaudoin, 1965; Hazard & Weiser, 1968
    Thelohania legeri: Chapman et al., 1966

        Host: Anopheles crucians Wiedemann, Anopheles quadrimaculatus Say

Parathelohania obscura (Kudo, 1929) Hazard & Anthony, 1974
    Thelohania obscura Kudo, 1929
    Thelohania obesa: Weiser, 1961

        Host: Anopheles varuna Iyengar (= Anopheles funestus of Kudo, 1929)

Parathelohania octolagenella Hazard & Anthony, 1974

        Host: Anopheles pretoriensis (Theobald)

Parathelohania periculosa (Kellen & Wills, 1962a) Hazard & Anthony, 1974
    Thelohania periculosa Kellen & Wills, 1962a
    Nosema chapmani Kellen et al., 1967

        Host: Anopheles franciscanus McCracken

<u>Parathelohania</u> sp. Hazard & Oldacre, 1975

      Host: <u>Anopheles albimanus</u> Wiedemann

<u>Parathelohania</u> sp. (Kudo, 1929) Hazard & Oldacre, 1975
    <u>Thelohania legeri</u>: Kudo, 1929

      Host: <u>Anopheles annularis</u> Van der Wulp

<u>Parathelohania</u> sp. (Kudo, 1929) Hazard & Oldacre, 1975
    <u>Thelohania legeri</u>: Kudo, 1929

      Host: <u>Anopheles barbirostris</u> Van der Wulp

<u>Parathelohania</u> sp. Hazard & Oldacre, 1975
    <u>Thelohania</u> sp. Chapman et al., 1966; Chapman et al., 1967

      Host: <u>Anopheles bradleyi</u> King

[a]<u>Parathelohania</u> sp. Chapman & Hazard, <u>hoc loco</u>

      Host: <u>Anopheles earlei</u> Vargas

<u>Parathelohania</u> sp. Hazard & Oldacre, 1975

      Host: <u>Anopheles funestus</u> Giles

<u>Parathelohania</u> sp. Hazard & Anthony, 1974
    <u>Thelohania legeri</u>: Tour et al., 1971

      Host: <u>Anopheles labranchiae atroparvus</u> Van Thiel

<u>Parathelohania</u> sp. Hazard & Oldacre, 1975

      Host: <u>Anopheles nili</u> (Theobald)

<u>Parathelohania</u> sp. Hazard & Oldacre, 1975

      Host: <u>Anopheles pharoensis</u> Theobald

<u>Parathelohania</u> sp. Hazard & Oldacre, 1975
    <u>Thelohania legeri</u>: Camey-Pacheco, 1965

      Host: <u>Anopheles pseudopunctipennis pseudopunctipennis</u> Theobald

<u>Parathelohania</u> sp. Chapman & Hazard, <u>hoc loco</u>

      Host: <u>Anopheles sinensis</u> Wiedemann

<u>Parathelohania</u> sp. Hazard & Anthony, 1974
    <u>Thelohania legeri</u>: Kudo, 1929

      Host: <u>Anopheles subpictus</u> Grassi

<u>Parathelohania</u> sp. Hazard & Oldacre, 1975

      Host: <u>Anopheles triannulatus</u> (Neva & Pinto)

<u>Parathelohania</u> sp. Hazard & Anthony, 1974
    <u>Thelohania legeri</u>: Sen, 1941

        Host: <u>Anopheles vagus</u> Donitz

<u>Parathelohania</u> sp. Hazard & Oldacre, 1975
    <u>Thelohania</u> sp. Laird, 1961

        Host: <u>Anopheles walkeri</u> Theobald

<u>Pilosporella chapmani</u> Hazard & Oldacre, 1975

        Host: <u>Aedes triseriatus</u> Say

<u>Pilosporella fishi</u> Hazard & Oldacre, 1975

        Host: <u>Wyeomyia vanduzeei</u> (Dyar & Knab)

[b]<u>Thelohania grassi</u> Missiroli, 1929

        Host: <u>Anopheles maculipennis</u> Meigen

[b]<u>Thelohania pyriformis</u> Kudo, 1924c

        Host: <u>Anopheles</u> sp.

[b]<u>Thelohania</u> sp. Kellen et al., 1965

        Host: <u>Culex apicalis</u> Adams

[b]<u>Thelohania</u> sp. Iturbe & Gonzales, 1921

        Host: <u>Culex pipiens pipiens</u> Linnaeus

[b]<u>Thelohania</u> sp. Kellen et al., 1965

        Host: <u>Culex territans</u> Walker

<u>Microsporidium</u> sp. Briggs, 1972

        Host: <u>Culex secutor</u> Theobald

ABSTRACTS

Mary Ann Strand

Alikhanov, S. G. (1972). /On the infestation of natural populations of <u>Aedes caspius caspius</u>
with microsporidians of the genus <u>Thelohania</u> in Azerbaidzhana.7   <u>Parazitologiya</u>, <u>6</u>: 381-384
(Russian, English summary).

   In the lowlands and foothills of Azerbaidzhana, infection incidence ranged from 0 to 90%.
   Seasonal peaks in infection level were observed.

Alikhanov, S. G. (1973). /Changes in the sex ratio of mosquitos <u>Aedes caspius caspius</u>
(Pall.) Edw. during infection of their natural populations with microsporidians, <u>Thelohania</u>
<u>opacita</u> Kudo, 1922./ <u>Parazitologiya</u>, <u>7</u>(2): 175-179 (Russian, English summary).

   The sex ratio (males:females) of uninfected populations was close to 1:1, while that of
   populations infected with an <u>Amblyospora</u> sp. (<u>T. opacita</u>) ranged from 1:2.5 to 1:8 over
   a 5 month period.   Apparently the altered sex ratio is due to higher mortality of males
   during the larval stages.

Anderson, J. F. (1968). Microsporidia parasitizing mosquitoes collected in Connecticut.
<u>J. Invertebr. Pathol.</u>, <u>11</u>: 440-455.

   <u>Amblyospora</u> and <u>Parathelohania</u> (<u>Thelohania</u>) species were found infecting 8 species of
   mosquitos and <u>Stempellia magna</u> was found in one species.   Diagnostic measurements of
   the microsporidia are given.   All Thelohaniidae infections were observed in adipose
   tissue.

Anthony, D. W. et al. (1972). Nosematosis: Its effect on <u>Anopheles albimanus</u> Wiedemann,
and a population model of its relation to malaria transmission. <u>Proc. Helminthol. Soc.</u>
<u>Wash., Special Issue</u>, <u>39</u>: 428-433.

   Infection by <u>Nosema algerae</u> (<u>Nosema stegomyiae</u>) significantly reduced the number of eggs
   laid by infected female <u>A. albimanus</u> and reduced their ability to produce eggs after the
   2nd gonotrophic cycle.   At a high dosage (5 x 10^4 spores/ml) the resulting infections
   reduced female longevity by half.   A population model is proposed to illustrate the
   effect of the infection on the ability of the population to transmit malaria.

Bailey, D. L. et al. (1967a).   A new Maryland record of <u>Thelohania</u> (Nosematidae: Microsporidia).   <u>J. Invertebr. Pathol.</u>, <u>9</u>: 354-356.

   A microsporidium, <u>Amblyospora canadensis</u> (<u>Thelohania inimica</u>), was isolated from larval
   <u>Aedes canadensis</u> collected in Maryland.   Statistical analysis of spore measurements of
   the isolate and those from California and Pennsylvania revealed that there were significant differences between spore measurements from different locations, but in no case
   was the spore size from one population significantly different from all others.   This
   exemplifies the difficulty in positive identification of this group by spore size and
   host alone.   Measurements and microphotographs of developmental stages are included.

Bailey, D. L. et al. (1967b).   <u>Stempellia magna</u> (Kudo) (Nosematidae: Microsporidia) in
<u>Culex restuans</u> Theobald from Virginia.   <u>Mosq. News</u>, <u>27</u>(1): 111-114.

   Late instar <u>Culex restuans</u> larvae collected in Virginia had a 35% incidence of infection.
   Later collections of mixed species from the same site showed the infections only in

C. restuans. Anopheles stephensi[*] larvae reared in water from the site also produced infections. The external appearance and activity of some larvae and pupae suggested they were healthy, but microscopic examination revealed the infection. The infection may not be fatal.

Bano, L. (1958). Partial inhibitory effect of Plistophora culicis on the sporogonic cycle of Plasmodium cynomolgi in Anopheles stephensi. Nature, 181: 430.

P. culicis invaded the fat bodies and the malpighian tubules and covered the surface of the mid-gut. In infected mosquitos, only a few oocysts of P. cynomolgi were found; the rate of the oocyst development was retarded and many degenerated.

Bresslau E. & Buschkiel, M. (1919). Die Parasiten der Stechmückenlarven. IV. Mitteilung der Beiträge zur Kenntnis der Lebensweise unserer Stechmücken. Biol. Zentralbl., 39: 325-336.

Nosema culicis is described as a new species. It was found infecting Culex pipiens larvae. The new species is compared to other Nosema species and is distinguished mainly on spore size. Also an Amblyospora sp. (Thelohania sp.) was seen infecting Culiseta annulata.

Briggs, J. D. (1972). 1971 Activities of the WHO International Reference Centre for diagnosis of diseases of vectors. Unpublished document WHO/VBC/72.408, 6 pp.

Camey-Pacheco, H. L. (1965). Encuesta para determinar la presencia de Coelomomyces (Blastocladiales) y Thelohania (Sporozoa: Microsporidia), parásitos de larvas de Anopheles, en Guatemala, como contribución a su posible control biológico. Univ. San Carlos, 67: 139-182.

Parathelohania sp. (Thelohania legeri) was found infecting Anopheles pseudopunctipennis in 2 of 9 localities sampled. Overall 44% of the larvae were infected with the peak incidence (95.2%) occurring in Sanerate during the dry season.

Canning, E. U. (1957a). Plistophora culicis Weiser (Protozoa, Microsporidia): Its development in Anopheles gambiae. Trans. R. Soc. Trop. Med. Hyg., 51: 8.

(See Canning, 1957b.)

Canning, E. U. (1957b). On the occurrence of Plistophora culicis Weiser in Anopheles gambiae. Riv. Malario., 36: 39-50.

Although considerable cell destruction was observed in infected mosquitos, their life span was not appreciably shortened.

Canning, E. U. & Hulls, R. H. (1970). A microsporidan infection of Anopheles gambiae Giles, from Tanzania, interpretation of its mode of transmission and notes on Nosema infections in mosquitoes. J. Protozool., 17(4): 531-539.

The development of Nosema algerae collected from A. gambiae, was traced in laboratory bred A. stephensi. Every tissue of the larvae, pupae and adults was attacked. Infected ova did not develop into viable larvae. Uninfected larvae hatching from unwashed eggs could have derived their infections by ingesting spores from outside of the eggshell, which in turn had picked up spores from ovarian connective tissue. N. algerae also attacks culicine mosquitos.

---

[*] Note: Recent unpublished data indicate that S. magna is not infectious to mosquitos other than C. restuans.

Chapman, H. C. (1966).   The mosquitoes of Nevada.   Univ. Nev. Bull., T-2, 43 pp.

Amblyospora spp. (Thelohania) were found in larvae of Culex tarsalis, Culiseta inornata, C. indicens, Aedes cinereus, A. melanimon, and A. dorsalis.

Chapman, H. C. (1974).   Biological control of mosquito larvae.   Ann. Rev. Entomol., 19: 33-59.

Literature concerning the biological control of mosquito larvae by pathogens is reviewed. The pathogen groups cited include: protozoa, bacteria, viruses, rickettsia, fungi, and nematodes.   The host pathogen-relationships, pathogen distribution and taxonomy, labora-tory studies, and field trials are discussed.   There is an apparent need for methods of mass production of pathogens to facilitate field studies.

Chapman, H. C. & Kellen, W. R. (1967).   Plistophora caecorum sp. n., a microsporidian of Culiseta inornata (Diptera: Culicidae) from Louisiana.   J. Invertebr. Pathol., 9: 500-502.

The microsporidium invades the gastric caecae of the mosquito and is transmitted trans-ovarially.   Pleistophora caecorum is not lethal to any stage of the mosquito nor does it affect the fecundity or life span of the adults.   Larvae of Culex salinarius, C. restuans, Anopheles bradleyi, A. crucians, Aedes sollicitans and Ae. vexans often occurred in the same pools with infected Culiseta inornata but no larvae of these companion species were infected.

Chapman, H. C. et al. (1966).   Host-parasite relationships of Thelohania associated with mosquitoes in Louisiana (Nosematidae: Microsporidia).   J. Invertebr. Pathol., 8: 452-456.

Microsporidia of the genera Amblyospora and Parathelohania (Thelohania) infected 13 species of mosquitos collected from 2 parishes in south-western Louisiana.   New host records were established for several mosquitos.   The host-parasite relationships of the mosquitos and associated Microsporida were classified into four types by the method described in Kellen et al. (1965).

Chapman, H. C. et al. (1967).   Pathogens and parasites in Louisiana Culicidae and Chaoboridae. Proc. N. J. Mosq. Exterm. Assoc., 54: 54-60.

A survey of pathogens and parasites was conducted from 1964 to 1967.   The field collec-tions are summarized by host species and pathogen genus.   Amblyospora and Parathelohania (Thelohania), Nosema, Pleistophora (Plistophora), and Stempellia species were found in one or more mosquito hosts.   The level of infection rarely exceeded 1%.   Several new host records are given.

Chapman, H. C. et al. (1969).   A two-year survey of pathogens and parasites of Culicidae, Chaoboridae, and Ceratopogonidae in Louisiana.   Proc. N. J. Mosq. Exterm. Assoc., 56: 203-212.

Twenty species of mosquitos collected in Louisiana were host to one or more species of microsporidan parasites.   Amblyospora (Thelohania) was the most common genus.   Levels of infection in the field rarely exceeded 1%.   The first record of Pleistophora culicis in the US is given.

Chapman, H. C. et al. (1970a).   Protozoans, nematodes, and viruses of anophelines.   Misc. Publ. Entomol. Soc. Am., 7: 134-139.

A review of the species of Anopheles reported as natural or experimentally induced hosts of various genera of Microsporida is given.

Chapman, H. C. et al. (1970b). A container for use in field studies of some pathogens and parasites of mosquitoes. Mosq. News, 30(1): 90-93.

A container was developed that allows pathogens to contact potential hosts but confines the host and excludes predators. The containers were tested in various locations where pathogens were known to be present. Larvae of various species of mosquitos were placed in the containers and fungi, Microsporida, nematodes, and viruses were successfully transmitted to them from the surrounding water.

Chapman, H. C. et al. (1973). Thelohania (Nosematidae: Microsporida) in Aedes mosquitoes of Alaska. Mosq. News, 33: 465-466.

Eight species of Aedes collected in Alaska were infected with a microsporidan, Amblyospora (Thelohania) sp. The Aedes species were: cataphylla, communis, excrucians, fitchii, hexodontus, pullatus, punctor, and riparius.

Chatton, E. (1911). Microsporidies considérées comme causes d'erreurs dans l'étude du cycle évolutif des trypanosomides chez les insectes. Bull. Soc. Pathol. Exot., 4: 662-664.

Chatton reports that the etiology of yellow fever was confused because of the presence of Pleistophora stegomyiae in the mosquito vector.

Clark, T. B. & Fukuda, T. (1967). Stempellia magna in the tree hole mosquito, Aedes sierrensis. J. Invertebr. Pathol., 9: 430-431.

Two fourth-stage Aedes sierrensis larvae infected with the microsporidan Stempellia magna, were found in California. This is the first report of Stempellia infecting the genus Aedes. It is evident that S. magna is common in A. sierrensis in an area of California but heavy infections are rare.

Clark, T. B. & Fukuda, T. (1971). Pleistophora chapmani n. sp. (Cnidospora: Microsporida) in Culex territans (Diptera: Culicidae) from Louisiana. J. Invertebr. Pathol., 18: 400-404.

In infected larvae, the parasite damages the cells of the mid-gut and the gastric caeca, but the larvae are able to pupate and emerge as apparently healthy adults. The microsporidium was not successfully transmitted to mosquitos by feeding them spores or cysts.

Fantham, H. B. et al. (1941). Some Microsporidia found in certain fishes and insects in eastern Canada. Parasitology, 33: 186-208.

Parathelohania illinoisensis (Thelohania legeri) is recorded from larvae of Anopheles punctipennis in Quebec Province and A. gambiae in Zululand. Some general remarks about zoogeographical distribution, habitat, host-specificity, and origin of species of Microsporida are given.

Fox, R. M. & Weiser, J. (1959). A microsporidian parasite of Anopheles gambiae in Liberia. J. Parasitol., 45: 21-30.

Nosema algerae (N. stegomyiae) was first observed with low frequency in natural populations of A. gambiae in Liberia. Later an outbreak of the disease occurred in a laboratory colony. Diseased mosquitos could not be recognized from external characteristics, but were evident from mid-gut dissections. The microsporidium invaded all tissues and organs except the air sacs, tracheae, and coelomic lumina. No infected larvae were seen.

Franz, D. R. & Hagmann, L. E. (1962).   A microsporidian parasite of <u>Aedes stimulans</u> (Walker).
<u>Mosq. News</u>, <u>22</u>: 302-303.

    This is the first record of <u>Amblyospora</u> sp. (<u>Thelohania</u>) in <u>Aedes stimulans</u> and first
microsporidan reported in a New Jersey mosquito.   Heavily infected larvae appear to
have a lumpy, whitish fat body resulting from the immense concentration of spores.   In
late stages of the infection, the fat body disintegrates and the larva becomes essen-
tially a sac of spores.

Garnham, P. C. C. (1957).   Microsporidia in laboratory colonies of <u>Anopheles</u>.   <u>Bull. W.H.O.</u>,
<u>15</u>: 845-847.

    The similarity of appearance between <u>Pleistophora culicis</u> pansporoblasts and <u>Plasmodium</u>
oocysts present difficulties when using <u>P. culicis</u> infected colonies for malaria
research.   Also, microsporidan infections may interfere with normal development of
oocysts and sporozoites.

Garnham, P. C. C. (1959).   Some natural protozoal parasites of mosquitoes with special
reference to <u>Crithidia</u>.   <u>Trans. 1st Int. Conf. Insect Pathol. Biol. Control (Prague)</u>,
pp. 287-294.   (E, r)

    <u>Pleistophora culicis</u> disappeared spontaneously from a mosquito colony previously 100%
infected.   No explanation is given.

Gol'berg, A. M. (1971).   /Microsporidia of <u>Culex pipiens</u> L. mosquitos.7   <u>Med. Parazitol</u>.
<u>Parazit. Bolezn</u>., <u>40</u>: 204-206.   (R, e)

    <u>Thelohania</u>* sp. was found infecting all stages of development in early spring while
<u>Weiseria spinosa</u> (a new species) was found in late summer-early autumn.   Infection
incidence of <u>W. spinosa</u> reached 93.4% of the larvae and 27.5% of the pupae.   Infection
of adults was rare.

Hagmann, L. E. (1972).   The occurrence of enzootic mosquito pathogens in New Jersey.   <u>Proc.</u>
<u>N. J. Mosq. Exterm. Assoc</u>., <u>59</u>: 175-176.

    Diseased mosquito larvae have been widely collected in New Jersey.   They are apparently
simultaneously infected with species of Thelohaniidae and the turquoise strain of
mosquito iridescent virus.   The attack rate is very low and stable from year to year,
fluctuating between 1 and 3% of the larval population.

Hazard, E. I. (1970).   Microsporidian diseases in mosquito colonies: <u>Nosema</u> in two <u>Anopheles</u>
colonies.   <u>Proc. 4th Int. Colloq. Insect Pathol</u>., pp. 267-271.   (E, f, g)

    Microsporidan epizootics have been found in mosquito colonies at the University of
Maryland, School of Medicine and at Walter Reed Army Institute of Research.   The
epizootic probably originated in colonies of <u>Anopheles gambiae</u> shipped from the London
School of Hygiene and Tropical Medicine.   Both American insectaries have colonized
mosquitos of the genera <u>Aedes</u>, <u>Anopheles</u>, and <u>Culex</u>;   however, <u>Nosema algerae</u> was found
in only <u>Anopheles</u> species.

Hazard, E. I. (1972).   Investigation of pathogens of anopheline mosquitos in the vicinity of
Kaduna, Nigeria.   Unpublished document WHO/VBC/72.384, 6 pp.

---

* Doubtful generic placement.

A survey of the pathogens and parasites of mosquitos in the savanna areas near Kaduna was conducted. Several species of Microsporida were found with low incidence infecting Anopheles funestus and A. gambiae.

Hazard, E. I. & Anthony, D. W. (1974). A redescription of the genus Parathelohania Codreanu 1966 (Microsporida: Protozoa) with a reexamination of previously described species of Thelohania Henneguy 1892 and descriptions of two new species of Parathelohania from anopheline mosquitoes. USDA Tech. Bull., 1505: 26 pp.

Taxonomic revision.

Hazard, E. I. & Fukuda, T. (1974). Stempellia milleri sp. n. (Microsporida: Nosematidae) from the mosquito Culex pipiens quinquefasciatus: Its morphological and behavioral characteristics as compared to other described Stempellia diseases. J. Protozool., 21(4): 497-504.

S. milleri was found in the blood cells and adipose tissue of field collected C. fatigans (= C. p. quinquefasciatus). Two types of spores were produced in the adipose cells and these were examined by electron microscopy. Host range experiments were conducted.

Hazard, E. I. & Lofgren, C. S. (1971). Tissue specificity and systematics of a Nosema in some species of Aedes, Anopheles, and Culex. J. Invertebr. Pathol., 18: 16-24.

The tissues invaded by Nosema algerae differ in different hosts. This variation in susceptibility illustrates the danger of differentiating Microsporida solely on the basis of the sites of infection.

Hazard, E. I. & Oldacre, S. (1973). Bimorphism in some Microsporida resulting in two generic types of vegetative development and two morphologically distinct spores. (Abstr.) Pap. 5th Int. Colloq. Insect Pathol. Microb. Control (Oxford), p. 12. (E, f, g)

Because of the recent discovery of developmental dimorphism in Microsporida in mosquitos, a possible dual infection of Nosema and Stempellia in larvae of Culex quinquefasciatus, among others, was examined. Both generic types could be morphological forms of the same species, particularly since neither type inhibits the growth of the other.

Hazard, E. I. & Oldacre, S. W. (1975). Revision of Microsporidia (Protozoa) close to Thelohania with descriptions of one new family, eight new genera, and thirteen species. USDA Tech. Bull. In press.

Taxonomic revision.

Hazard, E. I. & Savage, K. E. (1970). Stempellia lunata sp. n. (Microsporida: Nosematidae) in larvae of the mosquito Culex pilosus collected in Florida. J. Invertebr. Pathol., 15: 49-54.

A new species of Stempellia is described in C. pilosus collected from a flooded roadside ditch in Florida. Infected larvae were dull white, particularly in the thoracic region, but the infection is not easily diagnosed with the naked eye.

Hazard, E. I. & Weiser, J. (1968). Spores of Thelohania in adult female Anopheles: Development and transovarial transmission, and redescriptions of T. legeri Hesse and T. obesa Kudo. J. Protozool., 15(4): 817-823.

Two distinct spore types were found in Anopheles quadrimaculatus and A. crucians infected by Parathelohania anophelis and P. obesa, respectively. One type is found in the larvae and was used in the original descriptions of Kudo. The other type develops in the blood cells of adult females; one of which was previously described as <u>Nosema</u>. Confusion in microsporidan taxonomy has arisen because of the difference in appearance of spores in different stages of the mosquito.

Hesse, E. (1904a). <u>Thelohania legeri</u> n. sp. microsporidie nouvelle, parasite des larves d'<u>Anopheles maculipennis</u> Meig. <u>C. R. Hebd. Seances Mem. Soc. Biol. Ses. Fil.</u>, <u>57</u>: 570-571.

Hesse, E. (1904b). Sur le développement de <u>Thelohania legeri</u> Hesse. <u>C. R. Hebd. Seances Mem. Soc. Biol. Ses. Fil.</u>, <u>57</u>: 571-572.

Hesse, E. (1905). Microsporidies nouvelles des insectes. <u>C. R. Assoc. Fr. Avanc. Sci.</u>, <u>33</u>: 917-919.

Hesse, E. (1935). Sur quelques Microsporidies parasites de <u>Megacyclops viridis</u> Jurine. <u>Arch. Zool. Exp. Gèn.</u>, <u>75</u>: 651-661.

Iturbe, J. & Gonzalez, E. (1921). Contribución del Laboratorio Iturbe en el 3. Congreso Venezolano de Medicina, Caracas. Tip. Cultura Venezolana, 35 pp.

Iyengar, M. O. T. (1930). Microsporidian parasites of <u>Anopheles</u> larvae. <u>Trans. 7th Congr. Far East Assoc. Trop. Med. (Calcutta)</u>, <u>3</u>: 136-142.

Considerable mortality among anopheline larvae was caused by <u>Parathelohania</u> (<u>Thelohania</u> spp.) spp. in Lower Bengal. Serial sections of larvae revealed that the parasites were confined to the fat body.

Jenkins, D. W. (1964). Pathogens, parasites and predators of medically important arthropods. <u>Bull. W.H.O.</u>, <u>30</u> (Suppl.), 150 pp.

A review of the literature through 1963 is given.

Kellen, W. R. (1962). Microsporidia and larval control. <u>Mosq. News</u>, <u>22</u>: 87-95.

The species of <u>Amblyospora</u>, <u>Nosema</u> and <u>Parathelohania</u> found parasitizing mosquitos in California are listed along with their hosts and the appearance of infected larvae.

Kellen, W. R. (1963). Research on biological control of mosquitoes. <u>Proc. Calif. Mosq. Control Assoc.</u>, <u>31</u>: 23-25.

This is a report on the research activities of the Bureau of Vector Control in Fresno, California. The most common pathogens of mosquitos in California are species of Thelohaniidae. Infection levels usually range from 5% to less than 1% but occasionally more intense infections are found.

Kellen, W. R. (1965). Microsporida as parasites of mosquitoes. <u>Proc. 12th Int. Congr. Entomol. (London)</u>, pp. 728-729.

See Kellen et al., 1965.

Kellen, W. R. & Lipa, J. J. (1960). <u>Thelohania californica</u> n. sp., a microsporidian parasite of <u>Culex tarsalis</u> Coquillett. <u>J. Insect Pathol.</u>, <u>2</u>: 1-12.

The new species was found infecting larvae in California. It is highly pathogenic; less than 2% of the infected larvae observed in the laboratory emerged as adults.

/Note: This species has now been placed in the genus <u>Amblyospora.</u>/

Kellen, W. R. & Wills, W. (1962a).   New Thelohania from California mosquitoes (Nosematidae: Microsporidia).   J. Insect Pathol., 4: 41-56.

New species are described: Amblyospora gigantea, benigna, noxia, inimica, campbelli, unica, bolinasae, and Parathelohania (Thelohania) periculosa.

/Note: The generic placement of these species has been changed by Hazard & Oldacre, 1975./

Kellen, W. R. & Wills, W. (1962b).   The transovarian transmission of Thelohania californica Kellen and Lipa in Culex tarsalis Coquillett.   J. Insect Pathol., 4: 321-326.

Amblyospora californica is transovarially transmitted in C. tarsalis; however, sporogony does not occur in the female progeny.   The infections are lethal to the males but are benign in the females and do not interfere with their reproductive capacity.

Kellen, W. R. et al. (1965).   Host-parasite relationships of some Thelohania from mosquitoes (Nosematidae: Microsporidia).   J. Invertebr. Pathol., 7: 161-166.

By examining serial sections of infected mosquitos, it has been possible to determine relationships between sex, tissue specificity, suppression of sporogony, and the expression of patent infections.   From these observations, four basic types of infection are defined.

Kellen, W. R. et al. (1966a).   Transovarial transmission of some Thelohania (Nosematidae: Microsporidia) in mosquitoes of California and Louisiana.   J. Invertebr. Pathol., 8: 355-359.

Examination of larvae reared from field collected eggs confirmed that Amblyospora and Parathelohania (Thelohania) are frequently transmitted transovarially.   Progenies of Culex apicalis and Anopheles quadrimaculatus were most frequently infected with transovarially acquired Parathelohania 33.3 and 17.3%, respectively.   Peroral transmission may be a prerequisite for transovarial transmission.

Kellen, W. R. et al. (1966b).   Development of Thelohania californica in two hybrid mosquitoes. Exp. Parasitol., 18: 251-254.

Amblyospora californica was transovarially transmitted to Culex tarsalis x C. erythrothorax hybrids but did not develop and developed normally in C. tarsalis x C. peus hybrids.

Kellen, W. R. et al. (1967).   Two previously undescribed Nosema from mosquitoes of California (Nosematidae: Microsporidia).   J. Invertebr. Pathol., 9: 19-25.

Parathelohania periculosa which infects adult Anopheles franciscanus and Amblyospora californica which infects adult Culex tarsalis were mistakenly described as new species of Nosema.

Khaliulin, G. L. (1973).   /Microsporidia Thelohania sp., a parasite of larvae of Aedes excrucians Walker in the Mari-ASSR USSR./   Parazitologiya, 7: 180-182 (Russian, English summary).

The parasite was localized in the fat body and haemolymph of the larvae.   Three to eight spores per sporont were observed.

Khaliulin, G. L. & Ivanov, S. L. (1971). /Infection of larvae of <u>Aedes communis</u> DeG. with the microsporidia <u>Thelohania opacita</u> Kudo in the Mari ASSR./ <u>Parazitologiya</u>, <u>5</u>: 98-100 (Russian, English summary).

    The infection was localized in the fat body and haemolymph of the host. In nature 4-18% of the larvae are infected and 98-100% of those apparently infected die.

    /Note: T. opacita var. <u>mariensis</u> is now <u>Amblyospora khaliulini</u> Hazard & Oldacre./

Kramer, J. P. (1969). Parasite in laboratory colonies of mosquitoes. <u>Bull. W.H.O.</u>, <u>31</u>(4): 475-478.

    What is known concerning the biology of some typical parasites that are restricted to mosquitos is briefly outlined. Special reference is made to signs of infection in the host, transmission of the parasite, and effects on the host.

Kudo, R. (1920). On the structure of some microsporidian spores. <u>J. Parasitol.</u>, <u>6</u>: 178-182.

    <u>Stempellia magna</u> (<u>Thelohania magna</u>) is described as a new species. It was found infecting <u>Culex pipiens</u> larvae near Urbana, Illinois. Its spore shape and structure are compared with other Microsporida.

Kudo, R. (1921). Studies on microsporidia, with special reference to those parasitic in mosquitoes. <u>J. Morphol.</u>, <u>35</u>: 153-182.

    <u>Stempellia magna</u> (<u>Thelohania magna</u>) and <u>Parathelohania illinoisensis</u> (<u>Thelohania illinoisensis</u>) are described.

Kudo, R. (1922). Studies on microsporidia parasitic in mosquitoes. II. On the effect of the parasites upon the host body. <u>J. Parasitol.</u>, <u>8</u>: 70-77.

    Infections of <u>Stempellia magna</u> (<u>Thelohania magna</u>) in <u>Culex restuans</u> (= <u>C. territans</u> in Kudo) larvae, <u>Parathelohania illinoisensis</u> (<u>Thelohania illinoisensis</u>) in <u>Anopheles quadrimaculatus</u> larvae and <u>Amblyospora opacita</u> (<u>Thelohania opacita</u>) in <u>Culex territans</u> (= <u>C. apicalis</u> in Kudo) larvae are described. In all cases, infected individuals could be recognized by the white opaque appearance of their bodies. Infection experiments with <u>S. magna</u> were conducted.

Kudo, R. (1924a). Studies on microsporidia parasitic in mosquitoes. III. On <u>Thelohania legeri</u> Hesse (= <u>Th. illinoisensis</u> Kudo). <u>Arch. Protistenkd.</u>, <u>49</u>: 147-162.

    The process of schizogony in <u>Parathelohania illinoisensis</u> (<u>Thelohania illinoisensis</u>) is described. Kudo discounts reports of other authors that this parasite does not produce serious deleterious effects on the host larvae.

Kudo, R. (1924b). Studies on microsporidia parasitic in mosquitoes. VI. On the development of <u>Thelohania opacita</u>, a culicine parasite. <u>J. Parasitol.</u>, <u>11</u>: 84-88.

    The development of <u>Amblyospora opacita</u> (<u>Thelohania opacita</u>) in <u>Culex territans</u>, and <u>C. sp.</u> is given. The infection was fatal to the larvae concerned.

Kudo, R. (1924c). A biologic and taxonomic study of the Microsporidia. <u>Ill. Biol. Monogr.</u>, <u>9</u>: 76-344.

    Taxonomic revision.

Kudo, R. (1925a).   Studies on Microsporidia parasitic in mosquitoes.   IV. Observations upon the Microsporidia found in the mosquitoes of Georgia, U.S.A.   Zentralbl. Bakteriol. Parasitenkd. Infektionskr. Hyg. Abt. I Orig., 96: 428-440.

Several Microsporida were found infecting mosquito larvae in Georgia.   Although the intensity of infection was not great (about 4.5%), the infections were widespread. Also infections in adult anophelines were observed.

Kudo, R. (1925b).   Studies on microsporidia parasitic in mosquitoes.   V. Further observations upon Stempellia (Thelohania) magna Kudo, parasitic in Culex pipiens and C. territans. Biol. Bull. (Woods Hole), 48: 112-127.

C. restuans = C. territans in Kudo.

Kudo, R. (1929).   Studies on microsporidia parasitic in mosquitoes.   VII. Notes on microsporidia of some Indian mosquitoes.   Arch. Protistenkd., 67: 1-10.

Parathelohania obscura (Thelohania obscura) and P. indica (T. indica) are described as new species from larvae of Anopheles funestus and A. hyrcanus respectively, collected in India.   Parathelohania spp. were also observed in several Anopheles species.

Kudo, R. (1930).   Studies on microsporidia parasitic in mosquitoes.   VIII. On a microsporidan, Nosema aedis nov. spec., parasitic in a larva of Aedes aegypti of Porto Rico. Arch. Protistenkd., 69: 23-38.

The microsporidium was observed in the adipose tissue of the larva with the fat body of the thorax most heavily infected.   The new species was mainly distinguished from other Nosema species by spore size.

Kudo, R. R. (1960).   Protozoan parasites in certain insects of medical importance, pp. 46-66. In /Proceedings of a conference on the/ Biological Control of Insects of Medical Importance. Am. Inst. Biol. Sci., Washington, D.C.

Kudo, R. R. (1962).   Microsporidia in southern Illinois mosquitoes.   J. Insect Pathol., 4: 353-356.

Culex restuans and C. pipiens pipiens infected with Stempellia magna were collected from small pools.   Natural infection in more than 80% of the larvae originating in one egg raft, and the occurrence of infected adult host insects indicate that the infection is transmitted through eggs to larvae.

Kudo, R. R. & Daniels, E. W. (1963).   An electron microscope study of the spore of a microsporidian, Thelohania californica.   J. Protozool., 10(1): 112-120.

The ultrastructure of the spore is described and it is apparently a typical microsporidan spore.

/Note: T. californica is now considered Amblyospora californica./

Lainson, P. & Garnham, R. C. C. (1957).   Stages of Plistophora culicis encountered during dissections of Anopheles stephensi.   Trans. R. Soc. Trop. Med. Hyg., 51: 6.

P. culicis was found in the malpighian tubules, fat body and on walls of the mid-gut in larvae and adults.   In nature, infection incidence was about 1%, but can be up to 100% in laboratory colonies.

Laird, M. (1956). Studies of mosquitoes and freshwater ecology in the South Pacific.
Bull. R. Soc. N. Z., 6: 213 pp.

Details are given of the parasites associated with mosquitos (particularly anophelines)
in the South Pacific.

Laird, M. (1959a). Malayan protozoa. 1. Plistophora collessi n. sp. (Sporozoa: Microsporidia), an ovarian parasite of Singapore mosquitoes. J. Protozool., 6: 37-45.

Ovarian cysts were observed in 1% of field collected Culex tritaeniorhynchus and in 2
C. gelidus females. These cysts contained Pleistophora collessi spores. The parasite
caused degeneration of the oocyte, nurse cells, and epithelium of the invaded follicle.

Laird, M. (1959b). Parasites of Singapore mosquitoes, with particular reference to the
significance of larval epibionts as an index of habitat pollution, Ecology, 40: 206-221.

About 1% of Culex tritaeniorhynchus siamensis females had Pleistophora (Plistophora)
collessi in their ovaries.

Laird, M. (1961). New American locality records for four species of Coelomomyces (Blastocladiales: Coelomomycetaceae). J. Insect Pathol., 3: 249-253.

A larva of Anopheles walkeri containing a great number of Parathelohania (Thelohania)
spores was collected in Minnesota.

Laird, M. (1966). Some protozoa from Singapore insects. Proc. 1st Int. Congr. Parasitol.
(Rome), 1: 595-596.

A microsporidan parasitizing larval Aedes albopictus is reported. It may be a new
species of Amblyospora. The pansporoblasts are abundant in the fat body of infected
larvae.

Larson, L. V. (1970). Environmental factors and the natural occurrence of microsporidian-
infected mosquito larvae. Diss. Abstr. B. Sci. Eng., 31: 3460.

A field study determined that some environmental factors were correlated with the level
of infection in pools with mosquitos infected by Microsporida.   Temperature was
inversely correlated to percent infection of Culex tarsalis and C. restuans.   Carbon
dioxide, pH, and salinity, although limiting in their higher values, were not strictly
correlated in their lower ranges.   Infection range in this study was 0.6-15.8%.

Levchenko, N. G. & Dzerzhinskii, V. A. (1973). /Noxious effect of Thelohania opacita
(Microsporidia, Nosematidae) on larvae of blood-sucking mosquitos (Diptera, Culicidae)./
Izv. Akad. Nauk Kaz. SSR Ser. Biol. Nauk, 1: 58-61 (Russian).

Larvae of Aedes flavescens, A. caspius caspius, and Culex modestus collected from natural
reservoirs on the Ili river floodplain were infected by Amblyospora (T. opacita).
Laboratory experiments showed that no infected Aedes larvae survived to breed and only
1.4% of the Culex larvae did (compared to 5.1% of the controls).

Levchenko, N. G. et al. (1971a). /Detection of microsporidia in larvae of blood-sucking
mosquitos in the southeastern Kazakhstan./ Med. Parazitol. Parazit. Bolezn., 40: 619-620
(Russian, English summary).

Microsporidan infections have been discovered in mosquito larvae in south-eastern
Kazakhstan, SSR. Amblyospora spp. (Thelohania opacita) were found parasitizing Aedes

montchadskyi and A. flavescens. Parathelohania sp. (Thelohania legeri) was found in Anopheles hyrcanus.

Levchenko, N. G. et al. (1971b). /Detection of microsporidia in larvae of mosquitos and midges./ Alma Ata. Akad. Nauk Kazak, SSR Khabarshysy Vestnik, 9: 69-70.

Microsporida, possibly Amblyospora (Thelohania), have been found infecting larvae of Culex pipiens, Aedes cinereus, A. kasachstanicus, A. stramineus and Culicoides sp. in Kazakhstan.

Lom, J. & Weiser, J. (1972). Surface pattern of some microsporidian spores as seen in the scanning electron microscope. Folia Parasitol. (Praha), 19: 359-363.

Microsporida spores from several insect- and fish-invading species have been studied with the scanning electron microscope (SEM). This technique reveals minute details of their surfaces. This preliminary study indicates that the use of SEM may supply additional features for species differentiation.

Lutz, A. & Splendore, A. (1908). Uber Pebrine und verwandt Mikrosporidien. Zweite Mitteilung Zentralbl. Bakteriol. Parasitenkd. Infektionskr. Hyg. Abt. I Orig., 46: 311-315.

Nosema lutzi was observed in the intestines of adult Aedes aegypti (= Stegomyia fasciata).

/Note: N. lutzi may not be a valid species./

Marchoux, E. & Simond, P. L. (1906). Etudes sur la fièvre jaune. Deuxième Mémorie. Ann. Inst. Pasteur (Paris), 20: 16-40.

Marchoux, E. et al. (1903). La fièvre jaune. Rapport de la mission française. Ann. Inst. Pasteur (Paris), 17: 665-731.

Miller, F. M. et al. (1973). The biology of a new species of Stempellia, a microsporidium parasite of the mosquito Culex quinquefasciatus. Folia Entomol. Mex., 25-26.

Missiroli, A. (1928). Alcuni protozoi parassiti dell'Anopheles maculipennis. Riv. Malariol., 7: 1-3. (I, f, g, e)

Parathelohania legeri (Nosema sp.) and Sarcocystis anophelis were observed in adult A. maculipennis in Italy. S. anophelis is described as a new species.

Missiroli, A. (1929). Sui microsporidi parassiti dell'Anopheles maculipennis. Riv. Malariol., 8: 393-400. (I, f, e, g)

Thelohania grassi,* a new species infecting eggs of A. maculipennis, is described. Also, Parathelohania legeri was seen in the fat body of adult females.

Nöller, W. (1920). Kleine Beobachtungen an parasitischen Protozoan. (Zugleich vorläufige Mitteilung über die Befruchtung and Sporogonie von Lankesterella minima Chaussat.) Arch. Protistenkd., 41: 169-189.

Nöller reports seeing two Microsporida in mosquito larvae collected near Hamburg. These have been since determined to be Amblyospora khaliulini in Aedes communis and Amblyospora sp. in A. cantans.

---

* Doubtful generic placement.

Novák, D. (1969). Bemerkungen zum Auftreten infizierter Stechmückenlarven in Südmähren. Z. Tropenmed. Parasitol., 20: 229-231. (G, e)

Larvae of Aedes communis and A. cantans infected with the Microsporidan Amblyospora khaliulini and an Amblyospora sp. (Thelohania opacita) respectively were collected from pools in forests and the ditches passing through meadows.

Pillai, J. S. (1968). Thelohania barra n. sp., a microsporidian parasite of Aedes (Halaedes) australis Erichson, in New Zealand. Z. Angew. Entomol., 62: 395-398. (E, g)

The infection is easily recognized by the characteristic mottled appearance of infected larvae. The larva becomes sluggish and dies before pupating. At the type locality, about 3.5% of the larvae were infected. The infected larvae were found in a disused drain in brackish water.

/Note: T. barra has now been placed in the genus Parathelohania.7

Pillai, J. S. (1974). Pleistophora milesi, a new species of Microsporida from Maorigoeldia argyropus Walker (Diptera: Culicidae) in New Zealand. J. Invertebr. Pathol., 24(2): 234-237.

Diseased M. argyropus larvae were recognized by the presence of cysts in the abdominal segments. Some infected larvae were able to complete development and emerge as viable adults. Apparently the hosts are able to encapsulate and melanize the invading pathogens, but the reaction is variable. Larvae of Aedes notoscriptus were highly susceptible to the pathogen. All infected larvae of A. notoscriptus failed to pupate.

Reynolds, D. G. (1966). Infection of Culex fatigans with a microsporidian. Nature, 210: 967.

Larvae and adults of C. fatigans were experimentally infected with a microsporidan which was provisionally identified as Pleistophora culicis. No external signs of infection were noticed. The infection did not prevent larvae pupating or adults emerging from these pupae.

Reynolds, D. G. (1970). Laboratory studies of the microsporidian Plistophora culicis (Weiser) infecting Culex pipiens fatigans Wied. Bull. Entomol. Res., 60: 339-349.

At 27°, newly hatched larvae were exposed to 6000 spores/ml and 12.9% reduction in net reproductive rate occurred. Although the rate of the laboratory population was reduced by the pathogen, the effect is not sufficient to reduce a wild population, especially because of the slow action of the parasite.

Reynolds, D. G. (1971). Parasitism of Culex fatigans by Nosema stegomyiae. J. Invertebr. Pathol., 18: 429.

The reproduction rate (R, the number of offspring per female) was less in a population of Culex fatigans experimentally infected with Nosema algerae (N. stegomyiae) than the control population. The degree of reduction in R varied with the amount of inoculum to which the population was exposed. These reductions were caused by 2 methods: reduction in egg production, and increase in nonhatching eggs. Parasitism did not have a significant effect on the mortality rate.

Reynolds, D. G. (1972). Experimental introduction of a microsporidian into a wild population of Culex pipiens fatigans Wied. Bull. W.H.O., 46(6): 807-812.

Pleistophora culicis was introduced into a wild population on Nauru Island in the Pacific. Two years later the pathogen was still present, but its incidence was not high enough to affect the population adversely.

Savage, K. E. & Lowe, R. E. (1970).   Studies of _Anopheles quadrimaculatus_ infected with a _Nosema_ sp.   Proc. 4th Int. Colloq. Insect Pathol., pp. 272-278.   (E, f, g)

Mortality, percent of infection, and degree of infection of the mosquitos was correlated to the concentration of _Nosema_ spores to which the first instar larvae were exposed. Twenty-four hours after exposure to 7.2 x 10^6 spores/cc, 75% of the first instars had died.   A technique for detecting infection in adults is described.

Savage, K. E. et al. (1971).   Studies of the transmission of _Plasmodium gallinaceum_ by _Anopheles quadrimaculatus_ infected with a _Nosema_ sp.   Bull. W.H.O., 45: 845-847. (WHO/VBC/70.237, WHO/MAL/70.734)

When 52 female adult mosquitos infected with _Nosema_ sp. were fed on the blood of chicks infected with malaria, sporozoites were found in 15.4%.   However, sporozoites were found in 61.5% of the disease free mosquitos.   Malaria was transmitted by diseased mosquitos in only 3 tests out of 6 compared to 5 tests out of 7 for the disease free group.

Sen, P. (1941).   On the microsporidia infesting some anophelines of India.   J. Malaria Inst. India, 4: 257-261.

Parathelohania sp. (_Thelohania legeri_), _P. indica_ (_T. indica_), and _P. anomala_ (_T. anomala_) infected anopheline larvae collected near Calcutta.   The parasites were present throughout the year but were particularly prevalent between April and September.   Heavily infected larvae died within 2-6 days in the laboratory.

Simmers, J. W. (1974a).   _Parathelohania legeri_ infecting _Psorophora ciliata_.   J. Invertebr. Pathol., 23: 402.

One fourth instar larva infected by the microsporidan was collected from a pond overflow area in Illinois.   The infected larva had no visible external symptoms.   No microsporidan parasites have been previously identified from _P. ciliata_.   An infected _Anopheles punctipennis_ larva was found in the same area.

/Note: Hazard & Oldacre (1975) dispute Simmers' identification of the microsporidium and _P. ciliata_ as a host of _Parathelohania_./

Simmers, J. (1974b).   Microsporidia of the genus _Parathelohania_ occurring in three sympatric species of mosquitoes, _Culiseta inornata_, _Culex pipiens_, and _Culex territans_ in southern Illinois.   Trans. Ill. State Acad. Sci., 67(1): 16-19.

Simmers reports _Culiseta inornata_, _Culex pipiens_, and _C. territans_ as hosts of _Amblyospora opacita_ (_Parathelohania opacita_).

/Note: Hazard & Oldacre (1975) claim that _C. inornata_ and _C. pipiens_ had only ingested the spores of _A. opacita_ and were not infected after examining Simmers stained smears of these two mosquitos./

Simmers J. W. (1974c).   New host records of _Stempellia magna_, a microsporidan parasite of mosquitoes.   J. Parasitol., 60(4): 721-722.

In Illinois, the microsporidium was found infecting _Culiseta inornata_ and _Anopheles punctipennis_ for the first time as well as _Culex restuans_.   Sporogonic stages were present in the fat bodies and mature spores packed the haemocoels in all infections.   A concurrent infection of <u>Culiseta inornata</u> by <u>Parathelohania opacita</u> and <u>Stempellia magna</u> is also reported.

/Note: Hazard & Oldacre (1975) dispute Simmers' report of <u>C. inornata</u> as a host for <u>A. opacita</u> and <u>S. magna</u> and also his report of <u>A. punctipennis</u> as a host of <u>S. magna</u>./

Simond, P. L. (1903).   Note sur une sporozaire du genre <u>Nosema</u>, parasite du <u>Stegomyia fasciata</u>.   <u>C. R. Hebd. Seances Mem. Soc. Biol. Ses. Fil.</u>, <u>55</u>: 1335-1337.

Speer, A. J. (1927).   Compendium of the parasites of mosquitoes (Culicidae).   <u>Hyg. Lab. Bull. (Wash.)</u>, <u>146</u>: 1-36.

This is a list of mosquito parasites recorded throughout the world.

Tour, S. (1969).   Ecologie des microsporidioses chez les Culicides du "Midi" Mediterranean. Doctoral Thesis Université de Montpellier, France.

Tour, S. et al. (1971).   Systématique et écologie de microsporidies (Microsporidia-Nosematidae) parasites de larves de Culicides (Diptera-Culicidae).   Enquête sur les espèces du "Midi" Méditerranéen.   <u>Stempellia tuzetae</u> n. sp.   <u>Ann. Parasitol. Hum. Comp.</u>, <u>46</u>(3): 205-223.

A list of Microsporida and their hosts in southern France is given.   S. tuzetae, a new species, is described from a natural population of <u>Aedes detritus</u>.   The incidence of microsporidan disease in natural populations was low, less than 3%, and attempts to transmit the disease in the laboratory were unsuccessful.

Tsai, Y.-H. et al. (1969).   Parasites of mosquitoes in southwestern Wyoming and northern Utah. <u>Mosq. News</u>, <u>29</u>(1): 102-110.

The mosquito fauna were surveyed to determine kinds, distribution and prevalence of parasites present in Wyoming and northern Utah.   Occurrence of Microsporida was common, but infection rates were usually low, ranging from 0 to 12.3%.   Although the data are inconclusive about differences between controlled and uncontrolled areas, infections with Microsporida showed slightly higher incidence in pools where mosquitos were not controlled.

Undeen, A. H. & Alger, N. E. (1971).   A density gradient method for fractionating micro-sporidian spores.   <u>J. Invertebr. Pathol.</u>, <u>18</u>: 419-420.

A continuous density gradient made from Ludox was used to separate microsporidan spores from bacteria, mosquito tissue, and other contaminants.   Immature and mature spores form two peaks in early infections.

Undeen, A. H. & Alger, N. E. (1975).   The effect of the microsporidan, <u>Nosema algerae</u> on <u>Anopheles stephensi</u>.   <u>J. Invertebr. Pathol.</u>, <u>25</u>: 19-24.

A bioassay for determining the relative viability of <u>N. algerae</u> spores was devised. The viability of spores produced in <u>Heliothis zea</u> was about the same, and those from <u>Phormia regina</u> were significantly less viable, than those produced in <u>A. stephensi</u>. At low spore concentrations, the microsporidium had little effect on larval or pupal survival.

Undeen, A. H. & Maddox, J. V. (1973).   The infection of nonmosquito hosts by injection with spores of the microsporidan <u>Nosema algerae</u>.   <u>J. Invertebr. Pathol.</u>, <u>22</u>: 258-265.

The microsporidium developed in several different arthropods when spores were injected into their haemocoels. The spores produced from these infections were normal and infectious when fed to mosquito larvae. One corn ear worm larva yielded as many spores as 2000 mosquito larvae.

Vavra, J. & Undeen, A. H. (1970). Nosema algerae n. sp. (Cnidospora, Microsporida) a pathogen in a laboratory colony of Anopheles stephensi Liston (Diptera, Culicidae). J. Protozool., 17(2): 240-249.

The new species is highly pathogenic and infects a variety of tissues of both larvae and adults.

Walters, H. S. (1962). The morphology and life cycle of certain microsporidian parasites found in mosquitoes. Thesis, Pretoria, 126 pp.

Ward, R. A. & Savage, K. E. (1972). Effects of microsporidian parasites upon anopheline mosquitoes and malarial infection. Proc. Helminthol. Soc. Wash. Special Issue, 39: 434-438.

The susceptibility of Anopheles stephensi mosquitos to Plasmodium cynomolgi was reduced by their infection with Nosema algerae. Ninety to 95% of the exposed mosquitos died of the microsporidan infection within 14 days.

Weiser, J. (1946). /Studies on microsporidia from larvae of insects of our waters./ Vestn. Cesk. Spol. Zool., 10: 245-272. (Cz)

A key to Czechoslovakian species is given.

Weiser, J. (1947). /Key to identification of microsporidia./ Pr. Morav. Prirod. Spol., 18: 1-64. (Cz)

A lengthy key supplemented with photographs and drawings is given.

Weiser, J. (1959). Unterlagen der Taxonomie der Mikrosporidien. Trans. 1st Int. Conf. Insect Pathol. Biol. Control (Prague), pp. 277-285.

Weiser, J. (1961). Die Mikrosporidien als Parasiten der Insekten. Parey (Monographien zur angew. Entomol. Nr. 17) Hamburg and Berlin, 149 pp.

Weiser, J. & Coluzzi, M. (1964). Plistophora culisetae n. sp., a new microsporidian (Protozoa: Cnidosporidia) in the mosquito Culiseta longiareolata (Maquart 1838). Riv. Malariol., 43: 51-55. (E, i)

A new microsporidan, Pleistophora culiseta[*] is described from the muscles and the fat body of Culiseta longiareolata larvae. The infected larvae were collected from a rain-filled artificial basin in Italy. The larvae contained cysts in the 4th and 5th abdominal segments. Culex pipiens and Anopheles gambiae were also found infected in the same habitat.

Weiser, J. & Coluzzi, M. (1966). Plistophora culisetae[**] in the mosquito Culiseta longia-reolata Macqu. - further remarks. Proc. 1st Int. Congr. Parasitol. (Rome), 1: 596-597.

---

[*] Later Weiser & Coluzzi (1972) made P. culiseta a synonym of Pleistophora culicis.

[**] Now Pleistophora culicis, also see Weiser & Coluzzi (1972).

The pathogen was located in the 4th and 5th segments of the larval mosquito body. Only in the area of mycetomes formed by bacteroids are the gut cells invaded. Some cysts cross the gut wall and invade the connective tissue of the muscles and the fat body.

Weiser, J. & Coluzzi, M. (1972). The microsporidian Plistophora culicis Weiser, 1946 in different mosquito hosts. Folia Parasitol. (Praha), 19: 197-202. (E, r)

A microsporidan infection was found in a mosquito population living in a concrete basin (at Monticello, Italy). Three species appeared infected: Culex pipiens, C. fatigans, and Culiseta longiareolata. In view of the different spore sizes and location in the tissue, the presence of two different Microsporida was suggested, namely Plistophora* culicis and P. culisetae in Culex and Culiseta, respectively. However, based on cross-infection experiments using Anopheles stephensi, A. albimanus, and A. gambiae, and on changes in spore size during this transmission, and on general host tissue selection, the two species appear to be synonymous.

Welch, H. E. (1960). Effects of protozoan parasites and commensals on larvae of the mosquito Aedes communis (DeGeer) (Diptera: Culicidae) at Churchill, Manitoba. J. Insect Pathol., 2: 386-395.

Sporonts and spores of Amblyspora sp.** (Thelohania sp.) were found in fat bodies of A. communis larvae. Infected late instars were evident by the presence of creamy-white coloured masses in the thorax and abdomen. The disease is fatal to heavily infected larvae.

White, G. B. & Rosen, P. (1973). Comparative studies on sibling species of the Anopheles gambiae Giles complex (Dipt., Culicidae). II. Ecology of species A and B in savanna around Kaduna, Nigeria, during transition from wet to dry season. Bull. Entomol. Res., 62: 613-625.

During an intensive survey, numerous infections of mosquito larvae were observed. Parathelohania (Thelohania) was observed in 19% of gambiae-producing breeding sites. Infections were also found in adult females.

Wills, W. & Beaudoin, R. (1965). Microsporidia in Pennsylvania mosquitoes. J. Invertebr. Pathol., 7: 10-14.

Amblyospora canadensis (Thelohania inimica canadensis) infecting Aedes canadensis is the first microsporidan infection reported from that mosquito. Other host-parasite records from Pennsylvania are listed and confirm previous observations.

---

* Now Pleistophora.

** Hazard & Oldacre (1975) named this species Amblyospora khaliulini.

V. NON-MICROSPORIDAN PROTOZOAN PATHOGENS OF CULICIDAE (MOSQUITOS)[a]

Truman B. Clark

Plant Protection Institute
Bioenvironmental Bee Laboratory
Beltsville Research Center
Beltsville, MD 20705, USA

a The references in this bibliography are those that have appeared since Jenkins' 1964 bibliography and before 1975. Also included are some references that had been omitted from Jenkins' list. Papers dealing with facultative parasites such as *Tetrahymena* have been included only when dealing specifically with parasitism of mosquitos. Papers dealing with the fine structure or physiology of microorganisms have been included if they cover parasites specific to mosquitos. The author is indebted to Mrs Gayle Wentworth for her kind assistance with the organization and typing of the bibliography.

NON-MICROSPORIDAN PROTOZOAN PATHOGENS OF CULICIDAE (MOSQUITOS)

Host	Host stage infected	Pathogen	% Incidence	Locality	Lab. or field study	Reference
Aedes aegypti	Adults	Crithidia fasciculata	--	--	--	Kramer (1964)
"	"	"	--	--	--	Sinha (1972)
"	Larvae	Possible eugregarinid	--	India	Field	Briggs (1972)
"	"	Lankesteria culicis	Field inoculated, 42	USA (Texas)	Field & lab.	Barrett (1968)
"	"	"	18	USA (Texas)	Field	Barrett et al. (1971)
"	"	"	--	USA	Field	Briggs (1973)
"	"	"	60	USA (South-east)	Field	Gentile et al. (1971)
"	Larvae, adults	"	100	India	Lab.	Hati & Ghosh (1963)
"	Larvae, pupae	"	60	USA (Louisiana)	Field	Hayes & Haverfield (1971)
"	"	"	--	--	--	Kramer (1964)
"	Larvae, pupae, adults	"	--	--	Lab.	McCray et al. (1970)
"	Larvae	"	--	--	Lab.	Sheffield et al. (1971)
"	"	"	--	--	--	Sinha (1972)
"	"	"	0-100	USA (Florida)	Field	Stapp & Casten (1971)
"	Larvae, pupae, adults	Lankesteria culicis	95	USA (Texas)	Lab.	Walsh et al. (1969)
"	Larvae	Tetrahymena pyriformis			Lab.	Grassmick & Rowley (1973)

NON-MICROSPORIDAN PROTOZOAN PATHOGENS OF CULICIDAE (MOSQUITOS) (continued)

Host	Host stage infected	Pathogen	% Incidence	Locality	Lab. or field study	Reference
Aedes atlanticus	Larvae	Tetrahymena	--	USA (Louisiana)	Field	Chapman (1974)
Aedes dupreei	"	Internal ciliates	--	USA (Louisiana)	Field	Chapman et al. (1969)
Aedes nigromaculis	"	Vorticella sp.	--	USA (Iowa)	Field	Larson (1967)
Aedes sierrensis	Larvae, pupae, adults	Lankesteria clarki sp. n.	27.5	USA (California)	Lab. & field	Sanders et al. (1973)
"	Larvae	Lankesteria culicis	--	--	Lab. & field	Weiser (1968)
"	Larvae	Tetrahymena sp.	--	USA (California)	Field	Sanders (1972)
Aedes sinensis	"	Tetrahymena gelei	--	--	Lab. & field	Weiser (1968)
Aedes sollicitans	Adults	Crithidia fasciculata	--	USA (Louisiana)	Field	Chapman et al. (1967)
Aedes sticticus	Larvae	Tetrahymena	--	USA (Louisiana)	Field	Chapman (1974)
Aedes taeniorhynchus	"	Caulleryella	--	USA (Louisiana)	Field	Chapman (1974)
Aedes triseriatus	"	Lankesteria barretti n. sp.	--	USA (Texas)	Lab.	Vavra (1969)
Aedes trivittatus	"	Epistylis sp.	--	USA (Iowa)	Lab. & field	Larson (1967)
"	"	Vorticella sp.	72	USA (Iowa)	Lab. & field	Larson (1967)
"	"	Vorticella convallaria	--	USA (Iowa)	Lab. & field	Larson (1967)
"	"	Vorticella striata	--	USA (Iowa)	Lab. & field	Larson (1967)

NON-MICROSPORIDAN PROTOZOAN PATHOGENS OF CULICIDAE (MOSQUITOS) (continued)

Host	Host stage infected	Pathogen	% Incidence	Locality	Lab. or field study	Reference
Aedes vexans	Larvae	Internal ciliates	--	USA (Louisiana)	Field	Chapman et al. (1969)
"	"	"	--	USA (Louisiana)	Field	Chapman et al. (1967)
"	"	Epistylis sp.	1 to 100	USA (Iowa)	Field	Larson (1967)
"	"	Vorticella sp.	to 100	USA (Iowa)	Field	"
"	"	Vorticella convallaria	100	USA (Iowa)	Field	"
"	"	Vorticella striata	to 100	USA (Iowa)	Field	"
Anopheles freeborni	Larvae, pupae, adults	Crithidia fasciculata	--	USA (California)	Lab.	Clark et al. (1964)
Anopheles funestus	Not stated	Flagellate	--	Nigeria	Field	Hazard (1972)
Anopheles gambiae	Adults	Crithidia fasciculata	--	--	Lab.	Brooker (1971A)
"	"	"	--	--	Lab.	Brooker (1972)
"	Larvae	Epibionts	--	Kenya	Field	Briggs (1973)
"	"	Vorticellidae	--	Nigeria	Field	Briggs (1970)
Anopheles hackeri		Crithidia fasciculata	--	Malaya		Sandosham et al. (1962)
Anopheles quadrimaculatus	Culture only	"	--	--	Lab.	Brooker (1970)
"	"	"	--	--	Lab.	Brooker (1971b)
Anopheles punctipennis	Larvae	Epistylis sp.	100	USA (Iowa)	Lab. & field	Larson (1967)

NON-MICROSPORIDAN PROTOZOAN PATHOGENS OF CULICIDAE (MOSQUITOS) (continued)

Host	Host stage infected	Pathogen	% Incidence	Locality	Lab. or field study	Reference
Anopheles squamosus	Larvae	Vorticella sp.	--	Togo	Field	Briggs (1968)
Anopheles stephensi	"	Gregarina	--	England	Field	Briggs (1970)
Culex apicalis	Larvae, pupae, adults	Crithidia fasciculata	--	USA (California)	Lab.	Clark et al. (1964)
Culex boharti	Larvae, pupae, adults	"	--	USA (California)	Lab.	Clark et al. (1964)
Culex duttoni	Larvae	Epistylis sp.	--	Angola	Lab.	World Health Organization (1968)
Culex peus	Larvae, pupae, adults	Crithidia fasciculata	--	USA (California)	Lab.	Clark et al. (1964)
Culex pipiens	Larvae	Carchesium	100	USA (Iowa)	Lab. & field	Larson (1967)
"	Larvae, pupae, adults	Crithidia fasciculata	--	USA (California)	Lab.	Clark et al. (1964)
"	Adults	"	--	--	Lab.	Kramer (1964)
"	Larvae	Epistylis sp.	--	USA (Iowa)	Lab. & field	Larson (1967)
"	"	Vorticella sp.	to 100	USA (New York)	Lab. & field	Schober (1967)
"	"	Vorticella convallaria	to 100	USA (Iowa)	Lab. & field	Larson (1967)

NON-MICROSPORIDAN PROTOZOAN PATHOGENS OF CULICIDAE (MOSQUITOS) (continued)

Host	Host stage infected	Pathogen	% Incidence	Locality	Lab. or field study	Reference
Culex pipiens	Larvae	Vorticella striata	to 100	USA (Iowa)	Lab. & field	Larson (1967)
Culex pipiens fatigans	Adults	Crithidia fasciculata	--	--		Sinha (1972)
"	Larvae	Tetrahymena pyriformis	--	India	Field	Hati & Ghosh (1963)
Culex p. quinque-fasciatus	"	Helicosporidium parasiticum	--		Lab.	Kellen & Lindegren (1973)
Culex restuans	"	Internal ciliates	--	USA (Louisiana)	Field	Chapman et al. (1969)
"	"	Epistylis sp.	100	USA (Iowa)	Lab. & field	Larson (1967)
"	"	Vorticella convallaria	100	USA (Iowa)	Lab. & field	"
"	"	Vorticella striata	100	USA (Iowa)	Lab. & field	"
Culex salinarius	"	Vorticella convallaria	100	USA (Iowa)	Lab. & field	"
"	"	Vorticella striata	100	USA (Iowa)	Lab. & field	"
Culex tarsalis	Larvae, pupae, adults	Crithidia fasciculata		USA (California)	Lab.	Clark et al. (1964)
"	Larvae	Tetrahymena pyriformis	--	--	Lab.	Grassmick & Rowley (1973)
"	"	Vorticella convallaria	67	USA (Iowa)	Lab. & field	Larson (1967)
Culex territans	"	Helicosporida	--	USA (Louisiana)	Lab. & field	Chapman (1974)
"	"	Tetrahymena	--	USA (Louisiana)	Field	Chapman (1974)

NON-MICROSPORIDAN PROTOZOAN PATHOGENS OF CULICIDAE (MOSQUITOS) (continued)

Host	Host stage infected	Pathogen	% Incidence	Locality	Lab. or field study	Reference
Culex territans	Larvae	Vorticella striata	67	USA (Iowa)	Lab. & field	Larson (1967)
Culiseta incidans	Larvae, pupae, adults	Crithidia fasciculata	--	USA (California)	Lab.	Clark et al. (1964)
Culiseta inornata	Larvae, pupae, adults	Crithidia fasciculata	--	USA (California)	Lab.	Clark et al. (1964)
"	Larvae	Vorticella convallaria	--	USA (Iowa)	Lab. & field	Larson (1967)
Culiseta melanura	Larvae	Caulleryella	--	USA (Louisiana)	Field	Chapman (1974)
Mansonia titillans (Wik.)	Adults	Crithidia sp.	52	Colombia	Field	Page (1972)
Orthopodomyia signifera	Larvae	Caulleryella	--	USA (Louisiana)	Field	Chapman (1974)
"	"	Internal ciliates	--	USA (Louisiana)	Field	Chapman et al. (1969)
Culture only		Crithidia fasciculata	--	USA (Louisiana)	Lab.	Cowperthwaite et al. (1953)
"		"	--	USA (Louisiana)	Lab.	Dewey & Kidder (1973)
"		"	--	USA (Louisiana)	Lab.	Guttman (1963)
"		"	--	USA (Louisiana)	Lab.	Guttman & Eisenman (1965)
"		"	--	USA (Louisiana)	Lab.	Palmer (1974)
"		"	--	USA (Louisiana)	Lab.	Simpson & Simpson (1974)
"		Lankesteria barretti	--	USA (Louisiana)	Lab.	Vavra (1971)
"		Lankesteria culicis	--	USA (Louisiana)	Lab.	Vavra (1971)

ABSTRACTS

Mary Ann Strand

Barrett, W. L., jr (1968).   Damage caused by Lankesteria culicis (Ross) to Aedes aegypti (L.) Mosq. News, 28(3):   441-444.

Adult mosquitos, reared from field collected larve infected with L. culicis, are stunted and their Malpighian tubes are damaged by the development of gametocysts within the tubes.

Barrett, W. L., jr et al. (1971).   Distribution in Texas of Lankesteria culicis (Ross), a parasite of Aedes aegypti (L.).   Mosq. News, 31(1):   23-27.

Permanent water containers were found to yield larvae with higher infection rates. Where rates are high, a greater proportion in the container are infected with L. culicis.   Individual infection rates build up during the summer season, but little change was noted over winter.   Experimental infection trials were successful.

Bick, H. (1969).   An illustrated guide to ciliated Protozoa used as "biological indicators" in fresh water ecology.   Eighth fascicle.   Unpublished document WHO/VBC/69.139, 35 pp.

Tetrahymena is included in this key.

Briggs, J. D. (1968).   1967 activities of the WHO International Reference Centre for diagnosis of diseases of vectors.   Unpublished document WHO/VBC/68.97, 7 pp.

Briggs, J. D. (1970).   1969 activities of the WHO International Reference Centre for diagnosis of diseases of vectors.   Unpublished document WHO/VBC/70.250, 13 pp.

Briggs, J. D. (1973).   1970 activities of the WHO International Reference Centre for diagnosis of diseases of vectors.   Unpublished document WHO/VBC/73.422, 7 pp.

Briggs, J. D. (1972).   1971 activities of the WHO International Reference Centre for diagnosis of diseases of vectors.   Unpublished document WHO/VBC/72.408, 6 pp.

Tentative and final determinations of pathogens of disease vectors including protozoan parasites of mosquitoes are given.

Brooker, B. E. (1970).   Desmosomes and hemidesmosomes in the flagellate Crithidia fasciculata. Z. Zellforsch. Mikroskop. Anat., 105:   155-166.

The structure of the attachment mechanism between the flagellate and the gut wall of its host, Anopheles quadrimaculatus, was studied.

Brooker, B. E. (1971a).   Flagellar attachment and detachment of Crithidia fasciculata to the gut wall of Anopheles gambiae.   Protoplasma, 73(2):   191-202.

Experimental results indicate that flagellar adhesion is non-specific.   Not only is the flagellum able to adhere to the cuticular lining of the mosquito gut and the flagella of other cells, but also readily adheres to non-biological substrates.   In every case, the flagellar membrane follows the contour of the surface to which it is apposed.

Brooker, B. E. (1971b).   The fine structure of Crithidia fasciculata with special reference to the organelles involved in the ingestion and digestion of protein.   Z. Zellforsch. Mikroskop. Anat., 116(4):   532-563.

The organelles were observed in C. fasciculata infecting Anopheles quadrimaculatus. The process of ingestion was observed using an electron dense tracer (ferritin).

Brooker, B. E. (1972). Modifications in the arrangement of the pellicular microtubules of Crithidia fasciculata in the gut of Anopheles gambiae. Z. Parasitkd., 40(4): 271-280.

The presence and orientation of microtubules lying beneath the cell membrane is a typical feature of trypanosomatid flagellates. This study shows, in the case of C. fasciculata, that these microtubules are capable of adaptive modification in their arrangement. In heavy infections, groups of pellicular microtubules which have sunk beneath the surface are observed. This deviation from the normal distribution is probably due to physical forces exerted on the flagellates when they occupy a space whose dimensions differ considerably from their own.

Chapman, H. C. (1974). Biological control of mosquito larvae. Ann. Rev. Entomol., 19: 33-59.

The main genera of protozoans pathogenic to mosquito larvae are reviewed. Also, from previously unpublished studies, Caulleryella is reported to infect Culiseta melanura, Orthopodomyia signifera, and Aedes taeniorhynchus in Louisiana.

Chapman, H. C. et al. (1969). A two-year survey of pathogens and parasites of Culicidae, Chaoboridae, and Ceratopogonidae in Louisiana. Proc. N. J. Mosq. Exterm. Assoc., 56: 203-212.

A summary of larvae collected in the field with protozoan infections shows one species of mosquitoes infected by eugregarines, 3 with schizogregarines, and 5 with internal ciliates. Culex restuans with ciliatosis were the most frequently collected host-pathogen combination.

Chapman, H. C. et al. (1976b). Pathogens and parasites in Louisiana Culicidae and Chaoboridae. Proc. N. J. Mosq. Exterm. Assoc., 54: 54-60.

A survey of pathogens was conducted from 1964 to 1967. Crithidia fasciculata, internal ciliates, and Helicosporidia were found infrequently infecting mosquito larvae.

Clark, T. B. et al. (1964). The transmission of Crithidia fasciculata Leger 1902 in Culiseta incidens (Thomson). J. Protozool., 11(3): 400-402.

Larva-to-larva transmission can apparently account for the incidence of C. fasciculata in natural populations of adult mosquitoes. Infection of pupae occurred by contamination of the midgut of the prepupa with nectomondas while the peritrophic membrane was being shed prior to pupation. Of the adults that emerged from pupae of infected 4th instar larvae, 44% were infected. Adult-to-larvae transmission also occurred. After exposure to infected adults and their feces, 50% (10 larvae) of the 3rd instar larvae examined were infected.

Cowperthwaite, J. et al. (1953). Nutrition of Herpetomonas (Strigomonas) culicidarum. Ann. N. Y. Acad. Sci., 56: 972-981.

A synthetic medium was developed for the protozoan. The strain of H. culicidarum was isolated from Anopheles.

Dewey, V. C. & Kidder, G. W. (1973). The sites of action of allopurinol in Crithidia fasciculata. J. Protozool, 20(5): 678-682.

Purine and pyrimidine are required to reverse growth inhibition by allopurinol.   Allopurinol apparently interferes with the metabolism of hypoxanthine in inhibiting the growth of Crithidia.   Also, it acts as an inhibitor of pyrimidine biosynthesis.

Else, J. G. & Dangsupa, P. (1974).   Lankesteria, a gregarine protozoan, previously unreported in mosquitoes of Malaysia.   Southeast Asian J. Trop. Med. Pub. Health, 5(3):   454.

Gentile, A. G. et al. (1971).   The distribution, ethology and control potential of the Lankesteria culicis (Ross)-Aedes aegypti (L.) complex in southern United States.   Mosq. News, 31(1):   12-17.

The incidence of L. culicis follows the seasonal rise and fall of its host.   No correlation was found between incidence of the protozoan and water depth, sediment, detritus, and vegetation present.   Observations suggest that L. culicis is host specific.
Release of sporocysts in areas free from the parasite did result in establishment of the organism in Aedes aegypti populations.   However, symbiotic balance in the host-parasite complex was reestablished and the parasite had little apparent deleterious effect on the host population.

Grassmick, R. A. & Rowley, W. A. (1973a).   A ciliate infection in mosquito larvae caused by Colpoda sp. J. Protozool., 20(4):   509 (Abstract).

Naturally occurring infection caused by Colpoda sp. in larval stages of Culex tarsalis was observed in laboratory reared larvae.   Large numbers of active ciliates were seen in the larval-rearing medium and some were apparent in the dead larvae.

Grassmick, R. A. & Rowley, W. A. (1973b).   Larval mortality of Culex tarsalis and Aedes aegypti when reared with different concentrations of Tetrahymena pyriformis.   J. Invertebr. Pathol., 22(1):   86-93.

A significant difference in mortality was observed between the two mosquito species when exposed to identical concentrations of ciliates.   No significant deleterious effects occurred when A. aegypti larvae were exposed to up to 7000 ciliates/ml of larval medium. However C. tarsalis was extremely sensitive to the ciliates and at 7000/ml 76.3% died by the end of 11 days.

Grassmick, R. A. & Rowley, W. A. (1974).   Ciliated protozoa as possible mosquito-control agents.   (Abstr.) Proc. 3rd Int. Cong. Parasitol.   (Munich) 3:   1701.

In the laboratory, up to 90% of the Culex tarsalis larvae exposed to a clone of Tetrahymena pyriformis became infected and died.

Guttman, H. N. (1963).   Experimental glimpses at the lower Trypanosomatidae.   Exp. Parasitol. 14:   129-142.

A review article with 48 references.   It includes information on culture media and metabolism.

Guttman, H. N. & Eisenman, R. N. (1965).   Acriflavin-induced loss of kinetoplast deoxyribonucleic acid in Crithidia fasciculata (Culex pipiens strain).   Nature, 207:   1280-1281.

A kinotoplast is a DNA-containing, membrane-limited organelle peculiar to protozoa belonging to the Trypanosomatidae.   The apparent loss of the electron dense, highly polymerized DNA in the kinetoplast as a result of treatment with acriflavin is demonstrated. Variability in the effect of acriflavin may be related to the base composition of the DNA.

Hati, A. K. & Ghosh, S. M. (1961).   Vorticella infestation of mosquito larvae and its effect on their growth and longevity.   Bull. Calcutta Sch. Trop. Med., 9(4):   155.

An infestation of Vorticella sp. caused heavy mortality in Anopheles subpictus and A. stephensi larvae which were reared from a field collection.   Larval growth was retarded and most died during the 4th instar.   Attempts to infest larvae and eggs of Culex fatigans and Aedes aegypti were successful.

Hati, A. K. & Ghosh, S. M. (1963a).   On the gregarine (Lankesteria culicis Ross) infection in Aedes aegypti mosquito in Calcutta.   Bull. Calcutta Sch. Trop. Med., 11(1):   7-8.

Laboratory colonies of A. aegypti were found infected with L. culicis.   Heavy mortality was reported.

Hati, A. K. & Ghosh, S. M. (1963b).   Ciliated protozoal infection in Culex fatigans larvae. Bull. Calcutta Sch. Trop. Med., 11(3):   109.

Tetrahymena pyriformis was found in breeding places of mosquitoes and infecting larvae of C. fatigans.   Among the 200 larvae collected from a ditch near Calcutta only a few were infected.

Hayes, G. R. jr & Haverfield, L. E. (1971).   Distribution and density of Aedes aegypti (L.) and Lankesteria culicis (Ross) in Louisiana and adjoining areas.   Mosq. News, 31(1):   28-32.

The heaviest and most widespread distribution of L. culicis was coincident with flourish-ing populations of A. aegypti.   There was no evidence that the parasite exerts a limit-ing effect on the mosquito populations observed.   In areas observed, about 60% of the A. aegypti were infected with an average of 69.7 trophozoites per larva.

Hazard, E. I. (1972).   Investigation of pathogens of anopheline mosquitos in the vicinity of Kaduna, Nigeria.   WHO/VBC/72.384.   6 pp.

A survey of the pathogens and parasites of mosquitos in the savannah area near Kaduna was conducted.   Unidentified protozoans were found with low incidence infecting Anopheles funestus and Anopheles gambiae.

Kellen, W. R. & Lindegren, J. E. (1973).   New host records for Helicosporidium parasiticum. J. Invertebr. Pathol., 22:   296-297.

The susceptibility of various species of insects and mites to H. parasiticum was tested. Fresh sporocysts were suspended in water and applied to the mouth parts of the insects. Larvae of Culex pipiens quinquefasciatus were found to be susceptible to infection.

Kramer, J. P. (1964).   Parasites in laboratory colonies of mosquitos.   Bull. WHO, 31(4): 475-478.

What is known concerning the biology of some typical mosquito parasites and their effects on experimental results is reviewed.

Larson, L. V. (1967).   Association of Vorticella and Epistylis (Protozoa: Ciliata) with mosquito larvae.   WHO/VBC/67.20.   21 pp.

Laboratory studies are reported which demonstrate that heavy infestations of peritrichs have little effect on growth and survival rates of 4th instar larvae and pupae.

McCray, E. M. jr, Fay, R. W. & Schoof, H. F. (1970).   The bionomics of Lankesteria culicis and Aedes aegypti.   J. Invertebr. Pathol., 16(1):   42-53.

The protozoan L. culicis was used in laboratory studies to evaluate its effect on larval development, pupation, adult emergence, and egg production of A. aegypti. No significant pathological effects were observed, although the life span of infected females was reduced (43 days for infected and 50 days for uninfected).

Page, W. A. (1972).   Feeding behaviour and trypanosomatid infections of some tabanids and Culicidae in Colombia.   J. Entom. Ser A. Gen. Entomol., 47(1): 1-13.

Monsonia spp. mosquitos infected with Crithidia sp. were collected in Colombia.   The trypanosomatids were located mainly in the hindgut attached to the gut wall.

Palmer, F. B. (1974).   Biosynthesis of choline and ethanolamine phospholipids in Crithidia fasciculata.   J. Protozool. 21(1): 160-163.

The synthesis of phospholipids from ^{14}C labeled precursors (choline, ethanolamine, serine, and methionine) was examined in cultures of C. fasciculata.   All 4 precursors were incorporated into lipids to the extent of 1-2.5% of added radioactivity after 4 hours.

Sanders, R. D. (1972).   Microbial mortality factors in Aedes sierrensis populations.   Proc. Calif. Mosq. Control Assoc., 40: 66-68.

Tetrahymena sp. was found in the hemocoel of A. sierrensis larvae collected in California. The ciliates were often found jointly infecting with an unidentified mermithid nematode. Both parasites could grow at the same time in the body cavity of the host.

Sanders, R. D. & Poinar, G. O. jr (1973).   Fine structure and life cycle of Lankesteria clarki sp. n. (Sporozoa: Eugregarinida) parasitic in the mosquito Aedes sierrensis (Ludlow). J. Protozool., 20(5): 594-602.

A new parasite L. clarki is described from A. sierrensis in California.   The intra-cellular nature of the trophozoite and structural characteristics of the gamont separate it from L. culicis and L. barretti.   The midgut epithelial cells of the host are destroyed during the trophic stage and during gametogenesis and sporogony, the malpighian tubule cells may be also destroyed.

Sandosham, A. A., Coombs, G. L. & Hassan, A. (1962).   Crithidia fasciculata Leger, 1902 from the malpighian tubules of Anopheles hackeri Edwards, 1921.   Med. J. Malaya, 17: 95.

A new host record is reported from Malaya.

Schober, H. (1967).   Observations on Culex pipiens larvae infested with Vorticella sp. Mosquito News, 27(4): 523-524.

In late summer several breeding sites in New York were found where all or nearly all of the 4th instar larvae were infested with this ciliated protozoan.   In collection jars from infested sites, 32-38% of the larvae died; whereas in similar jars from uncon-taminated sites, 4% died.   The movements of the infested larvae were evidently impeded by the protozoan.   Larvae younger than 4th instar were not observed with infestations.

Service, M. W. (1970).   Studies on the biology and taxonomy of Aedes (Stegomyia) vittatus (Bigot) (Diptera: Culicidae) in Northern Nigeria.   Trans. R. Entomol. Soc. Lond., 122: 101-143.

Larvae from a rock pool had heavy infestations of Peritrichida and external fungal hyphae. No appreciable difference in the mortality of adults obtained from healthy and infected larvae and pupae was observed in the laboratory.   However, infected individuals appear less vigorous and significant mortality may be related to infection during times of stress.

Sheffield, H. G., Garnham, P. C. C. & Shiroishi, T. (1971).   The fine structure of the sporozoite of Lankesteria culicis.   J. Protozool., 18(1): 98-105.

Light and electron microscopy were used to observe the sporozoite of L. culicis after encystation in the intestine of 1st instar Aedes aegypti larvae.   Oocysts were fed to the mosquitoes and the sporozoite emerged, thereby becoming available for study within the gut.   The fine structure was compared with that reported for Stylocephalus africanus and other members of the sporozoa.

Simpson, A. M. & Simpson, L. (1974).   Labeling of Crithidia fasciculata DNA with $\sqrt{}^{-3}H\sqrt{}$ thymidine.   J. Protozool., 21(2): 379-382.

The continuous labeling of C. fasciculata DNA is hindered by the apparent presence of thymidine phosphorylase-like enzyme activity.   There is a cell-mediated conversion of thymidine to thymine in the medium.

Sinha, V. P. (1972).   Pathology of diseases of vectors of public health importance.   Patna. J. Med., 46(9): 253-258.

This is a general review paper which includes a table of host reactions to pathogens.

Stapp, R. R. & Casten, J. (1971).   Field studies on Lankesteria culicis and Aedes aegypti in Florida.   Mosq. News, 31(1): 18-22.

In Florida infestation with L. culicis in A. aegypti larvae was not high, averaging 66, 44, and 63 trophozoites per larvae in the 3 cycles observed.   Heavily infected larvae (200-800 trophozoites/larvae) did not appear adversely affected by the parasites.   DDT-resistant larvae were rendered slightly more susceptible to DDT by heavy infections.

Tuzet, O. & Rioux, J. A. (1965).   Les gregarines de Culicidae, Ceratopogonidae, Simuliidae, et. Psychodidae.   WHO/EBL/66.50.   18 pp.

A taxonomic review of the gregarines is given.

Vavra, J. (1969).   Lankesteria barretti n. sp. (Eugregarinida, Diplocystidae) a parasite of the mosquito Aedes triseriatus (Say) and a review of the genus Lankesteria Mingazzini. J. Protozool., 16(3): 546-570.

L. barretti was found parasitizing A. triseriatus from Texas.   The young cephalins occur within the midgut epithelial cells until these cells rupture and the cephalins are released into the space between the epithelium and the peritrophic membrane and become gamonts.   When the hosts pupate, they enter the lumen of the malpighian tubules.

Vavra, J. (1971).   Physiological, morphological, and ecological considerations of some microsporidia and gregarines.   pp. 92-103.   In A. M. Fallis (ed.).   Ecology and physiology of parasites - a symposium Univ. Toronto Press, Toronto.

Physiological studies on cultures of Lankesteria barretti and L. culicis are reported.

Walsh, R. D. jr & Callaway, C. S. (1969).   The fine structure of the gregarine Lankesteria culicis parasitic on the yellow fever mosquito Aedes aegypti.   J. Protozool., 16(3): 536-545.

The fine structure of the eugregarine was examined by light and electron microscope and compared with that of other gregarines.

Weiser, J. (1968).   Guide to field determination of major groups of pathogens and parasites affecting arthropods of public health importance.   WHO/VBC/68.59.   11 pp.

A key is given to facilitate the recognition of diseased vectors.

VI. *COELOMOMYCES* PATHOGENS OF CULICIDAE (MOSQUITOS)

Rand E. McNitt

and

John N. Couch

Department of Botany
University of North Carolina
Chapel Hill, NC 27514, USA

COELOMOMYCES PATHOGENS OF CULICIDAE (MOSQUITOS)

Host	Host stage infected	Parasite	% Incidence	Location	Lab. or field	Reference
Aedes aegypti	Larvae	Coelomomyces stegomyiae		Singapore	Field	Laird (1959a)
"		Coelomomyces stegomyiae		Ceylon	Field	Rajapaksa (1964)
Aedes albopictus		Coelomomyces stegomyiae		Singapore	Field	Keilin (1921)
"	Larvae	Coelomomyces stegomyiae		Ceylon	Field	Rajapaksa (1964)
"	Larvae	Coelomomyces quadrangulatus, var. lamborni		Malaya	Field	Couch & Dodge (1947)
Aedes atropalpus epactius	Larvae	Coelomomyces sp.		USA (Utah)	Field	Romney et al. (1971)
Aedes australis	Larvae	Coelomomyces tasmaniensis		Tasmania	Field	Laird (1956b)
"	Larvae	Coelomomyces opifexi	(55%)	New Zealand	Field	Pillai & Smith (1968) (Pillai (1969))
"	Larvae	Coelomomyces opifexi	3-100%	New Zealand	Lab.	Pillai & Woo (1973)
Aedes cantans	Adults	Coelomomyces sp.	low	England	Field	Service (1974)
Aedes caspius dorsalis		Coelomomyces sp.		USSR	Field	Khaliulin & Ivanov (1973)
Aedes cinereus		Coelomomyces psorophorae (?)		France (Strasbourg)		Eckstein (1922)
Aedes cyprius		Coelomomyces sp.		USSR		Khaliulin & Ivanov (1973)
Aedes excrucians	Adults	Coelomomyces sp.		USA (Alaska)	Field	Briggs (1969)
Aedes hebrideus		Coelomomyces sp.		Solomon Islands	Field	Genga & Maffi (1973)
Aedes melanimon		Coelomomyces sp.				Kellen et al. (1963)
Aedes multiformis		Coelomomyces stegomyiae		New Guinea	Field	Huang (1968)
Aedes notoscriptus		Coelomomyces finlayae		Australia		Laird (1959b)
Aedes polynesiensis	Larvae	Coelomomyces macleayae		Fiji	Field	Pillai & Rakai (1970)
Aedes polynesiensis		Coelomomyces stegomyiae		Tokelau	Field	Laird (1966)

COELOMOMYCES PATHOGENS OF CULICIDAE (MOSQUITOS) (continued)

Host	Host stage infected	Parasite	% Incidence	Location	Lab. or field	Reference
Aedes quadrispinatus		Coelomomyces stegomyiae		New Guinea	Field	Briggs (1967)
Aedes scatophagoides	Larvae	Coelomomyces psorophorae var.		Northern Rhodesia	Field	Muspratt (1946a)
Aedes scutellaris		Coelomomyces stegomyiae		Solomon Islands	Field	Laird (1956a)
Aedes simpsoni		Coelomomyces sp.		Africa (Bwamba)	Field	McCrae (1972)
Aedes sollicitans	Larvae	Coelomomyces sp.	<1%	USA (Louisiana)	Field	Chapman & Woodward (1966)
Aedes taeniorhynchus	Larvae, adults	Coelomomyces psorophorae var.		USA (Florida)	Field	Lum (1963)
Aedes togoi	Larvae	Coelomomyces sp.	ca. 1%	USSR		Fedder et al. (1971)
Aedes triseriatus	Larvae	Coelomomyces macleaya		Louisiana	Field	Couch (unpublished)
Aedes variabilis		Coelomomyces stegomyiae		New Guinea	Field	Briggs (1967)
Aedes vexans		Coelomomyces psorophorae (?)		France	Field	Eckstein (1922)
"	Larvae	Coelomomyces psorophorae var.		USA (Minnesota)	Field	Laird (1961)
"	Larvae	Coelomomyces psorophorae	3.4%	USSR	Field	Zharov (1973)
Ae. (Macleaya) sp.	Larvae	Coelomomyces macleayae		Australia	Field	Laird (1959b)
Ae. (Stegomyia) sp.	Larvae	Coelomomyces stegomyiae var. rotumae		Rotuma Island	Field	Laird (1959b)
Aedomyia catasticta	Larvae	Coelomomyces indicus		Australia	Field	Laird (1956a)
Anopheles aconitus	Larvae	Coelomomyces indicus		India (Bengal)	Field	Iyengar (1935)
Anopheles annularis	Larvae	Coelomomyces indicus		India (Bengal)	Field	Iyengar (1935)
Anopheles barbirostris	Larvae	Coelomomyces indicus		India (Bengal)	Field	Iyengar (1935)
Anopheles claviger	Larvae	Coelomomyces raffaelei		Italy	Field	Coluzzi & Rioux (1962)
Anopheles crucians	Larvae	Coelomomyces bisymmetricus		USA (Georgia)	Field	Couch & Dodge (1947)
"	Larvae	Coelomomyces cribrosus		USA (Georgia)	Field	Couch & Dodge (1947)
"	Larvae	Coelomomyces dodgei		USA (Georgia)	Field	Umphlett (1960) (unpublished)

COELOMOMYCES PATHOGENS OF CULICIDAE (MOSQUITOS) (continued)

Host	Host stage infected	Parasite	% Incidence	Location	Lab. or field	Reference
Anopheles crucians	Larvae	Coelomomyces keilini		USA (Georgia)	Field	Couch & Dodge (1947)
"	Larvae	Coelomomyces lativittatus		USA (Georgia)	Field	Couch & Dodge (1947)
"	Larvae	Coelomomyces quadrangulatus		USA (Georgia)	Field	Couch & Dodge (1947)
"	Larvae	Coelomomyces sculptosporus		USA (Georgia)	Field	Couch & Dodge (1947)
"	Larvae	Coelomomyces punctatus	12-67%	USA (Louisiana)	Field	Chapman & Glenn (1972)
"	Larvae	Coelomomyces dodgei	24-59%	USA (Louisiana)	Field	Chapman & Glenn (1972)
Anopheles earlei	Larvae	Coelomomyces lativittatus (var.?)		USA (Minnesota)	Field	Laird (1961)
Anopheles farauti	Larvae	Coelomomyces cairnsensis		Australia	Field	Laird (1956a)
"	Larvae	Coelomomyces sp.		British Solomon Is.	Field	Maffi & Genga (1970)
Anopheles funestus	Larvae, adults	Coelomomyces africanus		Kenya	Field	Haddow (1942)
"		Coelomomyces walkeri (Walker's type 1)		Sierra Leone	Field	Walker (1938)
"		Coelomomyces indicus		Northern Rhodesia	Field	Muspratt (1946a)
Anopheles gambiae		Coelomomyces africanus		Kenya	Field	Haddow (1942)
"	Adults	Coelomomyces walkeri		Sierra Leone	Field	Walker (1938) (validated by van Thiel (1962))
"	Larvae, adults	Coelomomyces ascariformis		Africa	Field	Walker (1938) (validated by van Thiel (1962))
"	Larvae	Coelomomyces indicus		Northern Rhodesia	Field	Muspratt (1946a,b)
"	Larvae	Coelomomyces indicus	95%+	Northern Rhodesia	Field	Muspratt (1962)
"	Larvae	Coelomomyces grassei		North Chad	Field	Rioux & Pech (1960, 1962)

COELOMOMYCES PATHOGENS OF CULICIDAE (MOSQUITOS) (continued)

Host	Host stage infected	Parasite	% Incidence	Location	Lab. or field	Reference
Anopheles gambiae	Larvae	Coelomomyces indicus			Lab.	Madelin (1968)
"	Larvae	Coelomomyces ascariformis		Upper Volta	Field	Rodhain & Gayral (1971)
Anopheles georgianus	Larvae	Coelomomyces quadrangulatus var.		USA (Georgia)	Field	Couch & Dodge (1947)
Anopheles hyrcanus var. nigerrimus	Larvae	Coelomomyces indicus		Bengal	Field	Iyengar (1935)
Anopheles jamesi	Larvae	Coelomomyces indicus		Bengal	Field	Iyengar (1935)
Anopheles minimus	Adults	Coelomomyces ascariformis	1-2%	Philippines	Field	Manalang (1930)
Anopheles pharoensis	Larvae	Coelomomyces indicus		Egypt	Field	Gad & Sadek (1968)
Anopheles pretoriensis	Larvae	Coelomomyces indicus		Northern Rhodesia	Field	Muspratt (1946a)
"	Larvae	Coelomomyces indicus	100%	Northern Rhodesia	Field	Muspratt (1962)
Anopheles punctipennis	Larvae	Coelomomyces cribrosus		USA (Georgia)	Field	Couch & Dodge (1947)
"	Larvae	Coelomomyces quadrangulatus		USA (Georgia)	Field	Couch & Dodge (1947)
"	Larvae	Coelomomyces quadrangulatus var. irregularis		USA (Georgia)	Field	Couch & Dodge (1947)
"		Coelomomyces sculptosporus		USA (Georgia)	Field	Couch & Dodge (1947)
Anopheles punctulatus		Coelomomyces solomonis		Solomon Islands & Guadalcanal	Field	Laird (1956a)
Anopheles quadrimaculatus		Coelomomyces punctatus		USA (Georgia)	Field	Couch & Dodge (1947)
"		Coelomomyces quadrangulatus var.		USA (Georgia)	Field	Couch & Dodge (1947)
Anopheles ramsayi	Larvae	Coelomomyces indicus		India (Bengal)	Field	Iyengar (1935)
Anopheles rivulorum	Larvae	Coelomomyces indicus		Northern Rhodesia	Field	Muspratt (1946a)
"	Larvae	Coelomomyces indicus	25%	Northern Rhodesia	Field	Muspratt (1962)
Anopheles rufipes	Larvae	Coelomomyces indicus		Northern Rhodesia	Field	Muspratt (1946a)

COELOMOMYCES PATHOGENS OF CULICIDAE (MOSQUITOS) (continued)

Host	Host stage infected	Parasite	% Inci-dence	Location	Lab. or field	Reference
Anopheles rufipes	Larvae	Coelomomyces indicus	100%	Northern Rhodesia	Field	Muspratt (1962)
"	Larvae	Coelomomyces africanus	100%	Upper Volta	Field	Rodhain & Gayral (1971)
Anopheles sinensis	Larvae	Coelomomyces raffaelei var. parvum		Japan	Field	Laird et al. (1975)
Anopheles squamosus	Larvae	Coelomomyces indicus		Northern Rhodesia	Field	Muspratt (1946)
"	Larvae	Coelomomyces indicus (?)	low	Northern Rhodesia	Field	Muspratt (1962)
Anopheles subpictus	Larvae	Coelomomyces anophelesica		India (Bengal)	Field	Iyengar (1935)
"	Larvae	Coelomomyces indicus		Cambodia	Field	Laird (1959b)
"	Larvae	Coelomomyces indicus		India (Bangalore)	Field	Iyengar (1961) (unpublished)
"	Larvae, pupae, adults	Coelomomyces indicus		India	Field	Gugnani et al. (1965)
Anopheles tesselatus	Adults	Coelomomyces walkeri		Java		van Thiel (1954)
Anopheles vagus	Larvae	Coelomomyces anophelesicus		India (Bengal)	Field	Iyengar (1935)
Anopheles varuna	Larvae	Coelomomyces indicus		India (Bengal)	Field	Iyengar (1935)
"		Coelomomyces anophelesicus		India (Bengal)	Field	Iyengar (1935)
Anopheles walkeri	Larvae	Coelomomyces quadrangulatus (var.?)		USA (Minnesota)	Field	Laird (1961)
"	Larvae	Coelomomyces sculptosporus		USA (Minnesota)	Field	Laird (1961)
Armigeres obturbans		Coelomomyces stegomyiae		Singapore	Field	Laird (1959b)
Culex annulirostris		Coelomomyces sp.		British Solomon Is.		Maffi & Genga (1970)
Culex erraticus		Coelomomyces pentangulatus		USA (Georgia)		Couch (1945)
Culex fraudatrix	Larvae	Coelomomyces cribrosus		North Borneo		Laird (1959b)

COELOMOMYCES PATHOGENS OF CULICIDAE (MOSQUITOS) (continued)

Host	Host stage infected	Parasite	% Incidence	Location	Lab. or field	Reference
Culex gelidus	Larvae	Coelomomyces n. sp.		Ceylon	Field	Rajapaksa (1964)
Culex modestus	Larvae	Coelomomyces iliensis		USSR (Kazakhstan)	Field	Dubitskii et al. (1970, 1973)
Culex orientalis	Larvae	Coelomomyces sp.		USSR (Vladivostock)	Field	Briggs (1969)
Culex peccator	Larvae	Coelomomyces sp.		USA	Field	Briggs (1969)
Culex pipiens fatigans	Larvae	Coelomomyces sp.		Ceylon	Field	Rajapaksa (1964)
"		Coelomomyces sp.		New Caledonia	Field	Lacour & Rageau (1957)
Culex portesi	Adults	Coelomomyces sp.		Trinidad	Field	Briggs (1969)
Culex restuans	Larvae, pupae, adults	Coelomomyces sp.	10-15%	USA (Louisiana)	Field	Chapman & Woodward (1966)
Culex salinarius	Larvae	Coelomomyces sp.	<0.1%	USA (Louisiana)	Field	Chapman & Woodward (1966)
Culex simpsoni	Larvae	Coelomomyces indicus		Northern Rhodesia	Field	Muspratt (1946a)
Culex tritaeniorhynchus summorosus	Larvae	Coelomomyces cribrosus		British North Borneo	Field	Laird (1959b)
"	Larvae	Coelomomyces omorii		Japan	Field	Laird et al. (1975)
Culex tritaeniorhynchus siamensis	Adults	Coelomomyces quadrangulatus var. parvus		Singapore	Field	Laird (1959a)
"	Larvae	Coelomomyces cribrosus		Singapore	Field	Laird (1959a)
Culiseta inornata	Larvae	Coelomomyces psorophorae var.	12%	Canada (Alberta)	Field	Shemanchuk (1959)
Psorophora ciliata	Larvae	Coelomomyces psorophorae		USA (Mississippi)	Field	Laird (1961)
Psorophora howardii	Larvae	Coelomomyces psorophorae var.		USA (South Carolina)	Field	Couch & Dodge (1947)

COELOMYCES PATHOGENS OF CULICIDAE (MOSQUITOS) (continued)

Host	Host stage infected	Parasite	% Inci- dence	Location	Lab. or field	Reference
Opifex fuscus	Larvae	Coelomomyces opifexi		New Zealand	Field	Pillai & Smith (1968)
Toxorynchites rutilus septentrionalis	Larvae	Coelomomyces macleayae		USA (Louisiana)	Field	Nolan et al. (1973)
Uranotaenia barnesi		Coelomomyces sp.		British Solomon Is.	Field	Maffi & Genga (1970)
Uranotaenia sappharina	Larvae	Coelomomyces uranotaeniae		USA (Georgia)	Field	Couch (1945)

ABSTRACTS

Mary Ann Strand & Donald W. Roberts

Anthony, D. W. et al. (1971). Scanning electron microscopy of the sporangia of species of Coelomomyces (Blastocladiales: Coelomomycetaceae). J. Invertebr. Pathol., 17: 395-403.

Surface microstructure differences were observed in scanning micrographs of sporangia from 6 species of Coelomomyces infecting 9 species of mosquito.

Aspöck, H. (1966). Parasitierung eines im Freiland aufgefundenen Intersexes von Aedes (Ochlerotatus) communis De Geer (Insecta, Culicidae) durch einen Pilz der Ordnung Blastocladiales. Z. Morph. Oekol. Tiere, 57: 231-243.

An intersex of A. communis was found in eastern Austria. The abdominal segments contained thousands of fungal sporangia which resembled Coelomomyces. The intersex may be caused by the fungal infection.

Bland, C. E. (1971). Scanning electron microscopy of Coelomomyces. Proc. Nat. Acad. Sci., 68: 2898a (Abstract).

See Bland & Couch (1973).

Bland, C. E. & Couch, J. N. (1973). Scanning electron microscopy of sporangia of Coelomomyces. Can. J. Bot., 51: 1325-1330.

The scanning electron microscopy (SEM) is used to examine sporangia of 18 described and 3 undescribed species of Coelomomyces. These species were divided into 8 groups based on surface structure of the sporangia. It is concluded that the SEM provides a graphic display of the sporangial wall structure, but it is not essential for species determination.

Briggs, J. D. (1966). 1965 Activities of the WHO International Reference Centre for diagnosis of diseases of vectors. WHO/EBL/66.73. 8 pp.

*Briggs, J. D. (1967). 1966 Activities of the WHO International Reference Centre for diagnosis of diseases of vectors. WHO/VBC/67.8. 8 pp.

Briggs, J. D. (1968). 1967 Activities of the WHO International Reference Centre for diagnosis of diseases of vectors. WHO/VBC/68.97. 7 pp.

*Briggs, J. D. (1969). 1968 Activities of the WHO International Reference Centre for diagnosis of diseases of vectors. WHO/VBC/69.171. 6 pp.

Briggs, J. D. (1970). 1969 Activities of the WHO International Reference Centre for diagnosis of diseases of vectors. WHO/VBC/70.250. 13 pp.

Briggs, J. D. (1972). 1971 Activities of the WHO International Reference Centre for diagnosis of diseases of vectors. WHO/VBC/72.408.

Note: The preceding table cites one or a few principal papers for each mosquito-Coelomomyces combination. These citations are marked with * in the following list.

Briggs, J. D. (1973). 1970 Activities of the WHO International Reference Centre for diagnosis of diseases of vectors. WHO/VBC/73.422. 7 pp.

Briggs, J. D. (1974). 1973 Activities of the WHO International Reference Centre for diagnosis of diseases of vectors. WHO, Mimeo. 19 pp.

Tentative and final determinations of pathogens including Coelomomyces infections of mosquitos are given.

Camey-Pacheco, H. L. (1965). Encuesta para determinar la presencia de Coelomomyces (Blastocladiales) y Thelohania (Sporozoa: Microsporidia), parásitos de larvas de Anopheles, en Guatemala, como contribución a su posible control biológico. Univ. San Carlos, 67: 139-182.

Coelomomyces fungi were not conclusively found infecting any of over 10 000 larvae examined.

Chapman, H. C. et al. (1969). A two-year survey of pathogens and parasites of Culicidae, Chaoboridae, and Ceratopogonidae in Louisiana. Proc. N. J. Mosq. Exterm. Assoc., 56: 203-212.

Larvae of 12 species and female adults of 1 species of mosquitos infected with Coelomomyces spp. were collected during the survey. Previously unrecorded hosts include: Aedes triseriatus, Anopheles bradleyi, and Culex peccator.

*Chapman, H. C. & Glenn, F. E., jr (1972). Incidence of the fungus Coelomomyces punctatus and C. dodgei in larval populations of the mosquito Anopheles crucians in two Louisiana ponds. J. Invertebr. Pathol., 19: 256-261.

The annual mean level of infection of A. crucians and C. punctatus ranged from 12% to 67% in one pond. The highest incidences of infection occurred in January, March, and October. In another pond, the annual level of infection of A. crucians by C. dodgei ranged from 24% to 59% with the highest levels occurring in January and September. No relationship between infection level and temperature was observed.

Chapman, H. C. & Woodard, D. B. (1966). Coelomomyces (Blastocladiales: Coelomomycetaceae) infections in Louisiana mosquitoes. Mosq. News, 26(2): 121-123.

Coelomomyces was found parasitizing larvae of Culiseta inornata, Psorophora ciliata, P. howardii, Aedes vexans, A. taeniorhynchus, A. sollicitans, Culex restuans, and C. salinarius in Louisiana. A. sollicitans and the two species Culex represent new host records for this fungus. The proportions of parasitized larvae are given for the different sampling locations. Anopheles, although found in Louisiana, was not infected.

Chapman, H. C. et al. (1967). Pathogens and parasites in Louisiana Culicidae and Chaoboridae. Proc. N. J. Mosq. Exterm. Assoc., 54: 54-60.

A survey of pathogens and parasites was conducted from 1964 to 1967. Levels of infection of mosquitos with Coelomomyces were usually less than 15%, often less than 1%, and in a few rare cases over 80%. Larvae of 11 species and female adults of Aedes sollicitans were infected.

*Coluzzi, M. & Rioux, J. A. (1962). Primo reperto in Italia di larve di Anopheles parassitate da funghi del genere Coelomomyces Keilin. Descrizione di Coelomomyces raffaelei n. sp. (Blastocladiales: Coelomomycetaceae). Riv. Malariol., 41: 29-37.

The presence of Coelomomyces in Europe was confirmed and a new species, C. raffaelei, was described from Anopheles claviger collected in Italy.

Couch, J. N. (1945). Revision of the genus Coelomomyces parasitic in insect larvae. J. Elisha Mitchell Sci. Soc., 61: 124-136.

Couch, J. N. (1962). Validation of the family Coelomomycetaceae and certain species and varieties of Coelomomyces. J. Elisha Mitchell Sci. Soc., 78: 135-138.

In accordance with the rules of botanical nomenclature, Latin diagnoses are given for the family and 11 previously described species of Coelomomyces.

Couch, J. N. (1968). Sporangial germination of Coelomomyces punctatus and the conditions favoring the infection of Anopheles quadrimaculatus under laboratory conditions. Proc. Joint U.S.-Japan Sem. Microb. Control Insect Pests (Fukuoka), pp. 93-105.

Rate of infection was highest when larvae and inoculum were incubated with soil from an infected area. Heavy infections also resulted when soil from uninfected areas was used and some infections were obtained without a soil addition.

Couch, J. N. (1971). Coelomomyces, a fungus for mosquito control. Proc. Nat. Acad. Sci., 68: 2898a (Abstract).

See Couch (1972).

Couch, J. N. (1972). Mass production of Coelomomyces, a fungus that kills mosquitoes. Proc. Nat. Acad. Sci., 69(8): 2043-2047.

Fourth-instar larvae of Anopheles quadrimaculatus infected with C. punctatus were used as inoculum in culturing experiments. If the inoculum is added when the sporangia are discharging spores and during the 1st, 2nd, or 3rd ecdyses, up to 100% infection occurs. Most larvae die as 2nd or 3rd instars. Field tests in North Carolina have averaged 60% infection.

Couch, J. N. & Dodge, H. R. (1947). Further observations on Coelomomyces, parasitic in mosquito larvae. J. Elisha Mitchell Sci. Soc., 63: 69-79.

Couch, J. N. & Umphlett, C. J. (1963). Coelomomyces infections, pp. 149-188. In E. A. Steinhaus (ed.), Insect Pathology, Vol. II, Academic Press, New York.

An extensive list of the species of Coelomomyces, the type host, location of first collection, and authors is given.

*Coz, J. (1973). Contribution a l'étude du parasitisme des Anopheles Ouest-Africain. Mermithidae et Coelomomyces. Cah. ORSTOM Ser. Entomol. Med. Parasitol., 11(4): 237-241.

Coelomomyces ascariformis and C. walkeri were observed infecting A. gambiae in Upper Volta. C. ascariformis parasitized larvae and C. walkeri infected the ovaries of adult females.

Dary, R. M. & Camey, L. (1966). Preliminary note on a survey for parasites of Anopheles larvae in Guatemala, as a contribution to relevant biological control research. WHO/EBL/66.75. 6 pp.

See Camey-Pacheco (1965).

134

*Dubitskii, A. M. et al. (1970).   ⟨The discovery in mosquito larvae of a fungus of the genus Coelomomyces in southeastern Kazakhstan (Preliminary communication)⟩.   Med. Parazitol. Parazit. Bolezni., 39: 737-738.   (R)

   Infected larvae of Culex modestus and C. pipiens were collected in the basin of the Ili River.   In some areas the incidence reached 100% and infected larvae were found throughout the basin.

Dubitskii, A. M. et al. (1973).   Pathogenicity of a fungus of the genus Coelomomyces for larvae of bloodsucking mosquitoes.   Izv. Akad. Nauk. Kaz. SSR Ser. Biol., 4: 1-6. (in Russian)

   The development of a Coelomomyces ⟨C. iliens⟩ in Culex modestus larvae was examined histologically and the effect of the fungus on adult emergence.

*Dubitskii, A. M. et al. (1973).   ⟨A new species of pathogenic fungus of the genus Coelomomyces, isolated from larvae of bloodsucking mosquitoes.⟩   Mikologiya i fitapatologiya, 7: 136-139.   (original in Russian)

   The Coelomomyces found in infected Culex modestus and C. pipiens larvae in Kazakhstan, USSR (see Dubitskii et al., 1970) is described as a new species, C. iliens.

*Ekstein, F. (1922).   Beiträge zur Kentniss der Stechmückenparasiten.   Zentr. Bakteriol. Parasitenk., 88: 128.

*Fedder, M. L. et al. (1971).   ⟨Finding of parasitic fungus of the genus Coelomomyces (Phycomycetes, Blastocladiales) in Aedes togoi Theobald mosquitoes in the Maritime territory.⟩ Med. Parazitol. Parazit. Bolezni., 40(2): 201-204.   (R)

   Twenty-four of 2123 female adult A. togoi examined were infected by Coelomomyces sp. Both nulliparous females and those with several gonotrophic cycles were infected.   A few infected larvae were also seen.

Federici, B. A. & Roberts, D. W. (1973).   Experimental laboratory infection of mosquito larvae with fungi of the genus Coelomomyces.   (Abstract)   Pap. 5th Int. Colloq. Insect Pathol. Microb. Control (Oxford), p. 82.   (E, F, G)

   Laboratory infections of larvae of Aedes taeniorhynchus with Coelomomyces psorophorae var. A. taeniorhynchus, Culiseta inornata with Co. psorophorae var. Cu. inornata and Anopheles quadrimaculatus with Co. punctatus were obtained repeatedly.   No infections occurred when larvae were exposed to large numbers of zoospores.   Units infective for mosquitos were not present prior to six days after zoospore introduction, indicating an unknown stage in the life cycle of Coelomomyces.

Federici, B. A. & Roberts, D. W. (1975).   Experimental laboratory infection of mosquito larvae with fungi of the genus Coelomomyces.   I. Experiments with Coelomomyces psorophorae var. in Aedes taeniorhynchus and Coelomomyces psorophorae var. in Culiseta inornata. J. Invertebr. Pathol., 26: 21-27.

   Laboratory infection of Aedes taeniorhynchus and Culiseta inornata was obtained with isolates of Coelomomyces psorophorae from the respective hosts.

Federici, B. A. & Roberts, D. W. (1976).   Experimental laboratory infection of mosquito larvae with fungi of the genus Coelomomyces.   II. Experiments with Coelomomyces punctatus in Anopheles quadrimaculatus.   J. Invertebr. Pathol., 27: 333-341.

   Experiments on the infection of Anopheles quadrimaculatus with Coelomomyces punctatus indicate that zoospores released from sporangia do not infect mosquitos.   Zoospores, or products from zoospores, apparently infect the copepod Cyclops vernalis.   The copepod is an alternate host for C. punctatus.

Federici, B. A. et al. (1975). Mosquito host range of Coelomomyces punctatus. Annals Entomol. Soc. Amer., 68: 669-670.

Anopheles albimanus, A. stephensi, Aedes aegypti, A. albopictus, A. taeniorhynchus, Culex pipiens pipiens, and Culiseta inornata were exposed to the fungus Coelomomyces punctatus, whose natural mosquito host is Anopheles quadrimaculatus. Of the test species only A. stephensi proved to be susceptible to infection.

Fulton, H. R. et al. (1974). A survey of North Mississippi mosquitoes for pathogenic micro-organisms. Mosq. News, 34: 86-90.

Coelomomyces punctatus, C. lativattus, and an unidentified species were detected in mosquitos during the survey.

*Gad, A. M. & Sadek, S. (1968). Experimental infection of Anopheles pharoensis larvae with Coelomomyces indicus. J. Egypt. Public Health Assoc., 43: 387-391.

Dead A. pharoensis larvae which were parasitized by C. indicus were introduced into rice fields. The introductions began after the crop had been cultivated and just prior to drying. The fields were re-examined the following season. Parasitized larvae were found in some fields. Introductions were again made. Although parasitized larvae could be found two weeks later, subsequent samplings, even after later introductions, failed to find infected larvae.

Gad, A. M. et al. (1967). The occurrence of Coelomomyces indicus Iyengar in Egypt, U.A.R. Mosq. News, 27: 201-202.

The fungus C. indicus was recorded for the first time in Egypt and the Middle East. The larvae and adults of A. pharoensis and the larvae of Culex antennatus were found infected in rice fields. Sporangia were found in all infected adults and larvae. Hyphae were, also, observed in the abdominal cavities of a few adults. The percentage of infection was 27.4 in C. antennatus larvae and 4.1 in A. pharoensis larvae for several sampling locations.

*Genga, R. & Maffi, M. (1973). Coelomomyces sp. infection in Aedes (Stegomyia) hebrideus Edwards on Nupani Island, Santa Cruz Group, Solomons. J. Med. Entomol., 10(4): 413-414.

Coelomomyces sp. was found infecting a 4th-instar larva of A. hebrideus in a collection of 17 larvae made in July 1970. The fungus appeared to present more features suggestive of C. solomonis than of C. stegomyiae but was not definitely identified. Sporangia were found packing the thorax, cephalic and abdominal segments, and ventral gills.

*Gugnani, H. C. et al. (1965). A note on Coelomomyces infection in mosquito larvae. Bull. Indian Soc. Malar. Commun. Dis., 2(4): 333-337.

During a survey of 11 different types of habitats, 7571 mosquito larvae of 6 different species were examined for Coelomomyces infection. Only Anopheles subpictus from a single site showed infection (C. indiana). The infection rate decreased with the decrease in water temperature from 61% to 12%. Sixty to seventy-five per cent. of the infections were fatal. The first record of a natural infection on a Crustacean (Daphnia) is reported. It appears likely that during seasons in which conditions are not favourable for the breeding of mosquitos, the fungus may survive on the microfauna of the pools.

*Haddow, A. J. (1942). The mosquito fauna and climate of native huts at Kisumu, Kenya. Bull. Entomol. Res., 33: 91-142.

Hazard, E. I. (1972).   Investigation of pathogens of anopheline mosquitoes in the vicinity of Kaduna, Nigeria.   WHO/VBC/72.384.   6 pp.

Coelomomyces was found infecting Anopheles funestus and A. gambiae with low incidence.

*Huang, Y. M. (1968).   Aedes (Verrallina) of the Papuan Subregion (Diptera: Culicidae).
Pacific Insects Monograph, 17: 1-74, Bishop Museum, Hawaii.

*Iyengar, M. O. T. (1935).   Two new fungi of the genus Coelomomyces parasitic in larvae of
Anopheles.   Parasitology, 27: 440-449.

Iyengar, M. O. T. (1962).   Validation of two species of Coelomomyces described from India.
J. Elisha Mitchell Sci. Soc., 78: 133-134.

In accordance with the rules of botanical nomenclature, Latin diagnoses are given for
C. indicus and C. anophelesicus.

Keilin, D. (1921).   On a new type of fungus: Coelomomyces stegomyiae, n.g., n. sp.,
parasitic in the body-cavity of the larva of Stegomyia scutellaris Walker (Diptera,
Nematocera, Culicidae).   Parasitology, 13: 225-234.

Kellen, W. R. (1963).   Research on biological control of mosquitoes.   Proc. Calif. Mosq.
Control Assoc., 31: 23-25.

This is a report on the research activities of the Bureau of Vector Control in Fresno,
California.   Coelomomyces psorophorae infecting adult female Aedes melanimon and
Culiseta incidens has been observed.   No diseased larvae were reported.

*Kellen, W. R. et al. (1963).   A new host record for Coelomomyces psorophorae Couch in
California (Blastocladiales: Coelomomycetaceae).   J. Insect Pathol., 5: 167-173.

C. psorophorae infections were found in 3% of the females of Aedes melanimon, although
larvae from the same area were not infected.   The development of sporangia may be
independent of the time of infection for some species of mosquitos.   Blood meals are
required for ovarial development of anautogenous mosquitos and may be necessary for
sporangial development in those females.

*Khaliulin, G. L. & Ivanov, S. L. (1973).   Parasitic fungus Coelomomyces sp. in mosquito
larvae in the Maryiskaya ASSR.   Med. Parazitol. Parazit. Bolezni., 42(4): 487.   (in Russian)

Coelomomyces sp. sporangia were found in larvae of Aedes cyprius, A. caspius dorsalis,
and Culex sp. collected in Maryiskaya ASSR.   The hyphae differed from C. dodgei and
C. pentangulatus in that it branched dichotomously.

Khaliulin, G. L. & Lavrentev, P. A. (1972).   /On diagnosing diseases of mosquito larvae./
Veterinariya, 5: 26-29.   (R)

The hemolymph of healthy larvae and those infected with Coelomomyces is compared.   This
comparison is suggested as a method for diagnosing diseases.

Kupriyanova, E. S. (1969).   The discovery of a parasitic fungus of the genus Coelomomyces in
mosquito larvae in the Maritime Province.   Med. Parazitol. Parazit. Bolezni., 38(4): 494-
495.   (R)

Larvae of Culex orientalis were found infected with Coelomomyces sp. on the outskirts of
Arsen'ev.   Although their behaviour appeared normal, infected larvae were bright orange.

Kuznetsov, V. G. & Mikheeva, A. I. (1970). /The occurrence of the fungus _Coelomomyces_ in larvae of _Aedes_ in the Soviet Far East./ _Parazitologiya_, 4(4): 392-393. (R)

C. psorophorae sporangia were found in larvae of A. vexans.

*Lacour, M. & Rageau, J. (1957). Enquête épidémiologique et entomologique sur la filariose de Bancroft en Nouvelle Calédonie et Dépendances. _Tech. Pap. South Pacific Comm._, 110: 1-24.

*Laird, M. (1956a). Studies of mosquitoes and fresh water ecology in the South Pacific. _Roy. Soc. New Zealand Bull._, 6: 1-213.

*Laird, M. (1956b). A new species of _Coelomomyces_ (fungi) from Tasmanian mosquito larvae. _J. Parasitol._, 42: 53-55.

*Laird, M. (1959a). Parasites of Singapore mosquitoes, with particular reference to the significance of larval epibionts as an index of habit pollution. _Ecology_, 40: 206-221.

*Laird, M. (1959b). Fungal parasites of mosquito larvae from the oriental and Australian regions, with a key to the genus _Coelomomyces_ (Blastocladiales: Coelomomycetaceae). _Can. J. Zool._, 37: 781-791.

*Laird, M. (1961). New American locality records for four species of _Coelomomyces_ (Blastocladiales, Coelomomycetaceae). _J. Insect Pathol._, 3: 249-253.

Laird, M. (1962a). _Coelomomyces_ fungi, an important group of mosquito parasites. WHO/EBL/1. 4 pp.

A general review of WHO research on the use of _Coelomomyces_ for mosquito control is given.

Laird, M. (1962b). Validation of certain species of _Coelomomyces_ (Blastocladiales, Coelomomycetaceae) described from the Australian and Oriental regions. _J. Elisha Mitchell Sci. Soc._, 78(2): 132-133.

In accordance with the rules of botanical nomenclature, Latin diagnoses are given for seven previously described species of _Coelomomyces_.

*Laird, M. (1966). Integrated control and _Aedes polynesiensis_: an outline of the Tokelau Islands project, and its results. WHO/EBL/66.69, WHO/FIL/66.63, WHO/Vector Control/66.204. 9 pp.

See Laird (1967).

Laird, M. (1967). A coral island experiment: a new approach to mosquito control. _WHO Chron._, 21: 18-26.

Methods of biological and chemical control of mosquitos were tested on the Tokelau Islands in the South Pacific. Infections of two mosquito species, _Aedes polynesiensis_ and _A. vexans nocturnus_, were observed after introduction of _Coelomomyces stegomyiae_ to the islands. Five years later, 37.1% of the sampled larvae were parasitized by the fungus. Apparently the fungus had become established and was self-perpetuating.

Laird, M. & Colless, D. H. (1962). A field experiment with a fungal pathogen of mosquitoes, in the Tokelau Islands. _Proc. 11th Cong. Entomol. (Vienna)_, 2: 867-868.

See Laird (1967).

*Laird, M. et al. (1975). _Coelomomyces omorii_ sp. n. and _C. raffaeli_ Coluzzi and Rioux var. _parvum_ var. n. from mosquitoes in Japan. J. Parasitol., 61(3): 539-544.

Lavits'ka, Z. G. et al. (1967). /A fungus _Coelomomyces quadrangulatus_ Couch - the parasite of mosquito larvae./ Dopov. Akad. Nauk Ukr. RSR Ser. B Geol. Geofiz. Khim. Biol., 12: 1116-1118. (Uk)

Larvae of _Aedes vexans_, _Culex pipiens molestus_, _A. rossicus_, and _A. geniculatus_ collected from a drying reservoir in the Ukraine were infected by _Co. quadrangulatus_.

*Lum, P. T. M. (1963). The infection of _Aedes taeniorhynchus_ (Wiedemann) and _Psorophora howardii_ Coquillett by the fungus _Coelomomyces_. J. Insect Pathol., 5: 157-166.

Dissection of mosquito larvae infected by _C. psorophorae_ revealed mycelium along the midgut, in muscle tissues and in the caeca, but sporangia were distributed throughout the body. In infected female adults, the fungus was confined to the ovaries, where each follicle was affected. Infected females failed to mature eggs and all infected larvae died.

Madelin, M. F. (1964). Laboratory studies on the infection of _Anopheles gambiae_ Giles by a species of _Coelomomyces_. WHO/EBL/17, WHO/Mal/435, WHO/Vector Control/64, 23 pp. Also WHO/EBL/17 Corr.1, 4 pp.

See Madelin (1968).

Madelin, M. F. (1965). Further laboratory studies of _Coelomomyces_ which infects _Anopheles gambiae_ Giles. WHO/EBL/52.65, 22 pp.

See Madelin (1968).

*Madelin, M. F. (1968). Studies on the infection by _Coelomomyces indicus_ of _Anopheles gambiae_. J. Elisha Mitchell Sci. Soc., 84(1): 115-124.

Eggs or 1st instar larvae were added to rearing basins containing resting sporangia and various tests were performed to determine the conditions conducive to infection. The following features were shared by the successful tests: (1) a freely evaporating body of water, (2) ca. pH 8, (3) ca. 28-32°C, (4) illumination for half of the day, (5) foundation of soil containing monpane clay, (6) inoculum of sporangia from Zambia, (7) mosquito population introduced as eggs or 1st instar larvae, and (8) presence of water fleas.

Madelin, M. F. & Beckett, A. (1972). The production of planonts by thin-walled sporangia of the fungus _Coelomomyces indicus_, a parasite of mosquitoes. J. Gen. Microbiol., 72(1): 185-200.

Sporangia of _C. indicus_ dehisced in water in 1-2 days. During dehiscence, the contents of the sporangia cleave into planonts. The cytological events surrounding this cleavage are described.

*Maffi, M. & Genga, R. (1970). Contributo alla conoscenza dell'infestazione da _Coelomomyces_ nei Culicidi delle Salomone Britanniche, Oceania. Parassitologia (Rome), 12: 171-178. (I, e)

A summary of sightings of _Coelomomyces_ in the British Solomon Islands is given.

Manalang, C. (1930). Coccidiosis in _Anopheles_ mosquitoes. Philippine J. Sci., 42: 279.

Martin, III, W. W. (1969).   A morphological and cytological study of the development of Coelomomyces punctatus parasitic in Anopheles quadrimaculatus.   J. Elisha Mitchell Sci. Soc., 85(2): 59-72.

   The development of the mycelium and resting sporangia is described using light, phase, and electron microscopy.   Spherical hyphagens are the earliest stage of the fungal infection.   These structures give rise to the mycelium.   Sporangia are produced at the tips or on lateral branches of hyphae and mature while still within the membrane of the parent hyphae.

Martin, W. W. (1971).   The ultrastructure of Coelomomyces punctatus zoospores.   J. Elisha Mitchell Sci. Soc., 87(4): 209-221.

   This is the first detailed account of the ultrastructure of the zoospore of a parasitic member of the Blastocladiales.   The shape of the nucleus differs significantly from that of other fungal zoospores.

*McCrae, A. W. R. (1972).   Age-composition of man-biting Aedes (Stegomyia) simpsoni (Theobald) (Diptera: Culicidae) in Bwamba County, Uganda.   J. Med. Entomol., 9(6): 545-550.

   Two of 224 adult female Ae. simpsoni dissected contained Coelomomyces sp. sporangia.

Morozov, V. A. (1967).   /The discovery of the parasitic fungus Coelomomyces in the larvae of Aedes in the vicinity of Krasnodar./   Med. Parazitol. Parazit. Bolezni., 36: 353.   (R)

   Larvae of A. vexans infected with C. psorophorae were found in a fresh water pond.   The water level in the pond periodically fluctuated.

Mullins, J. T. (1971).   Cultivation of Coelomomyces, a fungus parasite of mosquitoes.   Ann. Res. Rep. Instit. Food Agric. Sci. Univ. Florida (Gainesville), p. 70.

   Field collections of mosquito larvae infected with the fungus Coelomomyces in Florida showed that the extent of infection varied from 5% to 50%.   Attempts to establish a laboratory colony of infected mosquitos were unsuccessful.

*Muspratt, J. (1946a).   On Coelomomyces fungi causing high mortality of Anopheles gambiae larvae in Rhodesia.   Ann. Trop. Med. Parasitol., 40: 10-17.

*Muspratt, J. (1946b).   Experimental infection of the larvae of Anopheles gambiae (Dipt., Culicidae) with a Coelomomyces fungus.   Nature, 158: 202.

*Muspratt, J. (1962).   Destruction of the larvae of Anopheles gambiae Giles by a Coelomomyces fungus.   WHO/EBL/2, WHO/Vector Control/2, 2 July 1962.

Muspratt, J. (1963a).   Destruction of the larvae of Anopheles gambiae Giles by a Coelomomyces fungus.   Bull. Wld Hlth Org., 29: 81-86.

   In several areas of Northern Rhodesia, mortality of 95% has occurred in larvae of A. gambiae infected by the fungus.   During periods of high rainfall or where pools were kept filled by river water, the high level of mortality continued.   Successful introduction of Coelomomyces sp. into previously uninfected areas is reported.

Muspratt, J. (1963b).   Progress Report (May 1963) on investigations concerning three mosquito pathogens at Livingstone, Northern Rhodesia.   WHO/EBL/12, 7 pp.

   Coelomomyces indiana was found to be still active a year after introduction into a breeding pool inhabited by Anopheles gambiae.

Muspratt, J. (1964).   Parasitology of larval mosquitos, especially Culex pipiens fatigans Wied., at Rangoon, Burma.   WHO/EBL/18.   23 pp.

    In a one-month survey of mosquitos, Coelomomyces sp. was found in Aedes aegypti and A. albopictus.   The fungus was found in unpolluted sites.

Nolan, R. A. (1973).   Effect of plant hormones on germination of Coelomomyces psorophorae resistant sporangia.   J. Invertebr. Pathol., 21: 26-30.

    In germination tests, 41% germinated when incubated with indole-3-butyric acid (IBA), whereas 24% germinated in the no hormone test.   The effects of continuous incubation of resistant sporangia in 19 hormones and related substances were assessed.

Nolan, R. A. & Freake, G. W. (1974).   An electron microprobe analysis of the resistant sporangium wall of Coelomomyces psorophorae.   J. Invertebr. Pathol., 23: 121-122.

    A preliminary analysis of the elemental composition of the sporangium wall revealed that phosphorus, sodium and potassium were most abundant.

*Nolan, R. A. et al. (1973).   A mosquito parasite from a mosquito predator.   J. Invertebr. Pathol., 21: 172-175.

    A larva of Toxorhynchites rutilus septentrionalis collected from a treehole in Louisiana was found infected by Coelomomyces macleayae.   This treehole had been used previously for Aedes triseriatus infection experiments with the same fungus.

*Pillai, J. S. (1969).   A Coelomomyces infection of Aedes australis in New Zealand. J. Invertebr. Pathol., 14(1): 93-95.

    The identity of a fungus isolated from A. australis was confirmed by cross infection experiments as Coelomomyces opifexi.   This fungus had previously been identified from Opifex fuscus, an endemic New Zealand mosquito.

Pillai, J. S. (1971).   Coelomomyces opifexi (Pillai and Smith) (Coelomomycetaceae: Blastocladiales).   I. Its distribution and the ecology of infection pools in New Zealand. Hydrobiologia, 38: 425-436.

    The activity of C. opifexi is confined in New Zealand to the range of Aedes australis, although it also infects Opifex fuscus.   This fungus may have been derived from Australian sources.   Some correlation between pool salinity and the appearance of infected larvae was found.   Other factors, such as, temperature, pH, pool size, and biota were not correlated.

Pillai, J. S. & O'Loughlin, I. H. (1972).   Coelomomyces opifexi (Pillai and Smith). Coelomomycetaceae: Blastocladiales.   II. Experiments in sporangial germination. Hydrobiologia, 40(1): 77-86.

    Sporangia of this fungus, which is found in a superlittoral environment, did not germinate under salinity conditions equal to full sea water.   They germinate almost spontaneously in water with salinity less than 4.2%.   Those derived from living or moribund larvae germinated more readily than those from dead larvae.   Sporangia from dead larvae required a conditioning of 7 days at 23°C or 28°C to germinate.

*Pillai, J. S. & Rakai, I. (1970).   Coelomomyces macleayae (Laird), a parasite of Aedes polynesiensis (Marks) in Fiji.   J. Med. Entomol., 7(1): 125-126.

    The fungus was isolated from infected larvae found in tree-holes on Koro Island, Fiji. Apart from introduced Coelomomyces, only C. stegomyiae was previously known in this region.

*Pillai, J. S. & Smith, J. M. B. (1968).   Fungal pathogens of mosquitoes in New Zealand.
I. Coelomomyces opifexi sp. n., on the mosquito Opifexi fuscus Hutton.   J. Invertebr. Pathol.,
11: 316-320.

   Larvae of the endemic New Zealand mosquito O. fuscus collected on the Otago Peninsula,
   were infected with a previously undescribed species of Coelomomyces.   The mosquitos were
   collected from fresh to brackish water.   A table of characteristics is given to compare
   the new species of C. tasmaniensis and C. psorophorae.

*Pillai, J. S. & Woo, A. (1973).   Coelomomyces opifexi (Pillai and Smith) Coelomomycetaceae:
Blastocladiales.   III. The laboratory infection of Aedes australis (Erichson) larvae.
Hydrobiologia, 41(2): 169-181.

   Inoculum was prepared from field and laboratory infected larvae and was used 6-24 hours
   after preparation.   The main requirements for infection were (a) a suitable medium of
   low salinity, (b) inoculum containing a high concentration of active zoospores,
   (c) temperatures approaching summer conditions.   Infection rates ranged from 20-100%.

*Rajapaksa, N. (1963).   Records of a survey of Coelomomyces infections in mosquito larvae
carried out in the south-west coastal belt of Ceylon.   WHO/EBL/8, WHO/Vector Control/31,
23 April 1963.

   See Rajapaksa (1964).

*Rajapaksa, N. (1964).   Survey for Coelomomyces infections in mosquito larvae in the south-
west coastal belt of Ceylon.   Bull. Wld Hlth Org., 30: 149-151.

   During a survey conducted in urban and rural communities, 19 918 larvae representing
   18 species of mosquitos were screened for Coelomomyces infections.   Five species were
   parasitized: Aedes albopictus, Culex pipiens fatigans, C. gelidus, Ae. aegypti, and
   Armigeres abturbans.   The principal host was Ae. albopictus; 15.4% were infected.

Ramalingam, S. (1966).   Coelomomyces infections in mosquito larvae in the South Pacific.
Med. J. Malaya, 20(4): 334.

   None of the over 3000 adult female mosquitos examined from over 10 species was infected
   by Coelomomyces.   Larvae of five species were infected: Aedes polynesiensis,
   A. aegypti, A. oceanicus, A. tutuilae, and A. samoanus.   This is the first record of
   this fungus infecting the last three species.

*Rioux, J. A. & Pech, J. (1960).   Coelomomyces grassei n. sp. parasite d'Anopheles gambiae
(note préliminaire).   Acta Trop., 17: 179-182.

*Rioux, J. A. & Pech, J. (1962).   Validation of Coelomomyces grassei Rioux and Pech.
J. Elisha Mitchell Sci. Soc., 78: 134-135.

   In accordance with the rules of botanical nomenclature, Latin diagnosis is given for
   C. grassei.

Roberts, D. W. (1970).   Coelomomyces, Entomophthora, Beauveria, and Metarrhizium as parasites
of mosquitoes.   Misc. Publ. Entomol. Soc. Am., 7(1): 140-155.

   Coelomomyces spp. appear to be useful for mosquito control because they are largely
   restricted to the family Culicidae and they maintain themselves in the mosquito
   environment year after year.   Natural infections indicate that significant control can
   be achieved with these fungi.

Roberts, D. W. (1974).   Fungal infections of mosquitoes, p. 143-193.   In A. Aubin, J.-P. Bourassa, S. Belloncik, M. Pellissier, and E. Lacoursière (eds.), Le controle des moustiques/Mosquito contrôl, Univ. Quebec Press, Quebec.

The host range, distribution, life cycle, and possibilities as microbial control agents are discussed for fungi including Coelomomyces.   134 references.

Roberts, D. W. et al. (1973).   Dehiscence of Coelomomyces psorophorae sporangia from Aedes taeniorhynchus: Induction by amines and amino acids.   J. Invertebr. Pathol., 22: 175-181.

The ability to induce dehiscence of C. psorophorae sporangia would facilitate the use of stored sporangia for laboratory studies and biological control.   The effect of various chemicals on induction of dehiscence of nonsterile sporangia was examined.   The most active compound was Tris /tris(hydroxymethyl)amino-methane/.   It induced 50% dehiscence after 100 days of storage at 10°C; it was most effective at pH 8-9 and 1-20 mM.

Rodhain, F. (1969).   Sur la présence d'un champignon du genre Coelomomyces en République de Haute-Volta.   Ann. Parasitol. Hum. Comp., 44(3): 261-264.   (F, e)

A case of Coelomomyces africanus is reported in an Anopheles (probably A. squamosus) larva in Upper Volta.

Rodhain, F. & Brengues, J. (1974).   Présence de champignons du genre Coelomomyces chez des Anopheles en Haute-Volta.   Ann. Parasitol. Hum. Comp., 49(2): 241-246.   (F, e)

Four species of Coelomomyces: C. africanus, C. grassei, C. ascariformis, and C. walkeri, infect female A. gambiae in Upper Volta.   Descriptions of the parasites and the host's pathology are given for each species of parasite.   This report extends the geographical range of these species.   All were previously known from Africa south of the Sahara.

*Rodhain, F. & Gayral, P. (1971).   Nouveaux cas de parasitisme de larvaes d'Anopheles par des champignons du genre Coelomomyces en République de Haute-Volta.   Ann. Parasitol. Hum. Comp., 46(3): 295-300.   (F, e)

Coelomomyces infections of Anopheles larvae in Upper Volta are recorded.   A. rufipes larva was infected by C. africanus (tentative identification).   A. gambiae larvae were infected by C. ascariformis, C. indicus, and an undetermined species.

*Romney, S. V. et al. (1971).   Intergeneric transmission of Coelomomyces infections in the laboratory.   Utah Mosq. Abat. Assoc. Proc., 24: 18-19.

The discovery of a new species of Coelomomyces in rock pools in Utah is reported. Laboratory transmission of disease was accomplished by placing dissected infected larvae into water containing live larvae and debris from the infection site.   The fungus was freely transmitted in the field between three mosquito hosts (Aedes atropalpus nielseni, Culex tarsalis, and Culiseta incidens) and high rates of infection and mortality were found.   Examination of 90 species of animals representative of the rock pool fauna indicates that the parasite is specific for tissues of mosquito larvae.

Service, M. W. (1973).   Mortalities of the larvae of the Anopheles gambiae Giles complex and detection of predators by the precipitin test.   (WHO/VBC/72.360)   Bull. Entomol. Res., 62: 359-369.

A life table study was made on larvae of A. gambiae from ponds and ditches in Kenya. Daily survival rates and instar mortalities showed that there was high mortality among 4th instar larvae, due at least in part to infections of Coelomomyces.   Coelomomyces infections were common in some areas (77.1% and 74.8% of larvae in two areas) and very rare in others.

*Service, M. W. (1974). Further results of catches of Culicoides (Diptera: Ceratopogonidae) and mosquitoes from suction traps. J. Med. Entomol., 11(4): 471-479.

   Coelomomyces sporangia were found in a few adults of Aedes cantans.

Chapiro, M. & Roberts, D. W. (1976). Growth of Coelomomyces psorophorae mycelium in vitro. J. Invertebr. Pathol., 27: 399-402.

   Mycelium of Coelomomyces psorophorae from Aedes sollicitans was subcultured at 2- to 4-week intervals for two years in a modified mycoplasma medium. Three other varieties of C. psorophorae as well as C. dodgei and C. pentangulatus could not be subcultured.

Shcherban', Z. P. & Gol'berg, A. M. (1971). The pathogenic fungi Coelomycidium (Phycomycetes, Chytridiales) and Coelomomyces (Phycomycetes, Blastocladiales) in Culex and Aedes (Diptera, Culicidae) in Uzbekistan. Med. Parazitol. Parazit. Bolezni., 40(1): 110-111. (R, e)

   None of the female Aedes caspius collected while attacking animals was found infected with Coelomomyces. However, 5.8% collected on vegetation and with no blood in their guts were infected by C. psorophorae.

*Shemanchuk, J. A. (1959). Note on Coelomomyces psorophorae Couch, a fungus parasitic on mosquito larvae. Can. Entomologist, 91: 743-744.

Steinhaus, E. A. & Marsh, G. A. (1962). Report of diagnoses of diseased insects 1951-1961. Hilgardia, 33: 349-490.

   The location, host, submission date, and individual requesting diagnosis are given for each tentative diagnosis. For details see the host-parasite table.

Umphlett, C. J. (1962a). Comparative studies in the genus Coelomomyces Keilin. Diss. Abstr., 22: 2560.

   See Umphlett (1962b).

Umphlett, C. J. (1962b). Morphological and cytological observations on the mycelium of Coelomomyces. Mycologia, 54: 540-554.

   The structures of the mycelium and vegetative nuclei of Co. pentangulatus infecting Culex erraticus and Co. dodgei infecting Anopheles crucians were observed. Hyphal bodies in both species, apparently formed by fragmentation, may be the means by which the fungus is spread within the host's body. The stages of nuclear division were observed also.

Umphlett, C. J. (1964). Development of the resting sporangia of two species of Coelomomyces. Mycologia, 56: 488-497.

   The sporangia of C. pentangulatus parasitic on Culex erraticus and C. dodgei parasitic on Anopheles crucians were examined. The sporangia are initiated as swellings on the tips of hyphae within the coelom of the host larvae.

Umphlett, C. J. (1968). Ecology of Coelomomyces infections of mosquito larvae. J. Elisha Mitchell Sci. Soc., 84(1): 108-114.

   Samples of Anopheles quadrimaculatus larvae revealed that C. punctatus was present in larvae from one cove and absent from another cove in the same lake in North Carolina. Infection incidence varied from 6.3% to 80.8%. The environmental factors of the two coves are compared. Laboratory experiments showed that significantly more adults emerged from the uninfected area.

Umphlett, C. J. (1970).   Infection levels of Coelomomyces punctatus, an aquatic fungus parasite, in a natural population of the common malaria mosquito, Anopheles quadrimaculatus. J. Invertebr. Pathol., 15(3): 299-305.

The annual incidence of infection varied from 36.5% to 10.4% during three years of study. More larvae were produced in July than in August; however, a higher level of infection occurred in August.   Fourth-instar larvae had the highest incidence of infection.   The infection may cause the insects to remain in the larval stages for extended periods, thereby prolonging their vulnerability to other mortality factors.

*van Thiel, P. H. (1954).   Trematode, Gregarine and fungus parasites of Anopheles mosquitoes. J. Parasitol., 40(3): 271-279.

van Thiel, P. H. (1962).   Validation of two species of Coelomomyces.   J. Elisha Mitchell Sci. Soc., 78: 135.

In accordance with the rules of botanical nomenclature, Latin diagnoses are given for C. walkeri and C. ascariformis.

Walker, A. J. (1938).   Fungal infections of mosquitoes, especially of Anopheles costalis. Ann. Trop. Med. Parasitol., 32: 231-244.

Weiser, J. & Vávra, J. (1964).   Zur Verbreitung der Coelomomyces-Pilze in europäischen Insekten.   Z. Tropenmed. Parasitol., 15: 38-42.   (G, e)

Larvae of Aedes vexans infected by Coelomomyces psorophorae were found near Prague. This finding extends the northern limit of Coelomomyces.

Whisler, H. C. (1974).   Biology of Coelomomyces psorophorae in Culiseta inornata. (Abstract)   Proc. 3rd Int. Cong. Parasitol. (Munich), 3: 1713.

Time lapse cine-micrography was used to observe the developmental stages of the fungus.

Whisler, H. C. et al. (1972).   Germination of the resistant sporangia of Coelomomyces psorophorae.   J. Invertebr. Pathol., 19: 139-147.

The germination of C. psorophorae was recorded combining techniques of both light and electron microscopy.   Resting sporangia were collected from dead Culiseta inornata larvae in southern Alberta.   Germination was obtained by techniques outlined by Couch (1968).   The resistant sporangia readily germinated after reaching a fully cleaved stage of development.

Whisler, H. C. et al. (1974).   Alternate host for mosquito parasite Coelomomyces.   Nature, 251: 715-716.

Cyclops vernalis, a copepod, has been demonstrated to be the alternate host of Coelomomyces psorophorae which infects Culiseta inornata.   The copepod phase of the fungus resembles the Cyclops parasite Callimastix cyclopis.

Whisler, H. C. et al. (1975).   Life history of Coelomomyces psorophorae.   Proc. Nat. Acad. Sci. USA, 72: 963-966.

White, G. B. & Rosen, P. (1973). Comparative studies on sibling species of the _Anopheles gambiae_ Giles complex (Dipt., Culicidae). II. Ecology of species A and B in savanna around Kaduna, Nigeria during transition from wet to dry season. Bull. Entomol. Res., 62: 613-625.

Coelomomyces was found infecting larvae in 16% of the breeding sites examined. Infections in adult females were also observed.

*Zharov, A. A. (1973). Detection of a parasitic fungus _Coelomomyces psorophorae_ Couch (Phycomycetes: Blastocladiales) in _Aedes vexans_ Meigen mosquitoes in the Astrakhan region. Med. Parazitol. Parazit. Bolezni., 42(4): 485-487.

Seventy-five female _A. vexans_ (3.4% of those autopsied) were found with ovaries infected with _C. psorophorae_. In the ovaries, the development of follicles was inhibited.

VII. FUNGAL PATHOGENS, EXCEPT *COELOMOMYCES*, OF CULICIDAE (MOSQUITOS) [a]

Donald W. Roberts

Boyce Thompson Institute
Cell Physiology and Virology Program
Yonkers, NY 10701, USA

[a] Supported in part by USPHS research grant AI10010 from the National Institute of Allergy and Infectious Diseases.

FUNGAL PATHOGENS, EXCEPT <u>COELOMOMYCES</u>, OF CULICIDAE (MOSQUITOS)

Host	Host stage infected	Pathogen	% Incidence	Locality	Lab. or field study	Reference
Aedes sp.	Larvae	Culicinomyces sp.		Australia	Lab.	Sweeney et al. (1973)
"	Larvae	Saprolegnia sp.				Jettmar (1947)[b]
Aedes (Stegomyia) sp.	Larvae	Saprolegnia sp.				Martini (1920)[b]
Aedes sp.	Larvae	Smittium culisetae[a]		Japan	Field	Williams & Lichtwardt (1972)
Aedes aegypti	Larvae	Amoebidium parasiticum[a]		Puerto Rico	Lab. & field	Kuno (1973)
"	Larvae, adults	Beauveria bassiana	0 (L) 100 (A)	USA (California)	Lab.	Clark et al. (1967,1968)
"	Larvae	Beauveria tenellae	100	"	Lab.	Pinnock et al. (1973), Sanders (1972)
"	Larvae	Fusarium sp?		Ghana	Field	Macfie (1917)[c]
"	Larvae	Lagenidium giganteum (= L. culicidum)	100	USA (North Carolina, Georgia)	Lab.	Couch & Romney (1973), McCray et al. (1973a), Umphlett (1973)
"	Larvae	Metarrhizium anisopliae	100	USA (New York) France	Lab.	Roberts (1970,1974), Roberts & Mouchet (unpublished)
"	Larvae	Smittium culicis[a]	High	USA (Kansas)	Lab.	Williams & Lichtwardt (1972)
"	Larvae	Smittium culisetae[a]	High	USA (Kansas)	Lab.	Williams & Lichtwardt (1972)
"	Larvae	Smittium (= Rubetella) inopinata[a]	93-100	Italy	Lab.	Coluzzi (1966)
"	Larvae	Smittium simulii[a]	High	USA (Kansas)	Lab.	Williams & Lichtwardt (1972)
"	Adults	Yeast				Christophers (1952), Marchoux, Salimbeni & Simond (1903)[b]

[a] Trichomycete. Does not penetrate host cells. Attached to external exocuticle (<u>Amoebidium</u>) or commensal in gut (<u>Smittium</u>). Causes mortality in stressed individuals only.

[b] Cited in Jenkins (1964).

[c] Cited in Jenkins (1964).   Summary seen.

[d] Cited in Jenkins (1964).   Original seen.

FUNGAL PATHOGENS, EXCEPT COELOMYCES, OF CULICIDAE (MOSQUITOS) (continued)

Host	Host stage infected	Pathogen	% Incidence	Locality	Lab. or field study	Reference
Aedes albopictus	Larvae	Smittium culisetae[a]		USA (Hawaii)	Field	Williams & Lichtwardt (1972)
Aedes atropalpus epactius	Larvae	Culicinomyces clavosporus		USA (North Carolina)	Lab.	Couch et al. (1974)
Aedes atropalpus epactius	Larvae	Lagenidium giganteum		USA (North Carolina)	Lab.	Couch & Romney (1973)
Aedes atropalpus	Larvae	Metarrhizium anisopliae	100	USA (New York)	Lab.	Roberts (1970,1974)
Aedes australis	Larvae	Culicinomyces sp.		Australia	Lab.	Sweeney (1975a)
Aedes berlandi	Larvae	Saprolegnia declina		France	Lab.	Rioux & Achard (1956)[b]
"	Larvae	Smittium culicis[a]		France	Field	Manier (1969)
"	Larvae	Smittium culicis[a]	100	France	Lab.	Tuzet et al. (1961)
Aedes canadensis	Larvae, pupae	Entomophthora aquatica	0-80	USA (Connecticut)	Field	Anderson & Ringo (1969)
Aedes caspius	Larvae	Smittium culicis[a]		France	Field	Manier (1969)
Aedes communis (= Culex nemorosus)	Adults	Entomophthora conglomerata (= Empusa conglomerata, E. thaxteri)				Lakon (1919)[d], Sorokin (1877)[d]
Aedes detritus	Larvae	Amoebidium parasiticum[a]		Tunisia	Field	Manier et al. (1964)
"	Adults	Entomophthora culicis				Marshall (1938)[b]
"	Larvae	Fusarium oxysporum	>80 (lab.)	France	Lab. & field	Hasan & Vago (1972)
"	Larvae	Smittium culicis[a]		France	Field	Manier (1969)
"	Larvae	Smittium culicis[a]	100	France	Lab.	Tuzet et al. (1961)
Aedes dorsalis	Larvae	Beauveria tenella	100	USA (California)	Lab.	Pinnock et al. (1973), Sanders (1972)
Aedes geniculatus	Larvae	Saprolegnia sp.				Marshall (1938)[b]
"	Larvae	Smittium culicis[a]		France	Field	Manier (1969)

FUNGAL PATHOGENS, EXCEPT COELOMYCES, OF CULICIDAE (MOSQUITOS) (continued)

Host	Host stage infected	Pathogen	% Incidence	Locality	Lab. or field study	Reference
Aedes geniculatus	Larvae	Smittium culicis[a]	100		Lab.	Tuzet et al. (1961)
Aedes hexodontus	Larvae	Beauveria tenellae	100	USA (California)	Lab.	Pinnock et al. (1973)
Aedes mediovittatus	Larvae	Lagenidium giganteum (= L. culicidum)	100	USA (Georgia)	Lab.	McCray et al. (1973a), Umphlett (1973)
Aedes melanimon	Larvae	Smittium (= Rubetella) culicis[a]		USA (California)	Field	Clark et al. (1963)
Aedes nigromaculis	Larvae, adults	Beauveria bassiana	0 (L) 100 (A in lab.) 58 (A in field test)	USA (California)	Lab & field	Clark et al. (1967,1968)
"	Larvae	Lagenidium giganteum	100	USA (California)	Field	McCray et al. (1973b)
Aedes polynesiensis	Larvae	Lagenidium giganteum		USA (North Carolina)	Lab.	Couch & Romney (1973)
"	Larvae	Metarrhizium anisopliae		France	Lab.	Mouchet (unpublished)
Aedes rupestris	Larvae	Culicinomyces sp.	High	Australia	Field	Sweeney & Panter (1974)
Aedes rusticus	Larvae	Saprolegnia sp.				Marshall (1938)[b]
Aedes sierrensis	Larvae, adults	Beauveria bassiana	0 (L) 100 (A)	USA (California)	Lab. & field	Clark et al. (1967,1968)
"	Larvae	Beauveria tenellae	100 53-71 26-91	USA (California)	Lab. Field Field	Pinnock et al. (1973) Sanders (1972)
"	Larvae	Pythium sp. (near P. adhaerens)		USA (California)	Field & lab.	Clark et al. (1966)
Aedes sollicitans	Adults	Basidiomycete		USA (Louisiana)	Field	Chapman et al. (1967,1969)
"	Larvae	Lagenidium giganteum (= L. culicidum)	100	USA (Georgia)	Lab.	McCray et al. (1973a), Umphlett (1973)

FUNGAL PATHOGENS, EXCEPT COELOMYCES, OF CULICIDAE (MOSQUITOS) (continued)

Host	Host stage infected	Pathogen	% Incidence	Locality	Lab. or field study	Reference
Aedes sollicitans	Larvae	Metarrhizium anisopliae	100 (lab.) 98 field	USA (New York, Delaware)	Lab. & field	Roberts (1970,1974)
Aedes sticticus	Larvae	Smittium culicis[a]		USA (Wyoming)	Field	Williams & Lichtwardt (1972)
Aedes taeniorhynchus	Adults	Conidiobolus coronatus (= Entomophthora coronata)	76 (36% control mortality)	USA (Florida)	Lab.	Lowe & Kennel (1972)
"	Larvae	Lagenidium giganteum (= L. culicidum)	100	USA (Georgia)	Lab.	McCray et al. (1973a), Umphlett (1973)
"	Larvae	Metarrhizium anisopliae	100	USA (New York)	Lab.	Roberts (1970,1974)
Aedes triseriatus	Larvae	Lagenidium giganteum (= L. culicidum)	100	USA (Georgia)	Lab.	McCray et al. (1973a), Umphlett (1973)
"	Larvae	Lagenidium giganteum		USA (North Carolina)	Lab.	Couch & Romney (1973)
"	Larvae	Pythium sp. (near P. adhaerens)		USA (California)	Lab.	Clark et al. (1966)
"	Larvae	Smittium culisetae[a]		USA (Kansas)	Lab.	M. E. Chapman (cited in Williams & Lichtwardt, 1972)
"	Larvae	Smittium simulii[a]		USA (Kansas)	Lab.	M. E. Chapman (cited in Williams & Lichtwardt, 1972)
Aedes vexans	Larvae	Smittium culisetae[a]		USA (Hawaii)	Field	Williams & Lichtwardt (1972)
Aedes vittatus	Larvae	Smittium (= Rubetella) inopinata[a]	High	Italy	Field	Coluzzi (1966)
Anopheles sp.	Larvae	Aspergillus sp.				Christophers (1952)[b]
"	Larvae	Aspergillus glaucus	High		Lab.	Speer (1927),[b] Galli-Valerio & Rochaz de Jongh (1905)[d]
"	Larvae	Aspergillus niger	High		Lab.	Galli-Valerio & Rochaz de Jongh (1905),[d] Speer (1927)[b]

FUNGAL PATHOGENS, EXCEPT COELOMOMYCES, OF CULICIDAE (MOSQUITOS) (continued)

Host	Host stage infected	Pathogen	% Incidence	Locality	Lab. or field study	Reference
Anopheles sp.	Larvae	Lagenidium giganteum	5 (lab.) 8 (field)	USA (North Carolina)	Lab. Field	Umphlett & Huang (1972), Umphlett (1973), Umphlett & McCray (1975)
"	Larvae	Mucor stolonifera	Low	USSR	Lab.	Bačinskij (1926)[c]
"	Larvae	Oidium lactis	Low	USSR (Leningrad)	Lab.	Bačinskij (1926)[c]
"	Larvae	Penicillium glaucum	Low	USSR (Leningrad)	Lab.	Bačinskij (1926)[c]
"	Larvae	Saprolegnia sp.				Jettmar (1947)[b]
Anopheles albimanus	Larvae, adults	Beauveria bassiana	100	USA (California)	Lab.	Clark et al. (1967,1968)
"	Larvae	Metarrhizium anisopliae	100	USA (New York)	Lab.	Roberts (1970,1974)
Anopheles amictus hilli	Larvae	Culicinomyces sp.	High	Australia	Lab.	Sweeney (1975a), Sweeney et al. (1973)
Anopheles annulipes	Larvae	"		Australia	Lab.	Sweeney (1975a)
Anopheles atroparvus atroparvus	Larvae	Smittium culicis[a]		France	Field	Manier (1969)
Anopheles claviger	Larvae	"		France	Field	Manier (1969)
Anopheles coustani	Larvae	Trichophyton sp.				Dyé (1905)[b]
Anopheles freeborni	Larvae	Pythium sp. (near P. adhaerens)		USA (California)	Lab.	Clark et al. (1966)
Anopheles funestus	Larvae	Metarrhizium anisopliae	100	Nigeria	Field	Roberts (unpublished)
Anopheles gambiae	Larvae	"	100	Nigeria	Lab. & field	Roberts (unpublished)
Anopheles gambiae	Larvae	Smittium (= Rubetella) inopinata[a]	93-100	Italy	Lab.	Coluzzi (1966)

FUNGAL PATHOGENS, EXCEPT COELOMOMYCES, OF CULICIDAE (MOSQUITOS) (continued)

Host	Host stage infected	Pathogen	% Incidence	Locality	Lab. or field study	Reference
Anopheles hispaniola	Larvae	Entomophthora culicis				López-Neyra & Guardiola Mira (1938)[b]
Anopheles maculipennis	Larvae	Beauveria bassiana			Lab.	Roubaud & Toumanoff (1930)[b]
"	Larvae	Entomophthora culicis				López-Neyra & Guardiola Mira (1938)[b]
"	Adults	Fungus				Léon (1924)[c]
"	Larvae	Saprolegniaceae ?	100 (Lab.)	Yugoslavia	Lab. & field	Chorine & Baranoff (1929)[d]
"	Adults	Yeast				Laveran (1902)[b]
Anopheles plumbeus	Larvae	Smittium culicis[a]		France	Field	Manier (1969)
"	Larvae	" "		France	Lab.	Tuzet et al. (1961)
Anopheles punctipennis	Larvae	Culicinomyces clavosporus		USA (North Carolina)	Lab.	Couch et al. (1974)
"	Larvae	Lagenidium giganteum		USA (North Carolina)	Lab.	Couch & Romney (1973)
Anopheles quadri-maculatus	Larvae	Beauveria bassiana				Charles (1939)[d]
"	Larvae	Culicinomyces clavosporus		USA (North Carolina)	Lab.	Couch et al. (1974)
"	Larvae	Lagenidium giganteum		USA (North Carolina)	Lab.	Couch & Romney (1973)
"	Larvae	Metarrhizium anisopliae	100	USA (New York)	Lab.	Roberts (1970,1974)
"	Larvae	Spicaria sp.				Brown (1949)[b]
Anopheles rufipes	Larvae	Metarrhizium anisopliae	100	Nigeria	Field	Roberts (unpublished)
Anopheles stephensi	Larvae	Culicinomyces clavosporus		USA (North Carolina)	Lab.	Couch et al. (1974)

FUNGAL PATHOGENS, EXCEPT COELOMOMYCES, OF CULICIDAE (MOSQUITOS) (continued)

Host	Host stage infected	Pathogen	% Incidence	Locality	Lab. or field study	Reference
Anopheles stephensi	Larvae	Lagenidium giganteum		USA (North Carolina)	Lab.	Couch & Romney (1973)
"	Larvae	Metarrhizium anisopliae	100	USA (New York)	Lab.	Roberts (1967,1970,1974)
Anopheles subpictus	Adults	Aspergillus parasiticus		India	Field	Hati & Ghosh (1965)
Armigeres dentatus	Eggs	Lagenidium sp.		Malaya	Field	Mattingly (1972a,b)
Culex sp.	Larvae	Aspergillus sp.				Speer (1927)[b]
"	Larvae	Aspergillus glaucus			Lab.	Galli-Valerio & Rochaz de Jongh (1905)[d]
"	Larvae	Aspergillus niger	High		Lab.	Galli-Valerio & Rochaz de Jongh (1905),[d] Speer (1927)[b]
"	Larvae	Culicinomyces sp.		Australia	Lab.	Sweeney et al. (1973)
"	Larvae, adults	Entomophthora conglomerata (= Empusa conglomerata, Empusa thaxteri)				Thaxter (1888)[d]
"	Adults	Entomophthora culicis				Christophers (1952)[b]
"	Adults	Entomophthora culicis		Poland USA (Maine)		Nowakowski (1883)[b] Thaxter (1888)[d]
"	Larvae	Lagenidium giganteum	Low	USA (North Carolina)	Lab. & field	Couch (1935,1960)[d]
"	Larvae	Oidium lactis	Low	USSR (Leningrad)	Lab.	Bačinskij (1926)[c]
"	Larvae	Penicillium glaucum	Low	USSR (Leningrad)	Lab.	Bačinskij (1926)[c]
Culex apicalis	Larvae	Beauveria bassiana	100 (53% mortality in controls)	USSR (Byelorussia)	Lab.	Dyl'ko (1971)
Culex erraticus	Larvae	Culicinomyces clavosporus		USA (North Carolina)	Lab.	Couch et al. (1974)

FUNGAL PATHOGENS, EXCEPT COELOMOMYCES, OF CULICIDAE (MOSQUITOS) (continued)

Host	Host stage infected	Pathogen	% Incidence	Locality	Lab. or field study	Reference
Culex exilis	Larvae	Beauveria bassiana	100 (53% mortality in controls)	USSR (Byelorussia)	Lab.	Dyl'ko (1971)
Culex gelidus	Adults	Aspergillus parasiticus		India	Field	Hati & Ghosh (1965)
Culex hortensis	Larvae	Smittium (= Orphella) culicis[a]		France	Field	Tuzet & Manier (1947)[d]
"	Larvae	Smittium culicis[a]		France	Field	Manier (1969)
"	Larvae	Smittium (= Rubetella) inopinata[a]	High	Italy	Field	Coluzzi (1966)
Culex modestus	Adults	Coelomycidium sp.	2	USSR (Uzbekistan)	Field	Shcherban' & Gol'berg (1971)
"	Larvae	Smittium culicis[a]		France	Field	Manier (1969)
Culex nigripalpus	Larvae	Lagenidium giganteum (= L. culicidum)	100	USA (Georgia)	Lab.	McCray et al. (1973a), Umphlett (1973)
Culex pipiens	Larvae, adults	Beauveria bassiana (possibly Entomophthora culicis)				Dyé (1905)[b]
"	Larvae	Beauveria bassiana	"Some"		Lab.	Roubaud & Toumanoff (1930)[b]
"	Larvae, adults	" "	100 (70-95 in larval outdoor tests)	USA (California)	Lab., small scale outdoor tests against larvae	Clark et al. (1967, 1968)
"	Larvae	Beauveria tenella	100	USA (California)	Lab.	Pinnock et al. (1973), Sanders (1972)
"	Adults	Cephalosporium, possibly C. coccorum	Considerable mortality	England	Field	Service (1969)

FUNGAL PATHOGENS, EXCEPT COELOMYCES, OF CULICIDAE (MOSQUITOS) (continued)

Host	Host stage infected	Pathogen	% Incidence	Locality	Lab. or field study	Reference
Culex pipiens	Adults	Entomophthora?	High	Federal Republic of Germany	Field	Oda & Kuhlon (1973)
"	Adults, pupae, larvae	Entomophthora sp.	95 (adults in field) 65-100 (♀, lab.) 33-67 (♂, lab.) 63-88 (pupae, lab.) 25 (larvae, lab.)	USSR (near Moscow)	Field & lab.	Field & Gol'berg (1969,1970a, 1970b,1973)
"	Adults	Entomophthora spp.	97	Netherlands	Field	Teernstra-Eeken & Engel (1967)
"	Adults	Entomophthora conglomerata (= Empusa conglomerata, Empusa thaxteri				Brumpt (1941)[c] Lakon (1919)[d]
"	Adults	Entomophthora conglomerata	63 (man-holes) 47 (filtration fields) 2 (wooden barrels and wells 300 m from filtration fields) 30 (♀♀ on bank of water reservoir)	USSR	Field	Il'chenko (1968)

157

FUNGAL PATHOGENS, EXCEPT COELOMOMYCES, OF CULICIDAE (MOSQUITOS) (continued)

Host	Host stage infected	Pathogen	% Incidence	Locality	Lab. or field study	Reference
Culex pipiens	Adults	Entomophthora con-glomerata	16 (biting ♀♀ on bank of water reservoir) 0 (biting ♀♀ 700 m from water reservoir)	USSR	Field	Il'chenko (1968)
"	Adults	Entomophthora con-glomerata	40	USSR (near Moscow)	Field	Kupriyanova (1966a,b)
"	Adults	Entomophthora con-glomerata	95	USSR (near Moscow)	Field	
"		Entomophthora culicis				Braun (1855),[d] López-Neyra & Guardiola Mira (1938),[b] Marshall (1938),[b] Speer (1927)[b]
"	Adults	Entomophthora culicis	0-3	USSR	Lab.	Gol'berg (1973)
"	Adults	Entomophthora destruens	0-10	USSR	Lab.	Gol'berg (1973)
"	Adults	Entomophthora (Entomophaga) destruens	0-100	Czechoslovakia, England, France	Field	Novák (1965,1967,1971), Service (1969), Weiser & Batko (1966), Weiser & Novák (1964)
"		Entomophthora henrici			Lab. Field	Brumpt (1941)[c] Molliard (1918)[d]
"	Larvae	Smittium (= Rubetella) culicis[a]	100	France	Lab.	Tuzet et al. (1961)
"	Larvae	Smittium (= Rubetella) inopinata[a]	93-100	Italy	Field & lab.	Field & Coluzzi (1966)

FUNGAL PATHOGENS, EXCEPT _COELOMOMYCES_, OF CULICIDAE (MOSQUITOS) (continued)

Host	Host stage infected	Pathogen	% Incidence	Locality	Lab. or field study	Reference
Culex pipiens fatigans (= _C. pipiens quinquefasciatus_	Larvae	_Culicinomyces clavosporus_		USA (North Carolina)	Lab.	Couch et al. (1974)
"	Larvae	_Lagenidium giganteum_		USA (North Carolina)	Lab.	Couch & Romney (1973)
"	Larvae	_Lagenidium giganteum_ (= _L. culicidum_)	100	USA (Georgia)	Lab.	McCray et al. (1973a), Umphlett (1973)
"	Adults	_Conidiobolus coronatus_ (= _Entomophthora coronata_)	33 (lab.) (18% control mortality)	USA (Florida)	Lab.	Lowe & Kennel (1972)
"	Adults	_Conidiobolus coronatus_ (= _Entomophthora coronata_)		USA (Florida)	Field	Lowe et al. (1968)
"	Larvae	_Culicinomyces_ sp.		Australia	Lab.	Sweeney (1975a,b)
"	Adults	_Aspergillus parasiticus_		India	Field & lab.	Hati & Ghosh (1965)
"	Adults	_Cladosporium_ ?	One specimen	Singapore	Lab.	Laird (1959)[d]
"	Larvae	_Metarrhizium anisopliae_	100	Nigeria, France	Lab.	Roberts & Mouchet (unpublished)
"	Adults	_Saprolegnia monica_	Low	Australia	Lab.	Hamlyn-Harris (1932)[b]
C. pipiens pipiens	Adults	_Coelomycidium_ sp.	3	USSR (Uzbekistan)	Field	Shcherban & Gol'berg (1971)
"	Larvae	_Fusarium oxysporum_	>80 (lab.)	France	Lab.	Hasan & Vago (1972)
"	Larvae	_Metarrhizium anisopliae_	100 (lab.) 94 (field)	USA (New York, Delaware)	Lab. & field	Roberts (1967,1970,1974)
"	Larvae	_Smittium culicis_		France	Field	Manier (1969)

FUNGAL PATHOGENS, EXCEPT COELOMOMYCES, OF CULICIDAE (MOSQUITOS) (continued)

Host	Host stage infected	Pathogen	% Incidence	Locality	Lab. or field study	Reference
Culex restuans	Larvae	Culicinomyces clavosporus		USA (North Carolina)	Lab.	Couch et al. (1974)
"	Larvae	Lagenidium giganteum	>95 (lab.) 100 (field)	USA (North Carolina)	Lab. & field	Umphlett & Huang (1972), Umphlett (1973)
"	Larvae	Lagenidium giganteum		USA (North Carolina)	Lab.	Couch & Romney (1973)
"	Larvae	Metarrhizium anisopliae	100	USA (New York)	Lab.	Roberts (1970,1974)
Culex tarsalis	Larvae, adults	Beauveria bassiana	100	USA (California)	Lab.	Clark et al. (1967,1968)
"	Larvae	Beauveria tenellae	100	USA (California)	Lab.	Pinnock et al. (1973)
"	Larvae	Lagenidium giganteum (= L. culicidum)	100	USA (North Carolina, Georgia)	Lab.	Couch & Romney (1973), McCray et al. (1973a), Umphlett (1973)
"	Larvae	Lagenidium giganteum	8-100	USA (California)	Field	McCray et al. (1973b)
"	Larvae	Pythium sp. (near P. adhaerens)			Lab.	Clark et al. (1966)
Culex territans	Larvae	Culicinomyces clavosporus		USA (North Carolina)	Lab.	Couch et al. (1974)
Culex theileri	Larvae	Amoebidium parasiticum[a]		Tunisia	Field	Manier et al. (1964)
"	Larvae	Smittium (= Rubetella) culicis[a]		Tunisia	Field	
Culex tritaeniorhynchus summorosus	Larvae	Aspergillus sp.	High		Lab.	Laird (1959)[d]
Culiseta sp.	Larvae	Beauveria tenellae		USA (California)	Lab.	Sanders (1972)
Culiseta annulata	Larvae	Saprolegnia sp.				Marshall (1938)[b]
Culiseta (= Theobaldia) annulata	Larvae	Smittium culicisa		France	Field	Manier (1969)

FUNGAL PATHOGENS, EXCEPT COELOMOMYCES, OF CULICIDAE (MOSQUITOS) (continued)

Host	Host stage infected	Pathogen	% Incidence	Locality	Lab. or field study	Reference
Culiseta impatiens	Larvae	Smittium culisetae[a]		USA (Colorado)	Field	Lichtwardt (1964), Williams & Lichtwardt (1972)
Culiseta incidens	Larvae	Beauveria tenellae	100	USA (California)	Lab.	Pinnock et al. (1973)
"	Larvae	Lagenidium giganteum		USA (North Carolina)	Lab.	Couch & Romney (1973)
"	Larvae	Pythium sp. (near P. adhaerens)		USA (California)	Lab.	Clark et al. (1966)
"	Larvae	Smittium (= Rubetella) sp.[a]		USA (California)	Field	Clark et al. (1963)
"	Larvae	Smittium culisetae[a]		USA (Colorado)	Field	Lichtwardt (1964)
Culiseta inornata	Larvae	Metarrhizium anisopliae	100	USA (New York)	Lab.	Roberts (1970,1974)
"	Larvae	Pythium sp. (near P. adhaerens)		USA (California)	Lab.	Clark et al. (1966)
"	Larvae	Smittium culisetae[a]		USA (Colorado)	Field	Lichtwardt (1964)
Culiseta longiareolata	Larvae	Smittium (= Rubetella) inopinata[a]	High	Italy	Field	Coluzzi (1966)
Culiseta melanura	Larvae	Culicinomyces clavosporus		USA (North Carolina)	Lab.	Couch et al. (1974)
Culiseta morsitans	Larvae	Entomophthora aquatica	10	USA (Connecticut)	Field	Anderson & Ringo (1969)
"	Larvae	Saprolegnia sp.				Marshall (1938)[b]
Orthopodnyia californica	Larvae	Pythium sp. (near P. adhaerens)		USA (California)	Lab.	Clark et al. (1966)
Psorophora sp.	Larvae	Lagenidium giganteum	100	USA (North Carolina)	Field	Umphlett & Huang (1972), Umphlett (1973)
"	Larvae	Polyscytalum				Howard, Dyar & Knab (1912)[b]
Psorophora confinnis	Larvae	Culicinomyces clavosporus		USA (North Carolina)	Lab.	Couch et al. (1974)

FUNGAL PATHOGENS, EXCEPT <u>COELOMOMYCES</u>, OF CULICIDAE (MOSQUITOS) (continued)

Host	Host stage infected	Pathogen	% Incidence	Locality	Lab. or field study	Reference
Psorophora howardii	Eggs	Polyscytalum sp.				Martini (1920), Speer (1927)[b]
Psorophora lutzii		Polyscytalum sp.				Martini (1920)
Uranotaenia anhydor	Larvae	Pythium sp. (near P. adhaerens)		USA (California)	Lab.	Clark et al. (1966)
Uranotaenia sapphirina	Larvae	Culicinomyces clavosporus		USA (North Carolina)	Lab.	Couch et al. (1974)
Uranotaenia unguiculata	Larvae	Smittium culicis[a]		France	Field	Manier (1969)
Mosquitos		Beauveria bassiana (= B. cinerea from theobaldiae)				Morquer (1933)[b]
"		Entomophthora gracilis				Lakon (1919),[d] Picard (1914),[d] Thaxter (1888)[d]
"	Adults	Entomophthora papilliata (= Lamia apiculata)				Lakon (1919),[d] Marshall (1938)[b]
"	Adults	Entomophthora rhizospora	High		Field	Brumpt (1941),[c] Howard, Dyar & Knab (1912)[b]
"		Entomophthora schroeteri (= E. rimosa)				Brumpt (1941),[c] Schröter (1886)[b]
"		Entomophthora sphaerosperma (= Tarichium sphaerospermum, Empusa radicans, Entomophthora radicans, Entomophthora phytonomi)				Brumpt (1941),[c] Fresenius (1856),[d] Howard, Dyar & Knab (1912),[b] Thaxter (1888)[d]

FUNGAL PATHOGENS, EXCEPT <u>COELOMOMYCES</u>, OF CULICIDAE (MOSQUITOS) (continued)

Host	Host stage infected	Pathogen	% Incidence	Locality	Lab. or field study	Reference
Mosquitos		<u>Entomophthora variabilis</u>				Lakon (1919),[d] Thaxter (1888)[d]
"		<u>Trichoderma viride</u>				Steinhaus (1949)[d]
"	Larvae, adults	<u>Trichophyton</u> sp.?				Christophers (1952),[b] Liston (1901)[b]

ABSTRACTS

Mary Ann Strand & Donald W. Roberts

Anderson, J. F. & Ringo, S. L. (1969). Entomophthora aquatica sp. n. infecting larvae and pupae of floodwater mosquitoes. J. Invertebr. Pathol., 13: 386-393.

E. aquatica is a pathogen of larvae and pupae of Aedes canadensis and the larvae of Culiseta morsitans in woodland pools in Connecticut. No external changes are apparent in infected 4th instar larvae, although some changes in activity occur in larvae with extensive infections. Pupae killed by the fungus are identifiable by a white mat of conidiophores. Hyphal bodies are usually found in the parietal layer of the fat body, hemocoel, and nervous system.

Bačinskij, P. E. (1926). K biologii ličinok komarov anofeles i kuleks v svjazi s metodom biologičeskogo analiza Kol'kvica-Marssona i opytami zaraženija ličinok sporami plesnej. Gig. i Epidem., 5(4): 38-44. (Rev. appl. Ent. B, 15: 153.)

Braun, A. (1855). Algarum unicellularium genera nova et minus cognita, Leipzig, Engelmann, 111 pp.

Brown, E. S. (1949). Vorticellids (Protozoa: Ciliophora) epibiotic on larvae of the genus Aedes (Dipt., Culicidae). Ent. mth. Mag., 85: 31-34.

Brumpt, E. (1941). Les entomophthorées parasites des moustiques. Ann. Parasit. hum. comp., 18: 112-144. (Rev. appl. Ent. B, 30: 151.)

Chapman, H. C. et al. (1969). A two-year survey of pathogens and parasites of Culicidae, Chaoboridae, and Ceratopogonidae in Louisiana. Proc. N. J. Mosq. Exterm. Assoc., 56: 203-212.

Female adult Aedes sollicitans with many basidiomycete spores in their abdomens were found during the survey. They apparently acquire this fungus when feeding on exudates from Spartina spartinae plants. Identical fungal colonies have been isolated from the mosquitos and the exudate. The effect of this fungus on the mosquitos is not known.

Chapman, H. C. et al. (1967). Pathogens and parasites in Louisiana Culicidae and Chaoboridae. Proc. 54th Ann. Meet. N. J. Mosq. Exterm. Assoc. Atlantic City, March 15-17, pp. 54-60

Dissections of adult female mosquitos revealed 8.9 x 3.7 μ spores present as white opaque masses in their abdomens. Four mosquito species were involved, Aedes sollicitans, Culiseta inornata, Mansonia perturbans, and Psorophora confinnis. Incidence in Ae. sollicitans averaged 25% during one 2-month period. The identity of the fungus was unknown, but it is suspected to be a Basidiomycete. Its effect on the mosquito was unknown.

Charles, V. K. (1939). Notes on entomogenous fungi. Plant. Dis. Reptr., 23: 340. (Rev. appl. Ent. B, 28: 237.)

Chorine, V. & Baranoff, N. (1929). Sur deux champignons parasites d'Anopheles maculipennis Mg. C. R. Soc. Biol. (Paris), 101: 1025-1026. (Rev. appl. Ent. B, 17: 247.)

Christophers, S. R. (1952). The recorded parasites of mosquitos. Riv. Parassit., 13: 21-28.

Clark, T. B. et al. (1967). Experiments on the biological control of mosquitoes with the fungus Beauveria bassiana (Bals.) Vuill. Proc. Calif. Mosq. Control Assoc., 35: 99.

See Clark et al. (1968).

Clark, T. B. et al. (1968). Field and laboratory studies on the pathogenicity of the fungus Beauveria bassiana to three genera of mosquitoes. J. Invertebr. Pathol., 11: 1-7.

The susceptibility of Culex tarsalis, C. pipiens, Anopheles albimanus, Aedes aegypti, Ae. sierrensis, and Ae. nigromaculis to B. bassiana was tested. In field tests, conidial dust applied to the surface of the water was the most effective method for killing larvae. When applied at 3 lb./acre, 70-95% of the C. pipiens larvae died. Aedes spp. larvae were not susceptible in the field tests; however, adults of all species were susceptible in the laboratory. Eggs exposed to conidia hatched normally.

Clark, T. B. et al. (1963). Axenic culture of two trichomycetes from California mosquitoes. Nature, 197: 208-209.

Rubetella sp. and R. culicis were found inhabiting the rectum of mosquito larvae. Axenic culture was successfully accomplished using blood agar medium. Mosquitos were readily infected by exposing them to older cultures with conidia forming thalli.

Clark, T. B. et al. (1966). Pythium sp. (Phycomycetes: Phythiales) pathogenic to mosquito larvae. J. Invertebr. Pathol., 8: 351-354.

Motile zoospores of Pythium, the only infective stage, exhibited strongly positive chemotactic response to wounds of Aedes sierrensis larvae, but were apparently incapable of penetrating normal cuticle. Thus, wide variation of mortality rates in laboratory tests was a function of harshness of treatment and to zoospore concentration.

Coluzzi, M. (1966). Experimental infections with Rubetella fungi in Anopheles gambiae and other mosquitoes. Proc. 1st Int. Congr. Parasitol. (Rome), 1: 592-593.

Second instar larvae of laboratory colonies of Culex pipiens, Aedes aegypti, and Anopheles gambiae were exposed to 4th instar skins of larvae infected with Rubetella (inopinata?). Ninety-three to 100% of the prepupae became infected. No changes of vitality of C. pipiens and Ae. aegypti were observed. Twenty-seven per cent. (range = 12-87%) of the A. gambiae died. Death was the result of occlusion of the rectal ampulla.

Couch, J. N. (1935). A new saprophytic species of Lagenidium, with notes on other forms. Mycologia, 27: 376-387.

Couch, J. N. (1960). Some fungal parasites of mosquitoes. In /Proceedings of a conference on the/ Biological control of insects of medical importance, Washington, D.C., American Institute of Biological Sciences, pp. 35-48.

Couch, J. N. & Romney, S. V. (1973). Sexual reproduction in Lagenidium giganteum. Mycologia, 65: 250-252.

The description of L. giganteum is emended in light of more complete knowledge of the life cycle. Nine strains of Lagenidium from the United States and India were examined and all were determined to be L. giganteum. Larvae of 11 mosquito species were susceptible to L. giganteum in the laboratory.

Couch, J. N. et al. (1974). New fungus which attacks mosquitoes and related Diptera. Mycologia, 66(2): 374-379.

Culicinomyces clavosporus, a new genus and species, is described from Anopheles quadrimaculatus. The infections were discovered in a laboratory colony reared with lake water from North Carolina. The fungus is culturable on a wide variety of artificial media. All larval stages are susceptible. Infection probably starts by the ingestion of conidia and death usually occurs within 60 hours after exposure. Infection experiments revealed the susceptibility of 10 other species of mosquitos.

Dyé, L. (1905). Les parasites des culicides. Arch. Parasit. (Paris), 9: 5-77.

Dyl'ko, M. I. (1971). Tests of the suitability of entobacterin and beauverin for the biological control of mosquito larvae. Minsk Belaruskaya Akad. Navuk Vesti Ser. Biyalagichnykh Navuk, 4: 85-89. (Bull. Acad. Sci. Byelorussian SSR, Biol. Sci. Ser., 4: 85-89.)

Dosages of 1.2×10^8 and 2.4×10^8 Beauveria bassiana conidia per ml totally prevented production of adults from Culex apicalis and C. exilis larvae.

Fresenius, G. (1856). Notiz, Insekten-Pilze betreffend. Bot. Ztg. 14: 882-883.

Galli-Valerio, B. & Rochaz de Jongh, J. (1905-6). Über die Wirkung von Aspergillus niger und A. glaucus auf die Larven von Culex und Anopheles. Zbl. Bakt., I. Abt. Orig., 38: 174-177; 40: 630.

Gol'berg, A. M. (1969). The finding of entomophthoraceous fungi on mosquitoes (Family Culicidae) and midges (Family Ceratopogonidae). Med. Parazitol. Parazit. Bolezni, 38(1): 21-23. (R, e).

Entomophthora conglomerata and an unnamed Entomophthora sp. were found infecting Culex pipiens pipiens adults in filtration fields near Moscow. The maximum infection rate was 95% and it occurred in late July-early August.

Gol'berg, A. M. (1970a). [Experimental infection of mosquitoes of the family Culicidae with Entomophthora. Communication I. Species specificity of the fungus Entomophthora sp.] Med. Parazitol. Parazit. Bolezni, 39(4): 472-478.

Culex pipiens adults were frequently found infected by Entomophthora near Moscow. Experimental infection of the mosquitos was successful. After testing a number of mosquito species, including Aedes dorsalis, A. aegypti, Anopheles maculipennis messiae, A. m. atroparvus, the Entomophthora proved to be specific for Culex pipiens.

Gol'berg, A. M. (1970b). Experimental infection of mosquitoes of the family Culicidae with Entomophthora. Communication II. Susceptibility to entomophthorosis of preimaginal stages and adults Culex pipiens L. mosquitoes. Med. Parazitol. Parazit. Bolezni, 39(6): 694-698.

Conidia of Entomophthora sp. were used to experimentally infect Culex pipiens. Insects dying of entomophthorosis were capable of infecting healthy mosquitos for 6-7 days from onset of conidia elimination. Adults were found to be most susceptible with females dying more frequently (64.6-100%) than males (33.3-66.6%). Pupae (63.0-88.4%) and fourth instar larvae (24.6%) were also susceptible; however, younger instars could not be infected. From the artificially infected mosquitos, all stages of the fungus were obtained: hyphal bodies, conidia, and dormant spores.

Gol'berg, A. M. (1973). Experimental infestation of mosquitoes of the family Culicidae with Entomophthora. III. Use of fungal cultures of the Entomophthoraceae family. Med. Parazitol. Parazit. Bolezni, 42(5): 616-618.

Culex pipiens molestus was experimentally exposed to cultures of Entomophthora sp., E. culis, and E. destruens. These cultures caused less than 10% adult mortality and did not adversely affect larval development, even though a previous study (Gol'berg, 1970b) established that Entomophthora sp. conidia from cadavers were infective.

Hamlyn-Harris, R. (1932). Some further observations on Chara fragilis in relation to mosquito breeding in Queensland. Ann trop. Med. Parasit., 26: 519-524. (Rev. appl. Ent. B, 21: 52.)

Hasan, S. & Vago, C. (1972). The pathogenicity of Fusarium oxysporum to mosquito larvae. J. Invertebr. Pathol., 20: 268-271.

Aedes detritus larvae naturally infected with F. oxysporum were collected from marshy areas in southern France. In the laboratory, A. detritus and Culex pipiens pipiens larvae were infected by conidia obtained from cultures and diseased bodies. More than 80% died, and injured or young larvae were the most susceptible. Spores germinated in the intestinal lumen and the mycelium invaded all tissues.

Hati, A. K. & Ghosh, S. M. (1965). Aspergillus parasiticus infection in adult mosquitoes. Bull. Calcutta Sch. Trop. Med., 13(1): 18-19.

Growth of a fungus was noticed from thorax to abdomen of 3 species of adult mosquitos (Culex gelidus, C. fatigans, and Anopheles subpictus) collected near Calcutta. From cultures, the fungus was identified and subcultures were used to infect C. fatigans. The symptoms produced were similar to those of the naturally infected mosquitos.

Howard, L. O. et al. (1912). The mosquitoes of North and Central America and the West Indies Washington, Carnegie Inst., 1: 156-179.

Il'chenko, L. Ya. (1968). The infection of the mosquito Culex pipiens L. with the parasitic fungus Entomophthora conglomerata Sorok in the vicinity of Novocherkassk. Med. Parazitol. Parazit. Bolezni, 37: 613-615. (In Russian, Eng. sum.)

Adults and pupa C. pipiens were found infected with E. conglomerata from June to end of September in filtration fields, and other places with standing water. Although 15.6% of the females attacking near a reservoir were infected, no infected specimens were found in a village 700-8000 m away.

Jettmar, H. M. von. (1947). Mikrobien als Feinde von Stechmückenlarven. Acta trop. (Basel) 4: 193-209.

Kuno, G. (1973). Biological notes of Amoebidium parasiticum found in Puerto Rico. J. Invertebr. Pathol., 21: 1-8.

Under optimum rearing conditions, A. parasiticum does not cause mortality of Aedes aegypti larvae. However if starved, the mortality rate of infected larvae was significantly greater than the controls.

Kupriyanova, E. S. (1966a). Entomophthora fungus parasitizing mosquitoes of the Culex pipiens L. complex. Zool. Zhur., 45(5): 675-678. (Russian, Eng. sum.)

C. pipiens, breeding in filter beds of a sewage disposal system near Moscow, were found to be infected by E. conglomerata. The fungus infects mosquitos in adult stage or at the moment of eclosion from pupae. Healthy adults may become infected when they visit water bodies for oviposition.

Kupriyanova, E. S. (1966b).   Parasitization of mosquitoes of the Culex pipiens L. complex by fungi of the order Entomophthorales.   Unpublished World Health Organization document WHO/EBL/ 66.57, 10 pp.

An epizootic caused by Entomophthora conglomerata was found in a population of C. pipiens along the edges of sewage filter beds near Moscow.   About 40% of the recently emerged adults were infected.   Infections were not observed beyond the immediate vicinity of the larval habitat.

Laird, M. (1959b).   Parasites of Singapore mosquitoes, with particular reference to the significance of larval epibionts as an index of habitat pollution.   Ecology, 40: 206-221.

Lakon, G. (1919).   Die Insektenfeinde aus der Familie der Entomophthoreen.   Beiträge zu einer Monographie der insektentötenden Pilze.   Z. angew. Ent., 5: 161-216.

Laveran, A. (1902).   De quelques parasites des culicides.   C. R. Soc. Biol. (Paris), 54: 233-235.

Léon, N. (1924).   Action des ectoparasites sur les culicides.   Ann. Parasit. hum. comp. 2: 211-213.   (Rev. appl. Ent. B, 12:   143.)

Lichtwardt, R. W. (1964).   Axenic culture of two species of branched Trichomycetes.   Am. J. Bot., 51: 836-842.

Smittium culisetae was isolated from hind-guts of Culiseta impatiens larvae.   It was grown on 10% brain-heart infusion.

Liston, W. G. (1901).   A year's experience of the habits of Anopheles in Ellickpur.   Indian med. Gaz., 36:   361-366, 441-443.

López-Neyra, C. A. & Guardiola Mira, A. (1938).   Protofitos parásitos de los mosquitos y sus larvas en España.   Bol. Univ. Granada, 10:   105-114.   (Rev. appl. Ent. B, 27:   96.)

Lowe, R. E. & Kennel, E. W. (1972).   Pathogenicity of the fungus Entomophthora coronata in Culex pipiens quinquefasciatus and Aedes taeniorhynchus.   Mosq. News, 32(4):   614-620.

Third instar larvae of C. p. quinquefasciatus were not susceptible to the fungus; however, it did grow in one 3rd instar A. taeniorhynchus.   Pupae and adults of both species were also susceptible.   Histological examination revealed the penetration of the insect cuticle by the fungal germ tubes.   The fungus was detected in the connective tissue of an experimentally infected white mouse.

Lowe, R. E. et al. (1968).   Entomophthora coronata as a pathogen of mosquitoes. J. Invertebr. Pathol., 11:   506-507.

Infected adult Culex pipiens quinquefasciatus mosquitos were found in a laboratory colony in Florida.   One of the first signs of abnormality was a drastic reduction in egg production.   Histopathological examinations revealed that the fungus had invaded all parts of the infected bodies.

Macfie, J. W. S. (1917).   Fungal infections of mosquito larvae.   Rep. Accra Lab. (1916), 76-80.   (Rev. appl. Ent. B, 6:   16.)

Manier, J.-F. (1969).   Trichomycetes de France.   Ann. Sci. Nat. Bot. Biol. Veg. Ser., 12, 10(4):   565-672.

A list of insect hosts and Trichomycetes species is given.   Thirteen species of mosquitos were listed as hosts, infected mainly by Smittium culicis.

168

Manier, J.-F. et al. (1964).   Presence en Tunisie de deux Trichomycetes parasites de larves de Culicides.   Arch. Inst. Pasteur Tunis, 41: 147-152.

Amoebidium parasiticum was observed on the larvae of Aedes detritus and Culex theileri. Smittium (= Rubetella) culicis was also found on C. theileri.   Descriptions of the two fungi are given.

Marchoux, E. et al. (1903).   La fièvre jaune.   Rapport de la mission française.   Ann. Inst. Pasteur, 17: 665-731.

Marshall, J. F. (1938).   The British mosquitoes.   London, British Museum (Nat. Hist.), 341 pp.

Martini, E. (1920).   Über Stechmücken besonders deren europäische Arten und ihre Bekämpfung. Arch. Schiffs-u. Tropenhyg., 24: 1-267.

Mattingly, P. F. (1972a).   Mosquito eggs.   XVII. Further notes on egg parasitization in genus Armigeres.   Mosq. Syst., 4: 1-8.

Speculations as to the identity of parasites on the eggs of A. dentatus are made. (See Mattingly, 1972b.)

Mattingly, P. F. (1972b).   Mosquito eggs.   XX. Egg parasitism in Anopheles with further notes on Armigeres.   Mosq. Syst., 4: 84-86.

The egg parasite previously described by this author has been recognized as a Lagenidiales, probably Lagenidium sp.

McCray, E. M., jr (1973a).   Laboratory studies on a new fungal pathogen of mosquitoes. Mosq. News, 33(1): 54-60.

Motile zoospores of Lagenidium giganteum (see Umphlett, 1973) are normally ingested and probably enter the tissues of the mosquito larvae through the anterior portion of the digestive tract.   Sporangial formation and larval death are usually simultaneous, occurring about 60 hours after infection.   Larvae of several species of Aedes and Culex were susceptible but none of the Anopheles were.   There was no loss of infectivity after passage through different host species.   Younger larvae were more susceptible than older ones and no infected adults were found.

McCray, E. M. et al. (1973b).   Laboratory observations and field tests with Lagenidium against California mosquitoes.   Proc. and Papers of 41st Ann. Conf. Calif. Mosq. Control Assoc., p. 123-128.

Laboratory observations of the life cycle of L. giganteum revealed two modes of action. In a permanent body of water, the fungus produces asexual zoospores which are released from the infected larvae to infect other larvae and the cycle is repeated.   In inter-mittently dry and flooded areas, the fungus goes through a sexual cycle and produces resting oospores which germinate when rewetted.   Field tests in both types of locations resulted in dramatically reduced mosquito populations.   No infections were found in 1400 other aquatic organisms from the treated sites.

Molliard, M. (1918).   Sur la vie saprophytique d'une Entomophthora (E. Henrici n. sp.). C. R. Acad. Sci. (Paris), 167: 958-960.

Morquer, R. (1933).   Considérations biologiques sur les variations du Botrytis cinerea et spécialement sur une nouvelle forme (forma theobaldiae) pathogène pour les Culicides.   Bull. Soc. Hist. nat. Toulouse, 65: 603-617.   (Rev. appl. Ent. B, 22: 119.)

Novák, D. (1965).   Zum Auftreten der Mykosen bei Stechmücken in Mähren (Diptera: Culicidae).
Beitr. Entomol., 15: 135-137.   (G, e)

  Adults of Culex pipiens infected by Entomophthora conglomerata (see Weiser and Batko,
  1966) were collected in Czechoslovakia.   The infected mosquitos occurred in damp cellars
  or similar places in southern Moravia.   Infections occurred yearly in the same places
  but spread little.   The mosquitos entered the cellars in the fall and nearly all died
  during the winter.   C. annulata were not infected.

Novák, D. (1967).   Beobachtungen zur Verbreitung von Mykosen bei Stechmücken.   Zeitschrift
für Tropenmedizin und Parasitologie, 18: 488-491.

  Mortality of overwintering Culex pipiens pipiens adults due to Entomophthora destruens
  was observed at irregular intervals for 6 years at two sites.   Temperatures ranged from
  1 to 20°C and relative humidities from 45 to 100%.   Infected specimens were collected
  during all seasons of the year.   Infection levels, in most cases, were 50% or greater.
  Dead mosquitos were found on brick, wood, rubber, and iron substrates.   Spores survived
  6-7 months in cool, damp, humid environments.   Heat, dryness, and fresh paint destroyed
  the fungus.

Novák, D. (1971).   Weiterer Beobactungen zur Verbreitung von Mykosen bei Stechmücken von
Culex pipiens.   Biologia (Bratisl.), 28(8): 643-645.   (G, e, p)

  In two cellars in which mosquitos regularly overwintered, many were found to be infected
  by Entomophthora destruens (see Weiser & Batko, 1966).   Mortality due to this fungus
  was greatest in the autumn months when as many as 85% died.

Nowakowski, L. (1883).   Entomophthorae, przyczynek do znajomści pasożytnych grzbków,
sprawiajacych pomór owadów.   Pam. Akad. Umiejet. Krakow., Wydz. mat.-nat., 8: 153-183.

Oda, T. & Kuhlow, F. (1973).   Beobactungen über Sterblichkeit und Follikelgrösse bei Culex
pipiens pipiens L. im Verlauf der Überwinterung.   Z. Tropenmed. Parasitol., 24: 373-378.

  A high mortality rate from an unidentified fungus was observed in female mosquitos overwintering in cellars.   Many of the dead were covered with mould.

Picard, F. (1914).   Les Entomophthorées, leur parasitisme chez les insectes.   Bull. Soc.
Zool. agric., 13: 1-7, 25-30, 37-40, 62-65.

Pinnock, D. E. et al. (1973).   Beauveria tenella as a control agent for mosquito larvae.
J. Invertebr. Pathol., 22: 143-147.

  B. brongniartii (= tenella) was isolated from naturally infected Aedes sierrensis
  mosquitos captured in California.   Other species of Aedes, Culex, and Culiseta were
  also susceptible in laboratory tests.   The mortality rate of A. sierrensis was not
  correlated to inoculation concentration, but correlations were observed in tests with
  other species.   Mortality percentage was influenced by incubation temperature and
  early instars were the most susceptible stages.   In field trials, significant reductions
  of emerging A. sierrensis adults (53-71% over controls) were observed in treated tree
  holes.

Rioux, J. A. & Achard, F. (1956).   Entomophytose mortelle à Saprolegnia diclina Humphrey 1892
dans un élevage d'Aedes berlandi Seguy 1921.   Vie et Milieu, 7: 326-337.

Roberts, D. W. (1967).   Some effects of Metarrhizium anisopliae and its toxins on mosquito
larvae.   In Insect Pathology and Microbial Control (van der Laan, P. A., ed.) pp. 243-246.
North-Holland Publ. Co., Amsterdam.

Exposure to viable M. anisopliae conidia was fatal to several species of mosquitos. The fungus mycelium produces destruxins A and B in vitro. These compounds are also toxic to mosquito larvae.

Roberts, D. W. (1970). Coelomomyces, Entomophthora, Beauveria, and Metarrhizium as parasites of mosquitoes. Misc. Publ. Entomol. Soc. Am., 7(1): 140-155.

The use of fungi for mosquito control is discussed. Natural infections of Entomophthora occur in both larvae and adults and observations indicate that significant control can be achieved. Beauveria and Metarrhizium are not normally associated with mosquitos, so require repeated applications for use in control.

Roberts, D. W. (1974). Fungal infections of mosquitoes, p. 143-193. In Aubin, A. et al. (ed.), Le contrôle des moustiques/Mosquito control. Univ. Quebec Press, Quebec.

In this review paper, the host range, distribution, life cycle, and possibility as microbial control agents were discussed for fungal parasites, including Lagenidium, Entomophthora, Beauveria, and Metarrhizium.

Roubaud, E. & Toumanoff, C. (1930). Essais d'infection expérimentale de larves de culicides par quelques champignons entomophytes. Bull. Soc. Path. exot., 23: 1025-1027.

Sanders, R. D. (1972). Microbial mortality factors in Aedes sierrensis populations. Proc. Calif. Mosq. Control Assoc., 40: 66-68

Mosquitos infected with Beauveria brongniartii (= tenella) were found in 4 of 18 tree holes examined near Novato, California. Twenty-six to 91% of the larvae were infected.

Schröter, J. (1889). Pilze. In Cohn, F., ed., Kryptogamen-flora von Schlesien, Breslau, Kern, 3: 217

Service, M. W. (1969). Observations on the ecology of some British mosquitoes. Bull Entomol. Res., 59: 161-194.

Overwintering Culex pipiens adults were infected by Cephalosporium sp. (possibly C. coccorum) and Entomophthora sp. near conglomerata. Mortality levels exceeded 50% in November and December. Infection was virtually non-existent in populations resting on ceilings, which were drier than walls. In 11 overwintering sites, infection occurred only in those with damp walls.

Shcherban', Z. P. & Gol'berg, A. M. (1971). The pathogenic fungi Coelomycidium (Phycomycetes, Chytridiales) and Coelomomyces (Phycomycetes, Blastocladiales) in Culex and Aedes (Diptera, Culicidae) in Uzbekistan. Med. Parazitol. Parazit. Bolezni., 40(1): 110-111.

Coelomycidium sp. infected 2.4% of Culex modestus and 3.4% of C. pipiens pipiens mosquitos collected in the Fergana valley. In the laboratory infected females died within one week of taking a blood meal.

Sorokin, N. (1877). Über zwei neue Entomophthora-Arten. Beitr. Biol. Pflanzen, 2: 387-398.

Speer, A. J (1927). Compendium of the parasites of mosquitoes (Culicidae). Hyg. Lab. Bull. (Wash.), 146: 1-36.

Steinhaus, E. A. (1949b). Principles of insect pathology. New York, McGraw-Hill, 757 pp.

Sweeney, A. W. (1975a).   The insect pathogenic fungus Culicinomyces in mosquitoes and other hosts.   Australian Jour. Zool., 23: 59-64.

   The host range of an Australian Culicinomyces sp. was examined by testing against aquatic insects, shrimp, and fish.   Larval Culicidae (Anopheles annulipes, An. amictus hilli, Aedes australis and Culex pipiens fatigans), Chironomidae (Chironomus sp.), and Ceratopogonidae (Dasyhelea and Bezzia) were susceptible.   Another dipteran, Psychodidae (Telmatoscopus albipunctatus) was not susceptible.   Caddis-fly larvae (Trichoptera), dragonfly naiads (Zygoptera and Anisoptera), freshwater shrimp (Atyidae), and Gambusia fish were all nonsusceptible.

Sweeney, A. W. (1975b).   The mode of infection of the insect pathogenic fungus Culicinomyces in larvae of the mosquito Culex fatigans.   Australian Jour. Zool., 23: 49-57.

   The initiation and development of infection by an Australian Culicinomyces sp. in Culex pipiens fatigans larvae was followed by dissection and by histology.   The infection sites were the foregut and hindgut, and not the exterior integument.   After death conidia were produced on the external surface of the cadaver by conidiophores which penetrated from the hemocoel.

Sweeney, A. W. et al. (1973).   A fungal pathogen for mosquito larvae with potential as a microbial insecticide.   Search, 4(8): 344-345.   (WHO/VBC/73.444, WHO/MAL/73.805.)

   A fungal parasite was isolated from Anopheles amictus hilli.   It appears to be a new pathogen of aquatic Diptera, but it resembles the common terrestrial insect pathogenic fungus Metarrhizium anisopliae.   Mosquito larvae of three genera (Anopheles, Culex, and Aedes) are susceptible to infection.   Infection by ingestion is effective against mosquito larvae which feed non-selectively.   Prolonged subculturing did not lead to loss of pathogenicity.

Sweeney, A. W. & Panter, C. (1974).   The pathogenicity of the fungus Culicinomyces to mosquito larvae in a natural field habitat.   Unpublished World Health Organization document WHO/VBC/74.470, 2 pp.

   Spores of Culicinomyces sp. from Australia were introduced into two rock pools (6 and 40 litres, 106 and 105 spores/ml, respectively), which contained naturally occurring populations of Aedes rupestris.   Infected larvae were collected daily for 6 days, at which time the pools were flushed by heavy rain.   The percentage of larvae infected was not determined, but the experiment established that infection could occur in a natural mosquito habitat as well as in the laboratory.

Teernstra-Eeken, M. H. & Engel, A. (1967).   Notes on entomophthorous fungi on Heleomyzidae and Culicidae (Diptera).   J. Invertebr. Pathol., 9, 431-432.

   Entomophthora spp. were found parasitizing a population of Culex pipiens which was overwintering in caves in the Netherlands.   By the beginning of February 92-97% of the population was dead.

Thaxter, R. (1888).   The Entomophthoreae of the United States.   Mem. Boston Soc. nat. Hist., 4: 133-201.

Tuzet, O. & Manier, J. F. (1947).   Orphella culici n. sp., entophyte parasite du rectum des larves de Culex hortensis Fclb.   C. R. Acad. Sci. (Paris), 225: 264-265.

Tuzet, O. et al. (1961).   Rubetella culicis (Tuzet et Manier, 1947), trichomycète rameux, parasite de l'ampoule rectale des larves de culicides (morphologie et spécificité).   Viet et Milieu, 12: 167-187.

Umphlett, C. J. (1973).   A note to identify certain isolate of <u>Lagenidium</u> which kills mosquito larvae.   <u>Mycologia</u>, <u>65</u>(4):   970-972.

The synonomy of <u>L. giganteum</u> and <u>L. culicidum</u> is demonstrated.   The name is properly <u>L. giganteum</u>.

Umphlett, C. J. & Huang, C. S. (1972).   Experimental infection of mosquito larvae by a species of the aquatic fungus <u>Lagenidium</u>.   <u>J. Invertebr. Pathol.</u>, <u>20</u>:   326-331.

The susceptibility of <u>Culex restuans</u> to <u>L. giganteum</u> (Umphlett, 1973) was tested.   At low dosages, younger larvae were most susceptible, but over 80% of the larvae of all ages died at high dosages.

Umphlett, C. J. & McCray, E. M. jr (1975).   A brief review of the involvements of <u>Lagenidium</u>, an aquatic fungus parasite, with arthropods.   <u>Marine Fisheries Review</u>, <u>37</u>:   61-64.

Several species of <u>Lagenidium</u> have been reported as parasites of arthropods including mosquitos.   <u>L. giganteum</u> has been shown to be a virulent pathogen of several species of mosquitos.   However, more than 1400 acquatic non-target organisms (small crustaceans and insects) from sites where <u>L. giganteum</u> had been introduced were found not to be infected.   It has also been shown not to be pathogenic to small mammals.

Weiser, J. & Batko, A. (1966).   A new parasite of <u>Culex pipiens</u> L. <u>Entomophthora destruens</u> sp. nov. (Phycomycetes, Entomophthoraceae).   <u>Folia Parasitol. (Praha)</u>, <u>13</u>(2):   144-149.

A fungus previously reported to cause heavy mortality in some localities among hibernating adult mosquitos in Czechoslovakia is described.   It can be cultured on egg-yolk medium.   (See Novak, 1965, and Weiser & Novak, 1964.)

Weiser, J. & Novak, D. (1964).   Auftreten vom Mykosen bei Stechmücken.   <u>Entomophaga Mem.</u> <u>Hors Ser.</u>, <u>2</u>:   149-150.

A fungus infection caused by <u>Entomophthora destruens</u> (Weiser & Batko, 1966) is common in overwintering populations of <u>Culex pipiens</u> in Czechoslovakia.   More than 50% of natural shelters and basements are infected and a steadily increasing infection occurs in the hibernating population.   Although present in the shelters, <u>Culiseta Theobaldia</u> sp. mosquitos are not infected.

Williams, M. C. & Lichtwardt, R. A. (1972).   Infection of <u>Aedes aegypti</u> larvae by axenic cultures of the fungal genus <u>Smittium</u> (Trichomycetes).   <u>Am. J. Bot.</u>, <u>59</u>:   189-193

Spores produced by cultures isolated from various locations and dipteran hosts were fed to the mosquito larvae.   Some host but no geographical specificity was found.   This fungus apparently has little effect on larvae reared under optimum conditions.

VIII. NEMATODE PATHOGENS OF CULICIDAE (MOSQUITOS)

James J. Petersen

USDA, Agricultural Research Service
Gulf Coast Mosquito Research Laboratory
Lake Charles, LA 70601, USA

NEMATODE PATHOGENS OF CULICIDAE (MOSQUITOS)

Host	Host stage infected	Parasite	% Incidence	Location	Lab. or field study	Reference
Aedes aegypti (Stegomyia fasciatus)	Larvae	Mermis sp.	100	Guinea	Field	Gendre (1909)
"	"	Octomyomermis muspratti (Agamomermis sp.)	70-80	Zambia	"	Muspratt (1945)
"	Adults	Perutilimermis culicis (Agamomermis culicis)	6	USA (Louisiana)	Lab.	Petersen & Willis (1969a)
"	Larvae	Octomyomermis muspratti (Agamomermis sp.)	-	Nauru	(Field)[a]	Reynolds (1972)
"	"	Romanomermis culicivorax (Romanomermis sp.)	99	USA (Louisiana)	Lab.	Petersen et al. (1969)
"	"	Neoaplectana carpocapsae (DD136)	-	Canada (Ontario)	"	Welch & Bronskill (1962)
Aedes albopictus	"	Romanomermis culicivorax	-	USA (Louisana)	"	Unpublished data
Aedes atlanticus	"	Romanomermis culicivorax (Romanomermis sp.)	-	"	Field	Petersen et al. (1968)
"	"	Romanomermis culicivorax	50-62	"	(Field)	Petersen & Willis (1972b)
Aedes canadensis	"	Agamomermis sp.	-	North America	Field	Jenkins (1964)
"	"	Romanomermis culicivorax	-	USA (Louisania)	Lab.	Petersen et al. (1968)
Aedes cantans	Adults	Agamomermis sp.	83	USSR	Field	Shachov (1927)
Aedes calceatus	Larvae	Octomyomermis muspratti (Agamomermis sp.)	70-80	Zambia	Field	Muspratt (1945)
Aedes cinereus	"	Agamomermis sp.	-	USA (Alaska)	"	Frohne (1953)
"	"	"	-	North America	"	Jenkins (1964)
"	"	Romanomermis nielseni	53	USA (Wyoming)	"	Tsai et al. (1969)
"	"	Romanomermis culicivorax	-	USA (Louisiana)	Lab.	Unpublished data
Aedes communis (Culex nemeralis)	"	Agamomermis sp.	-	USA (Alaska)	Field	Frohne (1953)

[a] Denotes parasitism resulting from field releases.

NEMATODE PATHOGENS OF CULICIDAE (MOSQUITOS) (continued)

Host	Host stage infected	Parasite	% Incidence	Location	Lab. or field study	Reference
Aedes communis (Culex nemeralis)	Larvae, adults	Agamomermis sp.	-	Europe	Field	Stiles (1903)
"	Larvae	Hydromermis sp.	100	Canada (Manitoba)	"	Jenkins & West (1954)
"	"	Hydromermis churchillensis	-	"	"	Welch (1960a)
"	"	Hydromermis sp.	-	USA (Colorado)	"	Smith (1961)
"	Adults	Mermithid	-	USSR	"	Artyukhovsky & Kolycheva (1965)
"	Larvae	Romanomermis nielseni	17	USA (Wyoming)	Field	Tsai et al. (1969)
Aedes dorsalis	Adults	Agamomermis sp.	83	USSR	Field	Shachov (1927)
Aedes dupreei	Larvae	Romanomermis culicivorax	-	USA (Louisiana)	Lab.	Unpublished data
Aedes excrucians	Larvae	Agamomermis sp.	-	USA (Alaska)	Field	Frohne (1955a)
Aedes fitchii	Larvae	Romanomermis nielseni	50	USA (Wyoming)	"	Tsai et al. (1969)
Aedes flavescens	Adults	Mermithid ?	-	Canada (British Columbia)	"	Hearle (1929)
Aedes fulgens	Larvae	Octomyomermis muspratti (Agamomermis sp.)	70-80	Zambia	"	Muspratt (1945)
Aedes fulvus pallens	"	Romanomermis culicivorax	-	USA (Louisiana)	Lab.	Unpublished data
Aedes heischi	"	Agamomermis sp.	-	Kenya	Field	Lumsden (1955)
Aedes haworthi	"	"	70-80	Zambia	"	Muspratt (1945)
Aedes impiger (neararcticus)	"	"	-	USA (Alaska)	"	Frohne (1953)
"	"	Hydromermis sp.	-	Canada (Manitoba)	"	Jenkins & West (1954)
"	"	Hydromermis churchillensis	0-82	"	"	Welch (1960)
"	"	Mermithids	-	USA (Alaska)	"	Gorham (1972)
Aedes infirmatus	"	Romanomermis culicivorax	-	USA (Louisiana)	Lab.	Unpublished data
Aedes increpitus	"	Romanomermis nielseni	53	USA (Wyoming)	Field	Tsai et al. (1969)

NEMATODE PATHOGENS OF CULICIDAE (MOSQUITOS) (continued)

Host	Host stage infected	Parasite	% Incidence	Location	Lab. or field study	Reference
Aedes laguna	Larvae	Mermithid	-	Mexico	Field	Arnell & Nielsen (1972)
Aedes marshalli	"	Octomyomermis muspratti (Agamomermis sp.)	70-80	Zambia	"	Muspratt (1945)
Aedes metallicus	"	"	70-80	Zambia	"	"
Aedes michaelikati	"	Agamomermis sp.	-	Kenya	"	Lumsden (1955)
Aedes mitchellae	"	Romanomermis culicivorax	-	USA (Louisiana)	"	Petersen et al. (1968)
Aedes nigripes	"	Hydromermis sp.	-	Canada (Manitoba)	"	Jenkins & West (1954)
"	"	Hydromermis churchillensis	-		"	Welch (1960)
Aedes nigromaculis	"	Romanomermis culicivorax	0-51	USA (California)	(Field)	Hoy & Petersen (1973)
Aedes pionips	"	Agamomermis sp.	-	USA (Alaska)	Field	Frohne (1955b)
"	"	Hydromermis sp.	-	Canada (Manitoba)	"	Jenkins & West (1954)
Aedes polynesiensis	"	Romanomermis culicivorax	-	USA (Louisiana)	Lab.	Unpublished data
Aedes pullatus	"	Hydromermis sp.	-	USA (Colorado)	Field	Smith (1961)
"	"	Romanomermis nielseni	54	USA (Wyoming)	"	Tsai et al. (1969)
"	"	Mermithid	-	USA (Alaska)	"	Gorham (1972)
Aedes punctor	"	Agamomermis sp.	-	"	"	Frohne (1953)
"	"		-	North America	"	Jenkins (1964)
Aedes rusticus (maculatus)	Adults	Mermithid	-	USSR	"	Artyukhovsky & Kolycheva (1965)
Aedes scuttelaris	Larvae	Romanomermis culicivorax	-	USA (Louisiana)	Lab.	Unpublished data
Aedes sierrensis	"	"	53	"	"	Petersen et al. (1969)
"	"	Octomyomermis troglodytis (mermithid)	10	USA (California)	Field	Poinar & Sanders (1974)
Aedes soleatus	"	Agamomermis sp.	-	Kenya	"	Lumsden (1955)

NEMATODE PATHOGENS OF CULICIDAE (MOSQUITOS) (continued)

Host	Host stage infected	Parasite	% Incidence	Location	Lab. or field study	Reference
Aedes sollicitans	Adults	Perutilimermis culicis (Agamomermis culicis)	1-58	USA (New Jersey)	Field	Smith (1904)
	"	"	10-35	USA (Louisiana)	"	Petersen et al. (1967)
	Larvae	Romanomermis culicivorax	50	USA " "	Lab.	(1969)
	Adults	Perutilimermis culicis ?	-	USA (Florida)	Field	Savage & Petersen (1971)
	Larvae	Romanomermis culicivorax	4	USA (Louisiana)	"	Petersen & Willis (1971)
Aedes sticticus (aldrichii)	Adults	Paramermis canadensis	-	Canada (British Columbia)	"	Hearle (1926)
	Larvae	Romanomermis culicivorax	-	USA (Louisiana)	Lab.	Unpublished data
Aedes stimulans	Adults	Agamomermis sp.	85	USA (Maine)	Field	Anomymous (1970)
	Larvae	Neoaplectana carpocapsae (DD136)	-	Canada (Ontario)	Lab.	Bronskill (1962)
Aedes taeniorhynchus	"	Romanomermis culicivorax	49	USA (Louisiana)	Lab.	Petersen et al. (1969)
	Adults	Perutilimermis culicis (Agamomermis)	1	" "	"	Petersen & Willis (1969b)
Aedes thibaulti	Larvae	Romanomermis culicivorax	-	" "	"	Petersen et al. (1968)
	"	"	80	" "	"	Petersen et al. (1969)
Aedes tormentor	"	"	18	" "	"	"
	"	"	-	" "	Field	Petersen & Chapman (1972)
Aedes trichurus	"	Neoaplectana carpocapsae (DD136)	-	Canada (Ontario)	Lab.	Bronskill (1962)
Aedes triseriatus	"	Romanomermis culicivorax	2	USA (Louisiana)	"	Petersen et al. (1969)

NEMATODE PATHOGENS OF CULICIDAE (MOSQUITOS) (continued)

Host	Host stage infected	Parasite	% Incidence	Location	Lab. or field study	Reference
Aedes vexans	Adults	Paramermis canadensis	80	Canada (British Columbia)	Field	Steiner (1924)
"	Larvae	Agamomermis sp.	52	USA (Pennsylvania)	"	Stabler (1952)
"	Adults	Agamomermis sp. (Paramermis canadensis)	100	Canada (British Columbia)	"	Trpiš et al. (1968)
"	Larvae	Romanomermis culicivorax	-	USA (Louisiana)	"	Petersen et al. (1968)
"	"	"	50-62	" "	(Field)	Petersen & Willis (1972b)
Aedes zethus	"	Octomyomermis muspratti (Agamomermis sp.)	70-80	Zambia	Field	Muspratt (1945)
Anopheles spp.	"	Limnomermis aquatilis	-	France	"	Dujardin (1845)
"	"	Agamomermis sp.	-	USA (New Jersey)	"	Johnson (1903)
Anopheles albimanus	"	Romanomermis culicivorax	-	USA (Louisiana)	Lab.	Unpublished data
Anopheles annulipes	"	Agamomermis sp.	69	Australia	Field	Laird (1956)
"	"	"	17-80	India	"	Kalucy (1972)
Anopheles annularis (fulignosus)	Larvae, adults	Mermis sp.	-	India	"	Iyengar (1930)
Anopheles atropos	Larvae	Romanomermis culicivorax	-	USA (Louisiana)	Lab.	Chapman et al. (1970)
Anopheles barberi	"	"	-	"	"	Petersen et al. (1968)
"	"	Diximermis peterseni (Gastromermis sp.)	-	"	"	Petersen & Chapman (1970)
Anopheles barbirostris	"	Mermis sp.	-	India	Field	Iyengar (1930)
Anopheles bradleyi	"	Romanomermis culicivorax	-	USA (Louisiana)	Lab.	Petersen et al. (1968)
"	"	Diximermis peterseni (Gastromermis sp.)	-	"	"	Petersen & Chapman (1970)

NEMATODE PATHOGENS OF CULICIDAE (MOSQUITOS) (continued)

Host	Host stage infected	Parasite	% Incidence	Location	Lab. or field study	Reference
Anopheles crucians	Larvae	Romanomermis culicivorax	52	USA (Louisiana)	Field	Chapman et al. (1967a)
"	"	"	8-42	"	"	Petersen & Willis (1971)
"	"	"	1-100	"	(Field)	" (1972b)
"	"	"	1-53	USA (Florida)	Field	Savage & Petersen (1971)
"	"	Diximermis peterseni (Gastromermis sp.)	-	USA (Louisiana)	"	Chapman et al. (1969)
"	"	"	0-82	"	"	Petersen & Chapman (1970)
"	"	"	1-13	USA (Florida)	"	Savage & Petersen (1971)
Anopheles freeborni	"	Romanomermis culicivorax	50-85	USA (California)	(Field)	Petersen et al. (1972)
Anopheles funestus	Adults	Gastromermis sp.	10-49	Upper Volta	Field	Coz (1966, 1973)
"	"	Agamomermis sp.	9	Nigeria	"	Hanney (1960)
Anopheles gambiae	Larvae	"	-	Zambia	"	Muspratt (1945)
"	Adults	"	4	Nigeria	"	Hanney (1960)
"	Larvae	"	83	Kenya	"	Service (1973)
Anopheles hyrcanus	"	Mermis sp.	-	India	"	Iyengar (1930)
Anopheles leucosphyrus	"	"	-	Sumatra	"	Walandouw (1934)
Anopheles philippinensis	"	"	-	India	"	Iyengar (1930)
Anopheles philippinensis pseudopunctipennis	"	Romanomermis culicivorax	-	USA (Louisiana)	Lab.	Unpublished data
Anopheles punctipennis	"	Romanomermis culicivorax (mermithid)	33	"	Field	Chapman et al. (1967a)
"	"	Diximermis peterseni (Gastromermis sp.)	-	"	"	" (1969)
Anopheles quadri-maculatus	"	Romanomermis culicivorax (mermithid)	-	"	"	" (1967a)

NEMATODE PATHOGENS OF CULICIDAE (MOSQUITOS) (continued)

Host	Host stage infected	Parasite	% Incidence	Location	Lab. or field study	Reference
Anopheles quadri-maculatus	Larvae	Romanomermis culicivorax	1-100	USA (Louisiana)	(Field)	Petersen & Willis (1972)
"	"	"	17-84	"	"	Petersen et al. (1973)
"	"	Diximermis peterseni (Gastromermis sp.)	-	"	Field	Chapman et al. (1969)
"	"	"	1-89	"	"	Petersen & Chapman (1970)
Anopheles ramsayi (pseudojamesi)	"	Mermis sp.	-	India	"	Iyengar (1930)
Anopheles rufipes	"	Octomyomermis muspratti	-	Northern Rhodesia	"	Muspratt (1945)
Anopheles sinensis	"	Mermis sp.	-	India	"	Iyengar (1930)
Anopheles stephensi	"	Romanomermis culicivorax	-	USA (Louisiana)	Lab.	Unpublished data
Anopheles subpictus	Adults	Romanomermis iyengari (Mermis sp.)	-	India	Field	Iyengar (1930)
Anopheles tessellatus	Larvae	Mermis sp.	-	"	"	"
Anopheles varuna	"	"	-	"	"	"
Culex annulis	"	Romanomermis culicivorax	3-39	China (Province of Taiwan)	Lab., (field)	Mitchell et al. (1972)
Culex erraticus	"	Romanomermis culicivorax (mermithid)	20	USA (Louisiana)	Field	Chapman et al. (1967a)
"	"	Romanomermis culicivorax	37	USA (Louisiana)	"	Petersen & Willis (1971)
"	"	"	56	"	(Field)	(1972b)
Culex fatigans (pipiens quinque-fasciatus)	"	Agamomermis sp.	-	India	Field	Ross (1906)
"	"	Romanomermis culicivorax	11-97	USA (Louisana)	Lab.	Petersen et al. (1969)
"	"	"	0-21	Thailand	(Field)	Chapman et al. (1972)

NEMATODE PATHOGENS OF CULICIDAE (MOSQUITOS) (continued)

Host	Host stage infected	Parasite	% Incidence	Location	Lab. or field study	Reference
Culex fatigans (pipiens quinquefasciatus)	Larvae	Romanomermis culicivorax	98-100	China (Province of Taiwan)	Lab., (Field)	Mitchell et al. (1972)
"	Adults	Perutilimermis culicis (Agamomermis sp.)	1	USA (Louisiana)	Lab.	Petersen & Willis (1969a)
"	Larvae	Octomyomermis muspratti (Agamomermis sp.)	-	Zambia	Lab.	Muspratt (1965)
"	"	"	-	Nauru	(Field)	Reynolds (1972)
Culex fuscanus	"	Romanomermis culicivorax	-	China (Province of Taiwan)	Lab.	Mitchell et al. (1972)
Culex fuscocephalus	"	"	-	"	"	"
Culex nebulosus	"	Octomyomermis muspratti (Agamomermis sp.)	-	Zambia	Field	Muspratt (1945)
Culex peccator	"	Romanomermis culicivorax	-	USA (Louisiana)	Lab.	Petersen et al. (1968)
Culex pipiens	"	Neoaplectana carpocapsae	-	USA (California)	"	Poinar & Leutenegger (1971)
Culex pipiens pipiens	"	Mermithid	55	USA (Pennsylvania)	Field	Stabler (1952)
"	"	Romanomermis culicivorax	-	USA (Louisiana)	Lab.	Unpublished data
Culex pipiens molestus	Adults	Foleyella philistinae	-	Lebanon	Lab.	Schacher & Khalil (1968)
Culex restuans	Larvae	Romanomermis culicivorax	-	USA (Louisiana)	Field	Petersen et al. (1968)
"	"	"	33	"	Lab.	"
Culex rubithoracis	"	"	37	China (Province of Taiwan)	"	Mitchell et al. (1972)
Culex salinarius	"	"	56	USA (Louisiana)	"	Petersen et al. (1969)
"	"	Mermithid	8	USA (Pennsylvania)	Field	Stabler (1952)

NEMATODE PATHOGENS OF CULICIDAE (MOSQUITOS) (continued)

Host	Host stage infected	Parasite	% Incidence	Location	Lab. or field study	Reference
Culex tarsalis	Larvae	Romanomermis culicivorax	-	USA (Louisiana)	Lab.	Petersen et al. (1972)
Culex territans	"	Mermithid	-	USA (North Carolina)	Field	" (1968)
Culex tritaeniorhynchus summorosus	"	Romanomermis culicivorax	5-73	China (Province of Taiwan)	Lab. (field)	Mitchell et al. (1972)
Orthopodomyia signifera	"	"	-	USA (Louisiana)	Lab.	Petersen et al. (1968)
"	"	Mesomermis sp.	-	"	Field	Petersen & Willis (1969a)
Culiseta impatiens	"	Romanomermis nielseni	1	USA (Wyoming)	"	Tsai et al. (1969)
Culiseta inornata	"	Romanomermis culicivorax	30-86	USA (Louisiana)	Lab.	Petersen et al. (1969)
"	"	"	-	"	Field	Chapman et al. (1969)
Culiseta melanura	"	"	-	"	Lab.	Petersen et al. (1968)
Psorophora ciliata	"	"	-	"	Field	" (1968)
"	Adults	Mermithid	-	USA (Florida)	"	Savage & Petersen (1971)
Psorophora confinnis	Larvae	Romanomermis culicivorax (Mermithid)	-	USA (Louisiana)	"	Chapman et al. (1967a)
"	"	Romanomermis culicivorax	93	"	"	Petersen et al. (1968)
"	"	"	92	"	Lab.	" (1969)
"	"	"	50-62	"	(Field)	Petersen & Willis (1972b)
"	"	"	4-50	"	"	Petersen et al. (1973)
Psorophora cyanescens	"	"	-	"	"	" (1968)
Psorophora discolor	"	"	-	"	Field	Petersen et al. (1968)
Psorophora ferox	"	"	1	"	Lab.	" (1969)

NEMATODE PATHOGENS OF CULICIDAE (MOSQUITOS) (continued)

Host	Host stage infected	Parasite	% Incidence	Location	Lab. or field study	Reference
Psorophora horrida	Larvae	Romanomermis culicivorax	-	USA (Louisiana)	Lab.	Unpublished data
Psorophora howardii	"	"	-	"	"	"
Psorophora varipes	"	"	80	"	"	Petersen et al. (1969)
Uranotaenia lowii	"	Romanomermis culicivorax (Mermithid)	-	"	Field	Chapman et al. (1967a)
Uranotaenia sapphirina	"	"	35	"	"	"
"	"	Romanomermis culicivorax	93	"	"	Petersen et al. (1968)
"	"	"	7-65	"	"	Savage & Petersen (1971)
"	"	"	32	"	(Field)	Petersen & Willis (1972b)

ABSTRACTS

Mary Ann Strand

Anonymous. (1970). Michigan science in action. The biological control story. Mich. St. Univ. agric. Exp. Stn, East Lansing, 15: 16-17.

Anonymous. (1972). Mass-produced nematodes down mosquitoes. Agri. Res., U.S.D.A., 20(11): 3-4.

The work of J. J. Petersen & O. R. Willis at Lake Charles, La. in rearing the nematode Reesimermis nielseni for use as a biological control agent is presented.

Arnell, J. H. & Nielsen, L. T. (1972). Mosquito studies (Diptera, Culicidae). XXVII. The varipulpus group of Aedes (Ochlerotatus). Contr. Am. Entomol. Inst. (Ann Arbor), 8(2), 48 pp.

A mermithid nematode was found parasitizing Aedes laguna. About 25 larvae were infected among several hundred apparently uninfected ones. The nematodes leave the mosquito body through the siphon. The mosquito subsequently dies.

Artyukhovskii, A. K. & Kolycheva, R. V. (1965). /On the mermithosis of mosquitoes of the genus Aedes in the bottomland of the Khoper River./ Zool. Zhur., 44(3): 454-455. (Russian, English summary).

Mermithid larvae were found parasitizing 14% of the female A. communis and A. maculatus mosquitos. Ovary sterilization was observed in some infected females. Lightly infected individuals behaved normally, but ones that were heavily infected did not attack.

Bailey, C. H. & Gordon, R. (1973). Histopathology of Aedes aegypti (Diptera: Culicidae) larvae parasitized by Reesimermis nielseni (Nematoda: Mermithidae). J. Invertebr. Pathol., 22: 435-441.

Most serious effects on the hosts occurred between 4th and 6th day after infection. During this time, the nematodes developed most rapidly, depleting host storage tissues while accumulating storage material in their trophosomes. The severity of the effects depended on the intensity of infection.

Bronskill, J. F. (1962). Encapsulation of rhabditoid nematodes in mosquitoes. Can. J. Zool., 40(7): 1269-1275.

Juveniles of the rhabditoid nematode (DD136) enter the body cavity of larval mosquitos by penetrating the gut wall in the region of the proventriculus. A thick capsule develops around many of the immature adult nematodes. This defense reaction occurs in Aedes aegypti, A. stimulans and A. trichurus. The histological structure of the capsule is not affected during metamorphosis. Sometimes the encapsulated nematodes are partially or completely expelled during moulting.

Chapman, H. C. et al. (1970). Protozoans, nematodes, and viruses of anophelines. Misc. Publ. Entomol. Soc. Amer., 7: 134-139.

A review of the species of Anopheles reported as natural or experimentally induced hosts of various nematodes is given. Preliminary work of the Lake Charles laboratory on the biology of Reesimermis, Diximermis (Gastromermis), and Perutilimermis (Agamomermis) is reported.

Chapman, H. C. et al. (1969). A two-year survey of pathogens and parasites of Culicidae, Chaoboridae, and Ceratopogonidae in Louisiana. Proc. N. J. Mosq. Exterm. Assoc., 56: 203-212.

Sixteen species of mosquitos were host to one or more species of nematode. Reesimermis (Romanomermis) was the most frequently encountered nematode. It infected 14 species in the field and 34 others in laboratory tests.

Chapman, H. C. et al. (1972). Field releases of the nematode Reesimermis nielseni for the control of Culex p. fatigans in Bangkok, Thailand. WHO/VBC/72.412. 7 pp.

Field treatments of drains and ditches with preparasitic nematodes at dosages up to 2.52 x $10^5/m^2$ gave infection rates of 0 to 21%. The pupal population was not apparently reduced and no recycling of the parasite in nature was observed.

Chapman, H. C. et al. (1967a). Nematode parasites of Culicidae and Chaoboridae in Louisiana. Mosq. News, 27: 490-492.

During a periodic survey for pathogens, nematodes were found in 8 multivoltine species of mosquitos. The highest incidence (88.9%) was found in Aedes sollicitans infected with Perutilimermis (Agamomermis) culicis.

Chapman. H. C. et al. (1967b). Pathogens and parasites in Louisiana Culicidae and Chaoboridae. Proc. N. J. Mosq. Exterm. Assoc., 54: 54-60.

A survey of pathogens and parasites was conducted from 1964 to 1967. The field collections were summarized by host species and pathogen genus. Five species of mosquitos were infected by nematodes.

Coz, J. (1966). Contribution à l'étude du parasitisme des adultes d'Anopheles funestus par Gastromermis sp. (Mermithidae). Bull. Soc. Pathol. Exot., 59: 881-889. (French, English).

Infection of female A. funestus apparently inhibits ovarian development. The degree of inhibition is related to the age of the larvae when infected.

Coz, J. (1973). Contribution à l'étude du parasitisme des Anophèles Quest-Africane-Mermithidae et Coelomomyces. Cah. O.R.S.T.O.M. ser. Entomol. Med. Parasitol., 11(4): 237-241.

The use of Gastromermis sp. as a biological control agent against Anopheles funestus is considered. In Upper Volta, the natural infection incidence in adult females reached 17.6%.

Dadd, R. H. (1971a). Effects of size and concentration of particles on rates of ingestion of latex particulates by mosquito larvae. Ann. Entomol. Soc. Am., 64: 687-692.

Nematodes averaging 240 μ long readily fill the midgut of 4th instar mosquito larvae. However, nematodes swept toward the mouths of mosquito larvae are seen mostly to be rejected. Some presumably arrive end on, when their width, averaging 25 μ, allows entry.

Dadd, R. H. (1971b). Size limitations on the infectibility of mosquito larvae by nematodes during filter-feeding. J. Invertebr. Pathol., 18: 246-251.

The 1st and 2nd instar larvae of Culex pipiens cannot become infected by the pathogenic nematode Neoaplectana carpocapsae because the infective stage worm is too big to enter the larval mouth. Infective juvenile worms were readily ingested by the 4th instar and relatively poorly by the 3rd.   Very few of the nematodes were able to establish themselves in the hemocoel and those remaining in the peritrophic membrane disintegrate within a few hours.

Dujardin, F. (1845).   Histoire naturelle des helminthes ou vers intestinaux.   Paris Roret, 654 pp.

Frohne, W. C. (1953).   Mosquito breeding in Alaska salt marshes, with special reference to Aedes punctodes.   Mosq. News, 13: 96-103.

Frohne noted the infestation of Aedes punctor larvae by Agamomermis sp. in Alaska.

Frohne, W. C. (1955a).   News and notes.   Mosq. News, 15: 53.

Frohne, W. C. (1955b).   News and notes.   Mosq. News, 15: 125.

Gendre, E. (1909).   Sur des larves de Mermis parasites des larves du Stegomyia fasciata. Bull. Soc. Pathol. Exot., 2: 106-108.

Mermis sp. was found infesting Aedes aegypti (=Stegomyia fasciata) larvae in Guinea. In some locations all larvae were parasitized, however, these locations were rare.

Gordon, R., Bailey, C. H. & Barber, J. M. (1974).   Parasitic development of the mermithid nematode Reesimermis nielseni in the larval mosquito Aedes aegypti.   Can. J. Zool., 52: 1293-1302.

The most rapid growth phase of the nematode was from the 6th to 8th day of parasitism. During this time, the metabolites within the host's storage tissues were depleted and the development of the imaginal disc was suppressed.

Gorham, J. R. (1972).   Studies on the biology and control of arthropods of health significance in Alaska.   4.   Ecological studies of biting flies on the North Slope of Alaska: 1970.   Arctic Health Res. Center, P.H.S., H. E. W., 62 pp.

Hanney, P. W. (1960).   The mosquitoes of Zania Province, Northern Nigeria.   Bull. Entomol. Res., 51: 145-171.

During the wet season, 2 of the 48 adult Anopheles gambiae and 10 of 107 A. funestus were infested with nematodes (Agamomermis).   In many cases, ovarian development appeared impeded.

Hearle, E. (1926).   The mosquitoes of the lower Fraser Valley, British Columbia and their control.   Natl. Res. Counc. Rep. (Canada), 17, 82-83.

Hearle reports on the breeding places and general habits of the mosquitos.   He noticed a mermithid nematode (Paramermis canadensis) infesting adult Aedes sticticus (alrichii).

Hearle, E. (1929).   The life history of Aedes flavescens Müller.   A contribution to the biology of mosquitoes of the Canadian prairies.   Trans. R. Soc. Can. Biol. Sci., 23: 85-101.

During the survey, nematodes, possibly mermithids, were observed in the body cavities of A. flavescens adults.

Hooi, C. W. & Ramachandran, C. P. (1963).   (Mermithid?) (sic.) from two Malayan mosquitoes. Singapore Med. J., 4: 180.

Unknown nematodes were recovered from Anopheles letifer and Aedes albopictus.   The nematodes were found in the abdominal hemocoele.

Hoy, J. B. & Petersen, J. J. (1973).   Fish and nematodes - current status of mosquito control techniques.   Proc. Calif. Mosq. Control. Assoc., 41: 49-50.

Attempts to use Reesimermis nielseni to control Aedes nigromaculis have had varied results, but the nematode has been effectively used against Anopheles freeborni.

Ignoffo, C. M. et al. (1973).   Susceptibility of aquatic vertebrates and invertebrates to the infective stage of the mosquito nematode Reesimermis nielseni.   Mosq. News, 33: 599-602.

None of the nontarget organisms tested supported development of the nematode.

Iyenger, M. O. T. (1930).   Parasitic nematodes of Anopheles in Bengal.   Trans. 7th Cong. Far East. Assoc. Trop. Med. (Calcutta, 1927), 3: 128-135.

Larvae of several species of Anopheles were found to be parasitized by a Mermis sp. nematode.   A description of the worm and its development within the host is given. Although many nematodes may enter a larva, usually only one develops.   The worms leave the larva before pupation.   No infected pupae have been seen, but a few infected adults have been found.

Jenkins, D. W. (1964).   Pathogens, parasites, and predators of medically important arthropods. Bull. WHO 30 (Suppl.).   150 pp.

A review of the literature through 1963 is given.

Jenkins, D. W. & West, A. S. (1954) Mermithid nematode parasites in mosquitoes.   Mosq. News, 14: 138-143.

Mermithid nematodes (Hydromermis sp.) were found parasitizing Aedes communis larvae in Manitoba, Canada.   In some localities nearly all the late-developing larvae were infected and all infected larvae died.   A. excrucians larvae from the same pools were not infected.   Hydromermis nematodes were also seen in a few A. nearcticus and A. nigripes larvae.

Johnson, H. P. (1903).   A study of certain mosquitoes in New Jersey and a statement of the "Mosquito-malaria theory".   Ann. Rep. N. J. Agri. Exper. Sta., 23: 559-593.

Kalucy, E. C. (1972).   Parasitism of Anopheles annulipes Walker by a mermithid nematode. Mosq. News, 32: 582-585.

Larvae of A. annulipes were found parasitized by an unidentified mermithid nematode in Australia.   The nematode is apparently limited to A. annulipes because none was found parasitizing A. pseudostigmaticus larvae which were present in large numbers in the same location when A. annulipes was scarce.   The life history of the nematode in the laboratory is described.

Laird, M. (1956).   Studies of mosquitoes and freshwater ecology in the South Pacific. Bull. R. Soc. N. Z., 6: 126-129.

The characteristics of 457 potential mosquito breeding places were examined.   In some samples in Australia, 69% of the Anopheles annulipes larvae were infected with a nematode, Agamomermis sp.

Lumsden, W. H. R. (1955).   Entomological studies relating to yellow fever epidemiology, at Gede and Taveta, Kenya.   Bull. Entomol. Res., 46: 149-183.

During this extensive survey, Lumsden observed mermithid nematodes (Agamomermis sp.) parasitizing mosquito larvae collected from tree holes in one forest.

Mattingly, P. F. (1972). Mosquito eggs. XX. Egg parasitism in _Anopheles_ with further notes on _Armigeres_. Mosq. System, 4: 84-86.

An unidentified parasite of _Anopheles_ eggs is described in a paper by Pagayon (1963. Bull. Dept. Health Manila., 35: 41-45). It appears to be a nematode and is present in a large proportion of the mosquito specimens that Pagayon examined.

Mitchell, C. J., Chen, P. S. & Chapman, H. C. (1972). Exploratory trials utilizing a mermithid nematode as a control agent for _Culex_ mosquitos in Taiwan (China). Formosan Med. Assoc. J., 73(5): 241-254. (WHO/VBC/72.410.)

Several species of mosquitos were tested for susceptibility to _Reesimermis nielseni_. Among the species tested _C. P. fatigans_ is the most susceptible and _C. annulus_ the least. Infection rates were quite variable, ranging in one case from 5 to 47%. Attempts made to infect mosquito populations in their natural habitat met with mixed results. In no case was evidence obtained which would indicate that the parasite had become established in nature.

Muspratt, J. (1945). Observation on the larvae of the tree-hole breeding Culicini (Diptera: Culicidae) and two of their parasites. J. Entomol. Soc. South Afr., 8: 13-20.

A mermithid (later identified as _Octomyomermis muspratti_) was found infesting most species of tree-hole breeding mosquitos in Zambia. The nematode usually destroyed the host. Similar parasites were observed in _Anopheles_ larvae from swamp pools and stream bed pools in various parts of Africa.

Muspratt, J. (1947). The laboratory culture of a nematode parasite of mosquito larvae. J. Entomol. Soc. South Afr., 10: 131-132.

A method for culturing _Octomyomermis muspratti_ in _Aedes aegypti_ larvae is described. Sand from rearing jars containing naturally infected mosquitos was placed with newly hatched laboratory bred larvae. The sand contained nematode eggs which continued to hatch for 2-3 weeks.

Muspratt, J. (1963). Progress report (May 1963) on investigations concerning three mosquito pathogens at Livingstone, Northern Rhodesia. WHO/EBL/12. 7 pp.

Larval _Culex pipiens fatigans_ were found to be highly susceptible to _Reesimermis (Agamomermis) muspratti_ (Mermithidae).

Muspratt, J. (1965). Technique for infecting larvae of the _Culex pipiens_ complex with a mermithid nematode and for culturing the latter in the laboratory. Bull. WHO, 33(1): 140-144.

Second- or early third-instar larvae are placed in a jar containing the mermithid nematode (_Octomyomermis muspratti_) eggs. The parasitized larva is removed to a rearing vessel when it contains only one or two small worms because too large an infection may be fatal. The newly emerged nematodes are transferred to culture jars where they are kept in water and sand for 3-4 weeks. The water is decanted and the sand is allowed to dry. The dry jars may be left until required for reinfection. This mermithid matures within about 4 months.

Nasr, A. E. A. (1974). Nematode (family Mermithidae) infections in _Anopheles gambiae_ and _A. funestus_ in Nigeria. (Abstr.) Proc. 3rd Int. Cong. Parasitol. (Munich), 1: 325.

Mermithid nematodes were not further identified.

Nickle, W. R. (1972). A contribution to our knowledge of the Mermithidae (Nematoda). J. Nematol., 4: 113-146.

The genera of the Mermithidae are reviewed. Nickle makes an emended family diagnosis and redefines 16 of the genera. Romanomermis is rejected in favour of Reesimermis and Diximermis peterseni is described for the first time. The new genus Neomesomermis is proposed.

Nickle, W. R. (1973a). Identification of insect parasitic nematodes - a review. Exp. Parasitol., 33: 303-317.

Tables, photographs, and line drawings are given to help parasitologists identify nematode parasites.

Nickle, W. R. (1973b). Nematode parasites of insects. Proc. Ann. Tall Timbers Conf. Ecol. Animal Control Habitat Manage., 4: 145-163.

This general article gives basic information about the life histories and hosts of nematode parasites.

Obiamiwe, B. A. (1969). The life cycle of Romanomermis sp. (Nematoda: Mermithidae) a parasite of mosquitoes. Trans. R. Soc. Trop. Med. Hyg., 63(1): 18-19.

The life cycle of Reesimermis (Romanomermis) muspratti is described. Mosquitos infected for more than 8 days are pale and the parasites can be seen through the body wall.

Obiamiwe, B. A. & MacDonald, W. W. (1973). A new parasite of mosquitoes, Reesimermis muspratti sp. nov. (Nematoda: Mermithidae), with notes on its life-cycle. Ann. Trop. Med. Parasitol., 67: 439-444.

The nematode cultured by Muspratt (1965) is described as a new species. Aedes aegypti, Ae. polynesiensis, Culex pipiens molestus, C. p. fatigans, Anopheles stephensi, and An. albimanus are susceptible. Male nematodes were generally produced in underfed and females in well fed hosts.

Petersen, J. J. (1972). Factors affecting sex determination in a mermithid parasite of mosquitoes. J. Nematol., 4: 83-87.

The sex ratio (males:females) of Reesimermis nielseni increased as the number of parasites per host increased. Equal numbers of males and females were produced when the hosts averaged 3 nematodes per individual. Also, the species of host and the host's diet influenced the sex ratio.

Petersen, J. J. (1973a). Factors affecting mass production of Reesimermis nielseni a nematode parasite of mosquitoes. J. Med. Entomol., 10(1): 75-79.

Using the rearing procedures described by Petersen & Willis (1972), 7.2-7.5 x 10^5 post parasitic female nematodes were produced per 10^6 hosts exposed.

Petersen, J. J. (1973b). Relationship of density, location of hosts, and water volume to parasitism of larvae of the southern house mosquito by a mermithid nematode. Mosq. News, 33: 516-520.

Water volume was not found to be a major factor in determining the incidence of parasitism by Reesimermis nielseni of Culex pipiens quinquefasciatus. Preparasitic R. nielseni exhibited thigmotactic and negatively geotactic behaviour.

Petersen, J. J. (1973c) Role of mermithid nematodes in biological control of mosquitoes. Exp. Parasitol., 33(2): 239-247.

The factors which indicate that mermithid nematodes would be successful agents for biological control of mosquitos are reviewed.

Petersen, J. J. & Chapman, H. C. (1970). Parasitism of Anopheles mosquitoes by a Gastromermis sp. (Nematoda: Mermithidae) in southwestern Louisiana. Mosq. News, 30: 420-424.

The incidence of parasitism of anopheline larvae by Diximermis (Gastromermis) peterseni in Louisiana is apparently rare. However, large percentages of the larvae in some locales may be parasitized. The nematode appears to be specific to larvae of anophelines. It is easily cultured because it has a free living stage, and maturation, mating and oviposition take place readily in small containers of sand submerged in water. Infected larvae have not been observed to pupate and this may account for the limited distribution of the parasite.

Petersen, J. J. & Chapman, H. C. (1972). The development of a biological control agent of mosquitoes. Proc. 2nd Gulf Coast Conf. Mosq. Supp. Wildl. Manage., pp. 3-5.

An outline of the experiments and results leading to the development of Reesimermis nielseni (a mermithid nematode) as a control agent for mosquitos is given. Much of the information in this paper is contained in other works by Petersen.

Petersen, J. J., Chapman, H. C. & Willis, O. R. (1969). Fifteen species of mosquitoes as potential hosts of a mermithid nematode Romanomermis sp. Mosq. News, 29: 198-201.

The distribution of Reesimermis (Romanomermis) nielseni is apparently restricted because it prevents pupation and, thus, cannot be distributed by its host. Tests were made to evaluate 15 mosquito species as potential hosts. Psorophora confinnis was the most susceptible, while Culex territans, A. triseriatus, and P. ferox were judged highly resistant to attack and development of the parasite. The other species tested ranged between the two extremes with parasite development time, also, varying.

Petersen, J. J., Chapman, H. C. & Woodard, D. B. (1967). Preliminary observations on the incidence and biology of a mermithid nematode of Aedes sollicitans (Walker) in Louisiana. Mosq. News, 27: 493-498.

Sixteen locations in Louisiana were surveyed for the incidence of Perutilimermis (Agamomermis) culicis in adult female A. sollicitans from mid-August through December. Seventeen per cent. were found to be infected and no parasitism was observed after first prolonged cold weather. Death of the mosquito larva usually results when heavy infection occurs. Emergence of parasitized females was not retarded nor was their activity in seeking a blood mean. However, ovarian development appears to be affected and most infected females failed to produce eggs. A. sollicitans is the only species found in which this nematode can reach maturity.

Petersen, J. J., Chapman, H. C. & Woodard, D. B. (1968). The bionomics of a mermithid nematode of larval mosquitoes in southwestern Louisiana. Mosq. News, 28: 346-352.

The nematode Reesimermis (Romanomermis) nielseni was found in a few localities in Louisiana. Its limited distribution probably results from the failure of infected hosts to develop enough to emerge from their pond. Thirty-two species of mosquitos were found naturally infected or were infected in the laboratory by this nematode. Field observations showed the highest levels of parasitism occurred in pools where the water level fluctuates but never completely dries up.

Petersen, J. J., Hoy, J. B. & O'Berg, A. G. (1972). Preliminary field tests with Reesimermis nielseni (Mermithidae: Nematoda) against mosquito larvae in California rice fields. Calif. Vector Views, 19: 47-50.

First-, second-, and third-instar Anopheles freeborni were examined after preparasitic nematodes were introduced into rice paddies. When 500 nematodes/yd.^{2} were applied, 50% of the larvae became infected. Culex tarsalis larvae were not found to be parasitized.

Petersen, J. J., Steelman, C. D. & Willis, O. R. (1973). Field parasitism of two species of Louisiana rice field mosquitoes by a mermithid nematode. Mosq. News, 33: 573-575.

Psorophora confinnis and Anopheles quadrimaculatus larvae were exposed to preparasitic Reeseimermis nielseni in rice fields. From the results using various dosages, it was estimated that 3900 nematodes/yd.^{2} would be required for 95% parasitism of P. confinnis and 1300/yd.^{2} for 95% parasitism of A. quadrimaculatus.

Petersen, J. J. & Willis, O. R. (1969a). Incidence of Agamomermis culicis (Nematoda: Mermithidae) in Aedes sollicitans in Louisiana in 1967. Mosq. News, 29: 87-92.

Monthly samples of the adult female mosquito population were taken and overall 16.8% were found to be infected. In another survey, larvae were collected and reared in the laboratory; checks revealed that 10.6% of these were parasitized. Samples from 6 widely separated sites showed a fluctuating seasonal pattern. Parasitism generally increased when precipitation was heavy except when coupled with low temperatures. No parasitism was observed in other species except under laboratory conditions.

Petersen, J. J. & Willis, O. R. (1969b). Observations of a mermithid nematode parasitic in Orthopodmyia signifera (Coquillett) (Diptera: Culicidae). Mosq. News, 29: 492-493.

Larvae of O. signifera containing parasitic nematodes (tentatively identified as Mesomermis sp.) were collected from a tree hole in Louisiana. The genus Mesomermis had previously only been reported from blackflies. Many nematodes were found encapsulated and melanized which suggests that the mosquito may not be the parasite's primary host. Most of the infected mosquitos died before the nematodes matured.

Petersen, J. J. & Willis, O. R. (1970). Some factors affecting parasitism by mermithid nematodes in southern house mosquito larvae. J. Econ. Entomol., 63: 175-178.

In laboratory tests when the ratio of infective Reesimermis (Romanomermis) nielseni nematodes to Culex pipiens quinquefasciatus exceeded 3:1, almost 100% of the mosquitos became parasitized. The 2nd instar mosquito larvae were the most susceptible and the 4th instars were the least. All parasitized larvae failed to pupate.

Petersen, J. J. & Willis, O. R. (1971). A two-year survey to determine the incidence of a mermithid nematode in mosquitoes in Louisiana. Mosq. News, 31: 558-566.

The survey determined that Reesimermis nielseni occurred in 13 of 19 species of mosquito larvae in Louisiana. The parasite was active from April to November. The level of parasitism varied greatly, both for a given species and between species.

Petersen, J. J. & Willis, O. R. (1972a). Procedures for the mass rearing of a mermithid parasite of mosquitoes. Mosq. News, 32: 226-230.

The nematode, Reesimermis nielseni, was reared in the laboratory using larvae of Culex pipiens quinquefasciatus as hosts. It was most rapidly and economically reared with 0.35 cm^2 surface area per host, 1:12 ratio of hosts to preparasitic nematodes, and optimum feeding regime. Using this procedure, the infective stage nematodes cost 7-10 cents per million.

Petersen, J. J. & Willis, O. R. (1972b).   Results of preliminary field applications of
Reesimermis nielseni (Mermithidae: Nematoda) to control mosquito larvae.   Mosq. News, 32:
312-316.

Ten natural sites were treated 20 times with preparasitic nematodes.   When 1000
nematodes/m2 were added, 94% of the 2nd instar Anopheles larvae became infected and 64%
of all the Anopheles sampled.   The nematode appears to have become established in at
least 7 of the sites.

Petersen, J. J. & Willis, O. R. (1974a).   Diximermis peterseni (Nematoda: Mermithidae):
A potential biocontrol agent of Anopheles mosquito larvae.   J. Invertebr. Pathol., 24: 20-23.

About 300 1st instar A. quadrimaculatus larvae and a small culture of D. peterseni (about
60 adults and some eggs and preparasitic juveniles) were added to an artificial pond.
Initially 92% of the mosquitos became infected.   The parasite gradually disappeared and
then reappeared 8 months after introduction.   It has been observed for 40 months.
About 2300 mosquito larvae (85-90% infected) were released into a natural pond.   The
parasite has become established there with average parasitism of 92% between September
and March the first year and 88% during the same period the following year.

Petersen, J. J. & Willis, O. R. (1974b).   Experimental release of a mermithid nematode to
control Anopheles mosquitoes in Louisiana.   Mosq. News, 34(3): 316-319.

When Reesimermis nielseni preparasites were released in a natural site at dosage of
1000 nematodes/sq. yd., a mean of 76.3% of the Anopheles spp. larvae were infected
within 24 hours post treatment.   When twice the first dosage was used, a mean of 84.8%
was infected.   Vegetation and water depth did not influence the level of parasitism.

Poinar, G. O. jr & Leutenegger, R. (1971).   Ultrastructural investigation of the melanization
process in Culex pipiens (Culicidae) in response to a nematode.   J. Ultrastruct. Res., 36:
149-158.

A definitive capsule appeared 5-10 hours after the nematode Neoaplectana carpocapsae
entered the body cavity of the mosquito larvae.   This defense reaction killed the
nematodes in several instances, especially when a single parasite entered the host.
Multiple attacks usually resulted in the death of the host.   The encapsulation reaction
is not fast enough to halt the escape of the bacterium Achromobacter nematophilis,
which is liberated from the intestine of this nematode.   The bacterium causes a fatal
septicaemia.

Poinar, G. O. jr & Sanders, R. D. (1974).   Description and bionomics of Octomyomermis
troglodytis sp. n. (Nematoda: Mermithidae) parasitizing the western treehole mosquito
Aedes sierrensis (Ludlow) (Diptera: Culicidae).   Proc. Helminthol. Soc. Wash., 41: 37-41.

A new species of mermithid nematode is described from the larvae, pupae, and adults of
the mosquito in California.   The nematodes generally emerged from the 4th instar larvae
and entered the organic matter in the bottom of the treehole where they moult, mate,
and oviposit.   Preparasitic juveniles contact the resting mosquito larvae at the surface
of the water.   Parasitic development lasted from 20-22 days at 20°C.   All hosts died
soon after nematode emergence.   The incidence reached 38% in one treehole.

Reynolds, D. G. (1972).   Experimental introduction of a microsporidian into a wild population
of Culex pipiens fatigans Wied.   Bull WHO, 46(6): 807-812.

Eight culture jars were used to introduce Reesimermis muspratti into tree holes and a
bomb crater on Nauru Island.   Ten days after introduction parasitized larvae were found
in the tree holes.   Mermithids were not found in larvae from the bomb crater.

Ross, R. (1906).   Notes on the parasites of mosquitoes found in India between 1895-1899.
J. Hyg., 6: 101-108.

Sanders, R. D. (1972).   Microbial mortality factors in Aedes sierrensis populations.
Proc. Calif. Mosq. Control Assoc., 40: 66-68.

  Mosquitos infected with a mermithid nematode (later identified as Octomyomermis troglodyti
were found in 1 of 18 tree holes examined near Novato, California.

Savage, K. E. & Petersen, J. J. (1971).   Observations of mermithid nematodes in Florida
mosquitoes. Mosq. News, 31: 218-219.

  Anopheles crucians and Uranotaenia sp. were found to be parasitized by Reesimermis
(Romanomermis) nielseni and Diximermis (Gastromermis) peterseni.   Culex sp. was found in
the same sites with the infected species but none was infected.

Schacher, J. F. & Khalil, G. M. (1968).   Development of Foleyella philistinae Schacher &
Khalil, 1967 (Nematoda: Filariodae) in Culex pipiens molestus with notes on pathology in the
arthropod.   J. Parasitol., 54: 869-878.

  The morphogenesis of the nematode occurred in the fat body cells throughout the body
of the mosquito host.   In spite of large numbers of 3rd stage nematode larvae found in
the mosquitos, their survival was good.

Service, M. W. (1973).   Mortalities of the larvae of Anopheles gambiae Giles complex and
detection of predators by the precipitin test.   Bull. Entomol. Res., 62: 359-369.
(WHO/VBC/72.360.)

  A life table study was made on larvae of Anopheles gambiae from ponds and ditches in
Kenya.   In one pond, 82.8% of the 4th instar larvae were infected with nematodes
(Agamomermis) but in other areas no nematodes were found.

Shachov, S. D. (1927).   /On the parasite Agamomermis Stiles in mosquito Aedes dorsalis Mg.
and Aedes cantans Mg. in the vicinity of Kharkov./   Russ. Entomol. Obozr., 21: 27-32.

  Heavy infestations of the nematode were observed in adult A. dorsalis and A. cantans.
However, Culex pipiens and Anopheles maculipennis from the same areas were not infested.

Sinton, J. A. (1932).   Helminthic infections in Indian anopheline mosquitoes.   Rec. Malaria
Surv. India, 3(3): 347-351.   (Reference from Rev. Appl. Entomol., B 21: 83.)

Smith, J. B. (1904).   Report of the New Jersey State Agricultural Experiment Station upon the
mosquitoes occurring with the state, their habits, life history, etc.   MacCrellish & Quigley,
Trenton, N.J., pp. 81-84.

Smith, M. E. (1961).   Further records of mermithid parasites of mosquito larvae.   Mosq. News,
21: 344-345.

  H. E. Welch determined that the mermithid parasites collected from Aedes pullatus and
A. communis larvae in Colorado closely resembled Hydromermis churchillensis.

Stabler, R. M. (1945a).   Parasitism in mosquito control.   Entomol. Rep. Ann. Rep. Delaware
Co. Mosquito Ext. Comm. (1944), pp. 22-23.

Stabler, R. M. (1945b).   Mosquito transmission of disease.   Proc. N. J. Mosq. Exterm. Assoc.,
32: 119-129.

Stabler, R. M. (1952). Parasitism of mosquito larvae by mermithids (Nematoda). J. Parasitol.
38: 130-132.

Fourth instar mosquito larvae parasitized by mermithid nematodes (Agamomermis) were
collected in Pennsylvania. They had unusually white thoraces and their movements were
sluggish. The worms encircled the thoracic structures just inside the exoskeleton of
their hosts. No gross anatomical changes in the hosts were observed prior to nematode
emergence.

Steiner, G. (1924). Remarks on a mermithid found parasitic in the adult mosquito (Aedes
vexans Meigen) in B. C. Can. Entomol., 56: 161-164.

Steiner described and named the species Paramermis canadensis from material collected
by E. Hearle in British Columbia, Canada. Hearle noted that 80% of the adults dissected
were infected in one year and that no infected female mosquito had well developed
ovaries.

Stiles, C. W. (1903). A parasitic roundworm (Agamomermis culicis n.g., n. sp.) in American
mosquitoes (Culex sollicitans). Bull. U. S. Hyg. Lab., 13: 15-17.

Trpiš, M. (1969). Parasitical castration of mosquito females by mermithid nematodes.
Helminthologica, 10: 79-81.

Adult females of Aedes vexans were collected near Vancouver, B.C. Mermithid nematodes
(Agamomermis) were found infesting their body cavity along the gut. The infested
females appeared normal; however, they began dying 7 days after being fed a blood meal.
The ovaries of the parasitized females did not develop or were grossly underdeveloped.
When the worms began to leave their hosts, there were no eggs in the ovaries. The
mermithids caused 100% mortality.

Trpiš, M., Haufe, W. O. & Shemanchuk, J. A. (1968). Mermithid parasites of the mosquito
Aedes vexans Meigen in British Columbia. Can. J. Zool., 46: 1077-1078.

Newly emerged, apparently healthy female adult mosquitoes were collected. In the
laboratory they were found to be infected by mermithid nematodes (Agamomermis). The
nematodes suppressed ovarian development and nearly all died within 10 days.

Tsai, Y.-H. & Grundmann, A. W. (1969). Reesimermis nielseni gen. et. sp. n. (Nematoda:
Mermithidae) parasitizing mosquitoes in Wyoming. Proc. Helminthol. Soc. Wash., 36: 61-67.

A new species of mermithid nematode parasitizing larvae of Aedes communis is described.
It was also observed parasitizing A. cinereus, A. fitchii, A. increpitus, A. pullatus
and Culiseta impatiens in Wyoming. Seasonal fluctuations in sex ratio were observed
and the nematode undergoes one generation per year.

Tsai, Y.-H., Grundmann, A. W. & Rees, D. M. (1969). Parasites of mosquitoes in southwestern
Wyoming and northern Utah. Mosq. News, 29(1): 102-110.

The mosquito fauna were surveyed to determine kinds, distribution and prevalence of
parasites. The survey area represented a diversity of habitats and included areas of
intensive control and areas where no control had been practised. Reesimermis nielseni
infections were found in some areas.

Walandouw, E. K. (1934).   Nematoden als bestrijders von <u>Anopheles</u> larven.   <u>Geneesk.</u>
<u>Tijdschr. Ned.-Ind.</u>, <u>74</u>:   1219-1224.

   A preliminary description is given of a mermithid nematode found parasitizing <u>Anopheles</u>
<u>leucosphyrus</u> larvae from Sumatra.

Welch, H. E. (1960a).   <u>Hydromermis churchillensis</u> n. sp. (Nematoda: Mermithida) a parasite
of <u>Aedes communis</u> (DeG.) from Churchill, Manitoba, with observations on its incidence and
bionomics.   <u>Can. J. Zool.</u>, <u>38</u>:   465-474.

   The life cycle and description of the new species are presented.   Incidence of the
parasite was variable;   in general 10% of the larvae were killed, but in some pools the
percentage reached 80%.   No correlation between physical features of the pools and the
nematodes' distribution was found.

Welch, H. E. (1960b).   Notes on the identities of mermithid parasites of North American
mosquitoes, and a redescription of <u>Agamomermis culicis</u> Stiles, 1903.   <u>Proc. Helminthol. Soc.</u>
<u>Wash.</u>, <u>27</u>: 203-206.

Welch, H. E. (1964).   <u>Romanomermis iyengari</u>, species nov. (Nematoda: Mermithidae Braun,
1883).   <u>Pilot Register of Zoology</u> Card No. 4, Cornell University.

   This nematode was found parasitizing <u>Anopheles subpictus</u> in Bangalore, India, and has
been reclassified as <u>Reesimermis iyengari</u>.

Welch, H. E. & Bronskill, J. F. (1962).   Parasitism of mosquito larvae by the nematode,
DD-136 (Nematoda: Neoaplectanidae).   <u>Can. J. Zool.</u>, <u>40</u>:   1263-1268.

   The nematode becomes encapsulated after it penetrates the gut wall and invades the
prothoracic body cavity of <u>Aedes aegypti</u>.   Other <u>Aedes</u> species have a similar
encapsulation reaction to this parasite.   Bacteria transported by the nematode are
pathogenic to the mosquitos and encapsulation does not prevent host death.

IX. PATHOGENS OF
CERATOPOGONIDAE (MIDGES) [a]

Willis W. Wirth

Systematic Entomology Laboratory, IIBIII
Agricultural Research Service
US Department of Agriculture
Washington, DC 20560, USA

---

[a] Based on, and supplementary to, the reviews by Jenkins (1964), Bacon (1970), and Laird (1971).

PATHOGENS OF CERATOPOGONIDAE (MIDGES)

Host	Host stage infected	Parasite or pathogen	% incidence	Locality	Lab. or field study	Reference
VIRUSES						
Culicoides sp. prob. arboricola	Larva	Culicoides iridescent virus (CuIV)	over 100 larvae	USA (Louisiana)	Field	Chapman et al. (1968, 1969), Chapman (1973)
RICKETTSIAE						
Culicoides sanguisuga	Adult fat body	Cocci - ? Rickettsiae	33	USA (Massachusetts)	Lab., field	Hertig & Wolbach (1924)
BACTERIA						
Culicoides nubeculosus	Larva fat body	? Symbionts	-	Scotland	-	Lawson (1951)
Culicoides nubeculosus	Larva	Identification uncertain	-	-	-	Steinhaus (1946)
Culicoides salinarius	Larva	Pseudomonas sp.	-	Scotland	Field	Becker (1958)
Culicoides sp.	Larva	Bacterial symbiont	-	Germany	-	Mayer (1934)
Dasyhelea obscura	All stages	Hereditary bacterian symbiont	common	England	Field	Keilin (1921a, 1927)
Dasyhelea versicolor	Larva	Intracellular symbiont	-	Germany	-	Stammer, in Buchner (1930)
Dasyhelea sp.	Larva	"	-	"	-	"
FUNGI						
Culicoides furens	Adult (duct of crop of female)	"Fungal hyphae"	-	Jamaica	Field	Lewis (1958)

PATHOGENS OF CERATOPOGONIDAE (MIDGES) (continued)

Host	Host stage infected	Parasite or pathogen	% incidence	Locality	Lab. or field study	Reference
Culicoides nubeculosus	Adult (lumen of esophageal diverticulum;? fungus ingested with raisin juice)	"fungal hyphae"	-	Britain	Lab.	Megahed (1956)
Culicoides phlebotomus	Adult	Grubyella ochoterenai	experimental	Hispaniola	Lab.	Ciferri (1929)
Dasyhelea lithotelmatica	Larva hind-gut	Carouxella scalaris	-	France	-	Manier et al. (1961)
Dasyhelea lithotelmatica	Larva (rectal ampullae)	Rubetella inopinata	-	France	-	Manier et al. (1961), Coluzzi (1966)
Dasyhelea obscura	Larva	Monosporella unicuspidata	low	England	Lab., field	Keilin (1920a, 1921a, 1927)
Forcipomyia sp.	-	Laboulbeniales	-	Germany	-	Mayer (1934)
Ceratopogonidae	-	Entomophthora ovispora	epizootic	USSR (Moscow)	Field	Gol'berg (1969)
PROTOZOA						
Ceratopogon solstitialis	Larva-gut	Taeniocystis mira	to 20	France	Lab., field	Léger (1906)
Ceratopogon sp.	Larva-intestine	Schizocystis gregarinoides	to 50	France	-	Léger (1900, 1906), Weiser (1963a, b)
Ceratopogon sp.	Larva-fat body	Spiroglugea octospora	-	France	-	Léger & Hesse (1922), Kudo (1924), Jirovec (1936), Weiser (1961)
Ceratopogon sp.	Larva-	Toxoglugea vibrio	-	France	-	Léger & Hesse (1922), Kudo (1924), Jirovec (1936), Weiser (1961)

PATHOGENS OF CERATOPOGONIDAE (MIDGES) (continued)

Host	Host stage infected	Parasite or pathogen	% incidence	Locality	Lab. or field study	Reference
Culicoides arakawai	Adult	Akiba (Leucocytozoon) caulleryi	-	Japan	-	Akiba (1960), Bennett et al. (1965)
Culicoides austeni	Adult-haemocoel	Ciliates	3 in 540	Africa	Lab.	Sharp (1928)
Culicoides austeni	Adult-haemocoel	Flagellates (leptomonas type)	occasional	Africa	Lab.	"
Culicoides crepuscularis	Adult	Parahaemoproteus sp.	-	Canada (Ontario)	-	Fallis & Bennett (1961), Bennett et al. (1965)
Culicoides cubitalis	Larva	"Ectoparasitic cysts" c.f. Perezella	-	United Kingdom	-	Kettle & Lawson (1952)
Culicoides nanus	Larva-fat body	Nosema n. sp.	-	USA (Louisiana)	Field	Chapman (1973)
Culicoides sp. prob. nanus	Larva-fat body	Pleistophora spp.	-	USA (Louisiana)	Field	Chapman et al. 1967, 1968, 1969), Chapman (1973)
Culicoides sp. prob. nanus	Larva	Internal Ciliates		"	"	Chapman et al. (1969)
Culicoides odibilis	Larva	Perezella sp.	-	Scotland	-	Becker (1958)
Culicoides peregrinus	Adult haemocoel	Tetrahymena pyriformis (=Probalantidium knowlesi, Balantidium knowlesi Leptoglena knowlesi	-	India	Field	Ghosh (1925), Abe (1927), Grasse & Boissezon (1929), Jenkins (1964)
Culicoides near piliferus	Adult	Parahaemoproteus nettionis	-	Canada (Ontario)	-	Fallis & Bennett (1961), Bennett et al. (1965)

PATHOGENS OF CERATOPOGONIDAE (MIDGES) (continued)

Host	Host stage infected	Parasite or pathogen	% incidence	Locality	Lab. or field study	Reference
Culicoides pulicaris	Larva	Perezella sp.	-	Scotland	-	Becker (1958)
Culicoides riethi	Larva	Perezella sp.	-	"	-	"
Culicoides salinarius	Larva	Perezella sp.	-	"	-	"
Culicoides salinarius	-	Flagellates (? Strigomonas)	-	France	Field	Kremer et al. (1961)
Culicoides stilobezzioides	Adult	Parahaemoproteus sp.	-	Canada (Ontario)	-	Fallis & Bennet (1961) Bennett et al. (1965)
Culicoides sphagnumensis	Adult	Parahaemoproteus canachites	-	Canada (Ontario)		Bennett et al. (1965) " " "
Dasyhelea lithotelmatica	Larva	Stylocystis riouxi	-	France	Lab., field	Tuzet & Ormières (1964)
Dasyhelea obscura	Larva-fat body, nerve ganglia	Helicosporidium parasiticum	common	England	Lab., field	Keilin (1921a, b, 1927)
Dasyhelea obscura	Adult-mid-gut	Allantocystis dasyhelei	rare	England	Lab., field	Keilin (1920b, 1927)
Dasyhelea sp.	Larva	Glugea sp.	-	England	Field	Keilin (1927)
Forcipomyia sp.	Larva	Taeniocystis parva	-	Germany	-	Foerster (1938)
Sphaeromias sp.	Larva-fat body	Nosema sphaeromiadis	-	Czechoslovakia	Field	Weiser (1957, 1961, 1963)

PATHOGENS OF CERATOPOGONIDAE (MIDGES) (continued)

Host	Host stage infected	Parasite or pathogen	% incidence	Locality	Lab. or field study	Reference
NEMATODA						
Atrichopogon sp.	Adult	Mermis sp.	one	India	Field	Das Gupta (1964)
Culicoides albicans	Adult	Agamomermis sp.	-	France	-	Callot (1959)
Culicoides alatus	Adult	Mermis sp.	-	India	-	Sen & Das Gupta (1958)
Culicoides arboricola	Larva	Aproctonema chapmani	-	USA (Louisiana)	Field	Nickle (1969), Chapman et al. (1969), Chapman (1973)
Culicoides buckleyi	Adult	Mermithidae	-	Malaya	Field	Buckley (1938)
Culicoides circumscriptus	Adult, larva	Mermithidae	to 27	USSR	Field	Glukhova (1967)
Culicoides crepuscularis	Adult	Mermithidae	several	USA (Florida)	Field	Beck (1958)
Culicoides crepuscularis	Adult	Mermithidae	to 30-90	"	"	Smith & Perry (1967)
Culicoides grisecens	Adult, larva	Mermithidae	to 27	USSR	Field	Glukhova (1967)
Culicoides haematopotus	Adult	Mermithidae	to 30-90	USA (Florida)	Field	Smith & Perry (1967)
Culicoides nanus	-	Reesimermis (Romanomermis) sp.	-	USA (Louisiana)	Field	Chapman et al. (1969)
Culicoides nanus	Larva	Mermithidae	1-10	USA (Louisiana)	Field	Chapman et al. (1968)

PATHOGENS OF CERATOPOGONIDAE (MIDGES) (continued)

Host	Host stage infected	Parasite or pathogen	% incidence	Locality	Lab. or field study	Reference
Culicoides nubeculosus	Larva	Heleidomermis vivipara	-	USSR	-	Rubtsov (1970, 1972)
Culicoides nubeculous	Adult, larva	Mermithidae	to 27	USSR	Field	Glukhova (1967)
Culicoides obsoletus	Adult	Mermithidae	common	England	Field	Boorman & Goddard (1970)
Culicoides obsoletus	Adult	Mermithidae	5	England	Field	Service (1974)
Culicoides orientalis	Adult	Mermithidae	-	Malaya	Field	Buckley (1938)
Culicoides oxystoma	Adult	Mermithidae	-	Malaya	Field	Buckley (1938)
Culicoides peregrinus	Adult	Mermithidae	-	Malaya	Field	Buckley (1938)
Culicoides pictipennis	Adult	Mermithidae	5	England	Field	Service (1974)
Culicoides pulicaris	Adult	Agamomermis heleis	-	USSR	-	Rubtsov (1967), Mirzaeva (1971)
Culicoides pulicaris	Adult, larva	Mermithidae	to 27	USSR	Field	Glukhova (1967)
Culicoides puncticolis	Adult, larva	Mermithidae	to 27	USSR	Field	Glukhova (1967)
Culicoides pungens	Adult	Mermithidae	-	Malaya	Field	Buckley (1938)
Culicoides shortii	Adult	Mermithidae	-	Malaya	Field	Buckley (1938)

PATHOGENS OF CERATOPOGONIDAE (MIDGES) (continued)

Host	Host stage infected	Parasite or pathogen	% incidence	Locality	Lab. or field study	Reference
Culicoides stellifer	Adult	Mermithidae	one	USA (Florida)	Field	Smith (1966)
Culicoides stellifer	Adult	Mermithidae	to 30-90	"	"	Smith & Perry (1967)
Culicoides stigma	Larva	Heleidomermis vivipara	-	USSR	-	Rubtsov (1970, 1972)
Culicoides stigma	Adult, larva	Mermithidae	to 27	USSR	Field	Glukhova (1967)
Culicoides spp.	Larva	Mermithidae	to 67	Buryat ASSR	Field	Mirzaeva (1971)
Dasyhelea obscura	Larva	Mermithidae	few	England	Field	Keilin (1921a)
Leptoconops kerteszi	Adult	Mermithidae	one	Mexico	Field	Whitsel (1965)
Leptoconops sp.	Adult, larva	Mermithidae	to 27	USSR	Field	Glukhova (1967)
Biting midges	-	Mermithidae	-	general review	-	Weiser (1963)

ABSTRACTS

Mary Ann Strand & Willis W. Wirth

Abe, T. (1927). On the classification of <u>Balantidium</u> (Preliminary report). <u>Dobotugaku Zassi</u>, <u>39</u>: 191-196.

Generic synonomy.

Akiba, K. (1960). Studies on the Leucocytozoon found in the chicken, in Japan. II. On the transmission of <u>L. caulleryi</u> by <u>Culicoides arakawae</u>. <u>Jap. J. Vet. Sci.</u>, <u>22</u>: 309-317.

The zygotes, ookinetes, oocysts, and sporozoites of <u>L. caulleryi</u> were found in female <u>C. arakawae</u> which had fed on infected chickens. These stages are described and figured. Thirteen chickens were infected by injecting a suspension of <u>Culicoides</u> which had fed on infected chickens 2-7 days prior to preparation. About 14 days after injection, the parasites were seen in gametogeny in the peripheral blood. The transmission of <u>L. caulleryi</u> by <u>C. arakawae</u> to chickens in Japan was established.

Bacon, P. R. (1970). The natural enemies of the Ceratopogonidae - a review. <u>Commonw. Inst. Biol. Control Tech. Bull.</u>, <u>13</u>, 71-82.

Large summary table and list of references form an extensive review of the world literature.

Beck, E. (1958). A population study of the <u>Culicoides</u> of Florida (Diptera: Heleidae). <u>Fla. Entomol.</u>, <u>18</u>: 6-11.

Several adult <u>C. crepuscularis</u> were gynandromorphic with female heads and male genitalia. A mermithid nematode was found occupying the body cavity of each gynandromorph.

Becker, P. (1958). Some parasites and predators of biting midges, <u>Culicoides</u> Latreille (Dipt., Ceratopogonidae). <u>Entomol. Mon. Mag.</u> <u>94</u>: 186-189.

Symbiotic bacteria were found associated with <u>C. salinarius</u>. Also, <u>Perezella</u> sp. a ciliate, was found in the hemocoels of several <u>Culicoides</u> spp. collected in Scotland.

Bennett, G. F. et al. (1965). On the status of the genera <u>Leucocytozoon</u> Ziemann, 1898 and <u>Haemoproteus</u> Kruse, 1890 (Haemosporidiida: Leucocytozoidae and Haemoproteidae). <u>Can. J. Zool.</u>, <u>43</u>: 927-932.

The genera <u>Leucocytozoon</u> and <u>Haemoproteus</u> are revised by division of the former into the genera <u>Leucocytozoon</u> Ziemann and Akiba n. g., and the latter into <u>Haemoproteus</u> Kruse and <u>Parahaemoproteus</u> n. sp. In addition to morphological characters, the separation of the new genera is justified by their transmission by <u>Culicoides</u> midges, while <u>Haemoproteus</u> is transmitted by hippoboscid flies and <u>Leucocytozoon</u> by Simuliidae.

Boorman, J. & Goddard, P. A. (1970a).   _Culicoides_ Latreille (Diptera, Ceratopogonidae) from Pirbright, Surrey.   _Entomol. Gaz._ _21_: 205-216.

Female _C. oboletus_ adults are commonly parasitized by mermithid nematodes, most frequently during August and September.   Parasitized individuals can be identified by their yellowish abdomens.

Buchner, P. (1930).   _Tier und Pflanze in Symbiose._   900 pp.   Berlin.

Intracellular symbionts were observed in _Dasyhelea_ spp. by Stammer (p. 323-325).

Buckley, J. J. C. (1938).   On _Culicoides_ as a vector of _Onchocerca gibsoni_ (Cleland and Johnston, 1910).   _J. Helminthol._ _16_: 121-158.

Non-filarial namatode infections of _Culicoides_ are reported to occur in the abdomens of _C. oxystoma_, _pungens_, _peregrinus_, and _buckleyi_.

Callot, J. (1959).   Action d'un _Agamomermis_ sur les caractères sexuels d'un Cératopogonidé. _Ann. Parasitol. Hum. Comp._ _34_: 439-443.

About 5% of the _C. albicans_ collected near Strasbourg, France were infested by larvae of mermithids.   Parasitized females were nearly normal but males suffered a marked alteration of antenna and palpus.

Chapman, H. C. (1973).   Assessment of the potential of some pathogens and parasites of biting flies, pp. 71-77.   _In_ Biting fly control and environmental quality: Proceedings of a symposium.   Defense Research Board, DR 217, Ottawa, Canada.

Ceratopogonidae (_Culicoides_ spp.) are reported to be hosts of pathogens and parasites: 3 species with Microsporida, 1 species with a virus, and 5 species with nematodes.

Chapman, H. C., Woodard, D. B. & Petersen, J. J. (1967).   Pathogens and parasites in Louisiana Culicidae and Chaoboridae.   _Proc. N. J. Mosq. Exterm. Assoc._ _54_: 54-60.

A single _Culicoides_ species infected by _Pleistophora_ sp. was collected during the survey.

Chapman, H. C. et al. (1968).   New records of parasites of Ceratopogonidae.   _Mosq. News_, _28_: 122-123.

Ceratopogonid larvae from a number of habitats in Louisiana were examined for parasites. Several species hosted various _Pleistophora_ spp.   Infected larvae became opaque white to milky white and died before pupation.   A cytoplasmic, non-inclusion iridescent virus was found in 100 larvae.   Mermithid nematodes infested 1-10% of the _Culicoides nanus_ larvae.   All died before pupation.

Chapman, H. C. et al. (1969).   A two-year survey of pathogens and parasites of Culicidae, Chaoboridae, and Ceratopogonidae in Louisiana.   _Proc. N. J. Mosq. Exterm. Assoc._ _56_: 203-212.

During the survey _Culicoides_ spp. infected with _Culicoides_ iridescent virus, _Pleistophora_, internal ciliates, and _Reesimermis_ were collected.

Cifferri, R. (1929).   Sur un Grubyella parasite de simulidés, Grubyella ochoterenai, n. sp. Ann. Parasitol. Hum. Comp., 7: 511-523.

  G. ochoterenai was isolated from Simulium sp. in Mexico.   The pathogen was experimentally transferred to Culicoides phlebotomus.

Coluzzi, M. (1966). Experimental infections with Rubetella fungi in Anopheles gambiae and other mosquitos.   Proc. 1st Int. Congr. Parasit., Rome 1964.   1: 592-593.

  Coluzzi mentions the work of Manier et al. (1961) on Rubetella inopinata, a parasite of Dasyhelea lithotelmatica.

Das Gupta, S. K. (1964).   Mermithid infections in two nematoceran insects.   Current Sci., 33: 55.

  An adult Atrichopogon sp. was found infected by mermithid nematodes near Calcutta. Usually mermithid infections occur in larvae and infected larvae succumb before reaching the adult stage.

Fallis, A. M. & Bennett, G. F. (1961).   Ceratopogonidae as intermediate hosts for Haemoproteus and other parasites.   Mosq. News., 21: 21-28.

  A Table lists ceratopogonid species known to be vectors of pathogenic organisms throughout the world.

Foerster, H. (1938).   Gregarinen in schlesischen Insekten.   Z. Parasitenkd. 10: 157-209.

  Large numbers of Taeniocystis parva were found in the guts of Forcipomyia sp. larvae. The sporogonic stages of the gregarine are described.

Ghosh, E. (1925).   On a new ciliate, Balantidium knowlesii, sp. nov., a coelomic parasite in Culicoides peregrinus.   Parasitology, 17: 189

  Numerous specimens of the new species were observed.   Several were seen undergoing ordinary transverse fission.   A diagnosis of the species is given.

Glukhova, V. M. (1967).   /On parasitism in blood-sucking midges (Diptera: Ceratopogonidae) by nematodes of the superfamily Mermitoidae/.   Parazitologiya 1: 519-520. (R.e).

  In Karelyskoi ASSR several species of midges are reported to host mermithid nematodes. In one sample, 27% of Culicoides grisescens larvae were infected.

Gol'berg, A. M. (1969).   /The finding of entomophthoraceous fungi on mosquitos (Family Culicidae) and midges (Family Ceratopogonidae)/.   Med. Parazitol. Parazit. Bolezn., 38: 21-23 (R.e.).

  In a water reservoir near Moscow, an epizootic among midges caused by Entomophthora ovispora was observed.   Mosquitos infected with other Entomophthora species were found in the same location.

Grasse, P. & de Boissezon, P. (1929).   Turchiniella culicis n. g., n. sp., infusoire parasite de l'hemocoele d'un Culex adults.   Bull. Soc. Zool. Fr., 54: 187-191.

  The genus Leptoglena is proposed for Balantidium knowlesi Ghosh, which was described from Culicoides peregrinus.

Hertig, M. & Wolbach, S. B. (1924).   Studies on rickettsia-like micro-organisms in insects. J. Med. Res., 44: 329-374.

Rickettsia-like cocci were observed in smears of the abdomens of adult Culicoides sanguisuga collected in Massachusetts.   In a few cases lobes of the fat body apparently, also contained the cocci, but they were not seen in the intestine or malpighian tubules.

Jenkins, D. W. (1964).   Pathogens, parasites, and predators of medically important arthropods. Bull. Wld Hlth Org., 30 (suppl.): 150 pp.

A review of the literature through 1963 is given.

Jírovec, O. (1937).   Studien über Microsporiden.   Vestn. Cesk. Spol. Zool., 4: 5-79.

This paper contains a general taxonomic and biological review of described species and a review of the literature.   Nothing new on ceratopogonid hosts is given.

Keilin, D. (1920a).   On a new Saccharomycete Monosporella unicuspidata gen. n. nom., n. sp., parasitic in the body cavity of a dipterous larva (Dasyhelea obscura Winnertz).   Parasitology, 12: 83-91.

Parasitized larvae are recognized by the milky appearance of the body, especially the posterior segments.   Refactive cells of the pathogen completely fill the body cavity.   The fat body is apparently the only organ completely destroyed.   The incidence in nature is low.

Keilin, D. (1920b).   On two new gregarines, Allantocystis dasyhelei n. g., n. sp., and Dendrorhynchus systeni, n. g., n. sp., parasitic in the alimentary canal of the dipterous larvae, Dasyhelea obscura Winn. and Systenus sp.   Parasitology, 12: 154-158.

All stages of A. dasyhelei occur in the midgut of its host, D. obscura.   The incidence in nature is very low (12 infected of several hundred examined.   The growth stages of the gregarine are described.

Keilin, D. (1921a).   On the life-history of Dasyhelea obscura, Winnertz (Diptera, Nematocera, Ceratopogonidae) with some remarks on the parasites and hereditary bacterian symbiont of this midge.   Ann. Mag. Nat. Hist., ser., 9, 8: 576-590.

The parasites of D. obscura are reviewed.   Keilin lists 6, 3 of which he has named. The remaining include a microsporidian found in the alimentary tube and salivary glands of the larvae, a parasite (possibly a gregarine) in the perivisceral fat body of the larvae, and a nematode (maybe a mermithid) seen in the body cavity.

Keilin, D. (1921b).   On the life history of Helicosporidium parasiticum, n. g., n. sp., a new type of protist parasitic in the larva of Dasyhelea obscura Winn. (Diptera, Ceratopogonidae) and in some other arthropods.   Parasitology, 13: 97-113.

Patently infected individuals can be recognized by their milky white appearance.   In early infections the parasite is localized in the fat body and nerve ganglia.   Later all stages of H. parasiticum are found free in the body cavity.   Infection incidence is greatest during dry weather.

Keilin, D. (1927) Fauna of a horse-chestnut tree (Aesculus hippocastanum), dipterous larvae and their parasites. Parasitology, 19: 368-374.

  Keilin mainly reviews the parasites he has previously named. Also noted is a micro-sporidian, probably a Glugea, which invades the fat body and salivary glands of Dasyhelea larvae.

Kettle, D. S. & Lawson, J. W. H. (1952). The early stages of British biting midges Culicoides Latreille (Diptera: Ceratopogonidae) and allied genera, Bull. Entomol. Res. 43: 421-467.

  Ectoparasitic cysts were observed on the head and neck of C. cubitalis larvae.

Kremer, M., Vermeil, C. & Callot, J. (1961). Sur quelques Nematoceres vulnerants des eaux salees continentales de l'est de la France. Bull. Assoc. Philomathique d'Alsace et de Lorraine, 11: 1-7.

  Culicoides salinarius was found parasitized by a large number of small flagellates which resemble Strigomonas from Culex.

Kudo, R. (1924). A biologic and taxonomic study of the Microsporidia. Illinois Biol Monogr. 9: 1-268.

  Diagnoses are given of Spironema octospora and Toxonema vibrio which infect Ceratopogon sp.

Laird, M. (1971). A bibliography on diseases and enemies of medically important arthropods 1963-67 with some earlier titles omitted from Jenkins' 1964 list. Appendix 7, pp. 751-790. In Burges, H. D. & Hussey, N. W. (eds) Microbial Control of Insects and Mites. Academic Press, London, New York.

Lawson, J. W. H. (1951). The anatomy and morphology of the early stages of Culicoides nubeculosus Meigen (Diptera: Ceratopogonidae = Heleidae). Trans. R. Entomol. Soc. Lond., 102: 511-570.

  Bacterial symbionts were observed in the larval fat bodies.

Léger, L. (1900). Sur un nouveau Sporozoaire des larves de diptères. C. R. Hebd. Seances Acad. Sci. (Paris), 131: 722-724.

  A new protozoan, Schizocystis gregarinoides, is described from Ceratopogon larvae collected in Lake Luitel. It has the general characteristics of gregarines, but under-goes schizogony in the interior of the host.

Léger, L. (1906). Etude sur Taeniocystis mira Léger, Grégarine métamérique. Arch. Protistenkd., 7: 307-329.

  T. mira lives as a parasite in the midgut of Ceratopogon solstitialis larvae. About 20% of the larvae collected near Var, France were infected with this parasite and about 50% with Schizocystis gregarinoides.

Léger, L. & Hesse, E. (1922) Microsporidies bactériformes et essai de systématique du groupe. C. R. Hebd. Seances Acad. Sci. (Paris), 174: 327-330.

  Spirogluea (Spironema) octospora and Toxoglugea (Toxonema) vibrio are described as new genera and species. They were observed parasitizing the fatty tissue of Ceratopogon sp. larvae collected in Montessaux. Their effect on the host is not given.

Lewis, D. J. (1958).   Some observations on Ceratopogonidae and Simuliidae (Diptera) in Jamaica.   Ann. Mag. Nat. Hist. Ser., 13, 1: 721-732.

   The duct of the crop of a female Culicoides furens was packed with fungal hyphae. They appeared to be growing toward the esophagus.

Manier, J. F., Rioux, J. A. & Whisler, H. C. (1961).   Rubetella inopinata n. sp. et Carouxella scalaris n. g., n. sp., Trichomycètes parasites de Dasyhelea lithotelmatica Strenzke 1951 (Diptera: Ceratopogonidae). Nat. Monspeliensia, Ser. Bot., 13: 25-38.

   Two new species of Trichomycetes are described from the hind gut of D. lithotelmatica.

Mayer, K. (1934).   Die Metamorphose der Ceratopogonidae (Dipt.). Ein Beitrag zur Morphologie, Systematik, Ökologie und Biologie der Jugenstadien dieser Dipterenfamilie.   Arch. Naturgesh. (n.s.) 3: 205-288.

   Mayer first reviews the parasites observed by Keilin & Buchner.   He also reports observing a Laboulbeniales on Forcipomyia.   A large number of epiphytes and epizoans on midges are listed.

Megahed, M. M. (1956).   Anatomy and histology of the alimentary tract of the female of the biting midge Culicoides nubeculosus Meigen (Diptera: Heleidae = Ceratopogonidae). Parasitology, 46: 22-47.

   The lumen of the esophageal diverticulum of some females contained fungal hyphae. Megahed speculates that the spores were ingested with raisin sap and developed in the lumen.

Mirsaeva, A. G. (1971).   /Parasitism of nematodes of the superfamily Mermithoidea in larvae of the genus Culicoides./   Parazitologiya, 5: 455-457. (R.e.).

   Culicoides larvae infected with mermithid nematodes were collected from breeding sites in the Buryat, USSR.   The maximum incidence of the parasite was 66.6%.

Nickle, W. R. (1969).   Corethrellonema grandispiculosum n. gen., n. sp. and Aproctonema chapmani n. sp. (Nematoda: Tetradonematidae), parasites of the Dipterous insect genera, Corethrella and Culicoides in Louisiana.   J. Nematology, 1: 49-54.

   A. chapmani is described as a new species from Culicoides arboricola collected in Louisiana.   All stages including adult occur in the larval insect.   Death of the host occurs when the female nematode, replete with eggs, exits the host's body.

Rubtsov, I. A. (1967).   /A new species of Agamomermis from a biting midge/.   Parazitologiya, 1: 441-443. (R.e.).

   A new species, A. heleis, is described from a Culicoides pulicaris adult collected in Kazakhstan.   Only one female specimen was examined.   Differences between the new species and A. culicis are noted.

Rubtsov, I. A. (1970).   /New species and genera of mermithids from Mokresov.   New and poorly known species of the fauna of Siberia/.   3(CO) Nauka Novosibersk pp. 94-101 (R).

   A new genus of mermithid nematodes is described, Heleidomermis.   The type species is H. vivipara which infects Culicoides stigma and C. nubeculosus.

Rubtsov, I. A. (1972).   Aquatic Mermithids.   Vol. 1, Akad. Nauk S.S.S.R. Zool. Inst., Leningrad, 253 pp. (R).

  Heleidomermis is redescribed.   Six specimens were used to define this genus.

Sen. P. & Das Gupta, S.K. (1958).   Mermis (Nematode) as internal parasite of Culicoides alatus (Ceratopogonidae).   Bull. Calcutta Sch. Trop. Med., 6: 15.

Service, M. W. (1974).   Further results of catches of Culicoides (Diptera: Ceratopogonidae) and mosquitoes from suction traps, J. Med. Entomol., 11: 471-479.

  Service reports trapping 5 Culicoides adults parasitized with mermithid nematodes.

Sharp, N. A. D. (1928).   Filaria perstans: its development in Culicoides austeni, Trans. R. Soc. Trop. Med. Hyg., 21: 371-396.

  Ciliates and flagellates of the leptomonas type were observed infesting C. austeni adults. The incidence of these parasites was apparently low.

Smith, W. W. (1966).   Mermithid-induced intersexuality in Culicoides stellifer (Coquillett). Mosq. News, 26: 442-443.

  An intersexual gynandromorph-like specimen of Culicoides stellifer was collected in Florida.   It was parasitized by one or two mermithid nematodes.   The rate of natural parasitism is apparently very low.

Smith, W. W. & Perry, V. G. (1967).   Intersexes in Culicoides spp. caused by mermithid parasitism in Florida, J. Econ. Entomol., 60: 1026-1027.

  Parasitism by the mermithids resulted in genetically intended males being altered morphologically to resemble anterior-posterior type gynandromorphs while females are not noticeably changed.   In some areas incidence of parasitism varied from 30-90%.

Steinhaus, E. A. (1946).   Insect Microbiology, an account of the microbes associated with insects and ticks with special reference to the biological relationships involved, Comstock Pub. Co., Ithaca, N.Y., 763 pp.

Tuzet, O. & Ormières, R. (1964). Stylocystis riouxi n. sp. Grégarine parasite de Dasyhelea lithotelmatica Strenzke 1951 (Diptera Ceratopogonidae)., Bull. Soc. Zool. France 89: 163-166.

  A new species of gregarine is described from the midge D. lithotelmatica collected near Sète, France.   It is compared with S. praecox, a parasite of Tanypus (Dipt.).

Weiser, J. (1957).   /Parasites of some blood sucking insects/.   Cesk. Parasitol., 4: 355-358.   (Cz,g).

  Nosema sphaeromiadis is described as a new species.   It was found infecting the fat body of a larva Sphaeromias sp.

Weiser, J. (1961).   Die Microsporidien als Parasiten der Insekten, Monogr. Angew. Entomol., 17: 1-149.

  Nosema sphaeromiadis, Toxoglugea (Toxonema) vibrio, and Spiroglugea (Spironema) octospora, parasites of Ceratopogonids, are redescribed.

212

Weiser, J. (1963).   Diseases of insects of medical importance in Europe.   <u>Bull. Wld Hlth</u>
<u>Org.</u>, <u>28</u>: 121-127.

The protozoan and nematode parasites of the Ceratopogonidae are briefly reviewed.

Whitsel, R. H. (1965).   A new distribution record and an incidence of mermithid nematode
parasitism for <u>Leptoconops kerteszi</u> Kieffer (Diptera: Ceratopogonidae).   <u>Mosq. News</u>, <u>25</u>: 66-67

A single female from a collection of 10 <u>L. kerteszi</u> adults was parasitized by an immature
mermithid.   The collection was made in Baja California, Mexico.

X. PATHOGENS OF SIMULIIDAE (BLACKFLIES) [a]

Mary Ann Strand

Boyce Thompson Institute
Yonkers, NY 10701, USA

and

C. H. Bailey & Marshall Laird

Research Unit on Vector Pathology
Memorial University of Newfoundland
St. John's, Newfoundland, Canada A1C 5S7

---

[a] The literature for this table and bibliography covers the period 1962–1975.

PATHOGENS OF SIMULIIDAE (BLACKFLIES)

Host	Host stage infected	Pathogen	% Incidence	Locality	Lab. or field study	Reference
VIRUSES						
Cnephia mutata	Larvae	Cytoplasmic polyhedrosis	1-22	Canada	Field	Bailey et al. (1975)
Prosimulium mixtum	Larvae	"	<0.1	Canada	Field	"
Simulium ornatum	Larvae	Iridescent virus	rare	Czechoslovakia	Field, lab.	Weiser (1968)
BACTERIA						
Simulium damnosum	Larvae	Azotobacteraceae	--	Ghana	Field	Burton et al. (1973)
"	Larvae	Bacillaceae	--	Ghana	Field	"
"	Larvae	Micrococcaceae	--	Ghana	Field	"
"	Larvae	Pseudomonadaceae	--	Ghana	Field	"
FUNGI						
Australosimulium laticorne	Larvae	Harpella melusinae	--	New Zealand	Field	Crosby (1974)
Australosimulium longicorne	Larvae	H. melusinae	--	New Zealand	Field	"
Australosimulium multicorne	Larvae	H. melusinae	--	New Zealand	Field	"
Australosimulium stewartense	Larvae	H. melusinae	--	New Zealand	Field	"
Australosimulium tillyardianum	Larvae	H. melusinae	--	New Zealand	Field	"
Australosimulium ungulatum	Larvae	H. melusinae	--	New Zealand	Field	"
Australosimulium vexans	Larvae	H. melusinae	--	New Zealand	Field	Crosby (1974)
Australosimulium spp.	Larvae	Genistellaceae	--	New Zealand	Field	"
Cnephia mutata	Larvae	H. melusinae	--	Canada	Field	Frost & Manier (1971)

PATHOGENS OF SIMULIIDAE (BLACKFLIES) (continued)

Host	Host stage infected	Pathogen	% Incidence	Locality	Lab. or field study	Reference
Prosimulium sp.	Larvae	H. melusinae	--	Canada	Field	Frost & Manier (1971)
	Larvae	Penella hovassi	--	Canada	Field	" " "
Simulium argus	Larvae	Smittium simulii	--	USA	Field, lab.	Lichwardt (1964)
Simulium argyreatum	Larvae	Amoebidium sp.	--	France	Field	Manier (1964)
Simulium aureum	Larvae	H. melusinae	--	France	Field	" "
"	Larvae	S. simulii	--	France	Field	" "
Simulium bezzii	Larvae	H. melusinae	--	France	Field	Manier (1969)
"	Larvae	S. simulii	--	France	Field	" "
"	Larvae	Stipella vigilans	--	France	Field	" "
Simulium damnosum	Larvae	Aspergillus sp.	0.3	Cameroons	Field	Lewis (1965)
"	Larvae	Entomophthora sp.	2.8	Cameroons	Field	" "
"	Adults	Phycomycete	--	Liberia	Field	Briggs (1970, 1973)
"	Larvae	Fungus?	7.6	Cameroons	Field	Lewis (1965)
"	Adults	Fungus?	--	Liberia	Field	Briggs (1970, 1973)
Simulium equinum	Larvae	H. melusinae	75	England	Field	Moss (1970)
"	Larvae	H. melusinae	--	France	Field	Manier (1969)
"	Larvae	Paramoebidium chattoni	40	England	Field	Moss (1970)
"	Larvae		--	France	Field	Manier (1969)
"	Larvae	Stipella sp.	20	England	Field	Moss (1970)
"	Larvae	Smittium sp.	25	England	Field	Moss (1970)
Simulium euryadminiculum	Larvae	H. melusinae	--	Canada	Field	Frost & Manier (1971)
Simulium fasciatum	Larvae	Amoebidium sp.	--	France	Field	Manier (1971)
"	Larvae	H. melusinae	--	France	Field	" "
Simulium monticola	Larvae	Penella hovassi	--	France	Field	Frost & Manier (1971), Manier (1969)
Simulium ornatum	Larvae	H. melusinae	--	France	Field	Manier (1969)
"	Larvae	P. chattoni	--	France	Field	" "
"	Larvae	S. simulii	--	France	Field	" "
"	Larvae	Stipella vigilans	--	France	Field	" "

PATHOGENS OF SIMULIIDAE (BLACKFLIES) (continued)

Host	Host stage infected	Pathogen	% Incidence	Locality	Lab. or field study	Reference
Simulium o. nitidifrons	Pupae	Coelomomyces sp.	--	Morocco	Field	Briggs (1967)
Simulium rostratum	Larvae	Fungus?	rare	USSR	Field	Bobrova (1971)
Simulium variegatum	Larvae	H. melusinae	--	France	Field	Manier (1969)
"	Larvae	S. vigilans	--	France	Field	Manier (1969)
"	Larvae	S. simulii	--	France	Field	"
Simulium venustum	Larvae	Coelomycidium simulii	2.2	Canada	Field	Ezenwa (1974d)
Simulium vittatum	Larvae	C. simulii	1.4	Canada	Field	"
"	Larvae	H. melusinae	--	Canada	Field	Frost & Manier (1971)
"	Larvae	P. hovassi	--	Canada	Field	" " "
"	Larvae	Fungus?	--	USA	Field	Frost & Manier (1971)
Simulium sp.	Larvae	Coelomycidium simulii	--	--	Lab.	Maurand & Manier (1968), Loubes & Manier (1970)
Simuliidae	Larvae, pupae, adults	C. simulii	freq.	USSR	Field	Rubtsov (1969)
PROTOZOA						
Boophthora erythrocephala	Larvae	Pleistophora simulii	--	USSR	Field	Rubtsov (1966b)
Cnephia mutata	Larvae	Caudospora brevicauda	3.7	Canada	Field	Ezenwa (1974c)
			3.0	Canada	Field	Frost & Nolan (1972)
			--	USA	Field	Jamnback (1970)
Simulium (Eusimulium) latipes	Larvae	Caudospora simulii	rare	France	Field, lab.	Vavra (1968)
			--	--	Lab.	Briggs (1972)

PATHOGENS OF SIMULIIDAE (BLACKFLIES) (continued)

Host	Host stage infected	Pathogen	% Incidence	Locality	Lab. or field study	Reference
Simulium (Eusimulium) latipes	Larvae	P. debaiseuxi	--	USSR	Field	Rubtsov (1966b)
"	Larvae	P. simulii	0.6	Canada	Field	Ezenwa (1974c, d)
"	Larvae	Thelohania bracteata	1.2	Canada	Field	"
"	Larvae	T. fibrata	--	USSR	Field	Rubtsov (1966b)
"	Larvae	T. varians	--	USSR	Field	"
Eusimulium sp.	Larvae	T. bracteata	--	USSR	Lab.	Briggs (1972)
"	Larvae	T. fibrata	--	Canada	Field	Steinhaus & Marsh (1962)
Gymnopais sp.	Larvae	Weiseria sommeranae	--	USA	Field	Jamnback (1970)
Odagmia caucasica	Larvae	Stempellia rubtsovi	--	USSR	Field	Issi (1966)
Odagmia ornata	Larvae	T. varians	--	USSR	Field	Rubtsov (1966b)
"	Adults	Microsporida	10, 2.1	USSR	Field	Shipitsina (1963)
Prosimulium alpestre	Larvae	C. alaskansis	--	USA	Field	Jamnback (1970)
Prosimulium flaveantennus	Larvae	Thelohania sp.	--	USA	Field	"
Prosimulium fuscum	Larvae	C. simulii	2.5	Canada	Field	Frost & Nolan (1970)
Prosimulium hirtipes	Larvae	Microsporida	max. 95	USA	Field	Anderson & Dicke (1960)
Prosimulium inflatum	Larvae	Weiseria laurenti	--	France	Field	Doby & Saquez (1964)
Prosimulium magnum	Larvae	C. pennsylvanica	--	USA	Field	Beaudoin & Wills (1965)
Prosimulium mixtum	Larvae	Caudospora simulii	rare	Canada	Field	Frost & Nolan (1970)

PATHOGENS OF SIMULIIDAE (BLACKFLIES) (continued)

Host	Host stage infected	Pathogen	% Incidence	Locality	Lab. or field study	Reference
Prosimulium mixtum/fuscum	Larvae	C. simulii	2.3	Canada	Field	Ezenwa (1974d)
			1/4	Canada	Field	Frost (1970)
Prosimulium sp.	Larvae	protozoan	--	USA	Field	Briggs (1969)
Simulium adersi	Larvae	Thelohania sp.	--	Ghana	Field	"
"	Larvae	C. nasiae	--	Ghana	Field	Jamback (1970)
Simulium arakawae	Larvae	Microsporida	--	Japan	Field	Steinhaus & Marsh (1962)
Simulium arcticum	Larvae	Thelohania sp.	--	USA	Field	"
Simulium bezzii	Larvae	Pleistophora simulii	--	France	Field	Maurand (1967)
"	Larvae	Stempellia simulii	--	France	Field	"
Simulium canadensis	Larvae	Thelohania sp.	--	USA	Field	Steinhaus & Marsh (1962)
Simulium corbis	Larvae	P. simulii	0.9	Canada	Field	Ezenwa (1974d)
"	Larvae	P. multispora	3.8	Canada	Field	"
Simulium damnosum	Larvae	Haplosporidia	0.3	Cameroons	Field	Lewis (1965)
"	Larvae	Ciliate	1.0	Cameroons	Field	"
"	Adults	Tetrahymena-like ciliate	rare	Ghana	Field	Marr & Lewis (1964)
				Ivory Coast	Lab.	Berl et al. personal communication, 1977
"	Larvae	Microsporida	--	Tanzania	Field	Briggs (1968)
Simulium decorum	Larvae	P. multispora	(14/497)	Canada	Field	Ezenwa (1973)
"	Larvae	T. bracteata	(4/402)	Canada	Field	"
Simulium equinum	Larvae	P. simulii	--	France	Field	Maurand & Manier (1967)
"	Larvae	T. bracteata	--	France	Field	" "
Simulium monticola	Larvae	T. bracteata	--	France	Field	Maurand (1967)
"	Larvae	P. simulii	--	France	Field	"
Simulium morsitans	Larvae	P. simulii	--	France	Field	Rubtsov (1966b)
"	Larvae	P. multispora	--	USSR	Lab.	Briggs (1972)
Simulium ornatum	Larvae	P. leasi	--	England	Field	Gassouma (1972)
"	Larvae	P. simulii	--	France	Field	Maurand (1967)
"	Larvae	P. tillingbournei	--	England	Field	Gassouma (1972)
"	Larvae	P. tillingbournei	--	England	Lab.	Gassouma & Ellis (1973)

PATHOGENS OF SIMULIIDAE (BLACKFLIES) (continued)

Host	Host stage infected	Pathogen	% Incidence	Locality	Lab. or field study	Reference
Simulium ornatum	Larvae	T. avacuolata	--	England	Field	Gassouma (1972)
"	Larvae	Thelohania bertrami	--	England	Field	Gassouma (1972)
"	Larvae	T. bracteata	--	France	Field	Maurand (1967)
"	Larvae	T. canningi	--	England	Field	Gassouma (1972)
"	Larvae	T. canningi	--	England	Lab.	Gassouma & Ellis (1973)
"	Larvae	T. minuta	--	England	Field	Gassouma (1972)
"	Larvae	T. minuta	--	England	Lab.	Gassouma & Ellis (1973)
"	Larvae	T. simulii	--	England	Field	Gassouma (1972)
"	Larvae	Thelohania sp.	--	England	Lab.	Briggs (1968)
Simulium piperi	Larvae	Thelohania sp.	--	USA	Lab.	Steinhaus & Marsh (1962)
Simulium reptans	Adults	Microsporida	11, 2.4	USSR	Field	Shipitsina (1963)
Simulium taylori	Larvae	Protozoa	--	Tanzania	Field	Briggs (1968)
Simulium tuberosum	Larvae	P. simulii	0.5	Canada	Field	Ezenwa (1974d)
Simulium variegatum	Larvae	T. bracteata	--	France	Field	Maurand (1967)
Simulium venustum	Larvae	Haplosporidium simulii	--	USA	Field	Beaudoin & Wills (1968)
"	Larvae	P. multispora	3.4	Canada	Field	Ezenwa (1974d)
"	Larvae	P. multispora	9.5	Canada	Field	Ezenwa (1973)
"	Larvae	P. simulii	1.6	Canada	Field	Ezenwa (1974d)
"	Larvae	P. simulii	(7/11)	Canada	Field	Frost (1970)
"	Larvae	P. simulii	(13/726)	Canada	Field	Ezenwa (1973)
"	Larvae	P. simulii	--	France	Field	Maurand (1967)
"	Larvae	P. simulii	0.5	Canada	Field	Ezenwa (1974d)
"	Larvae	T. bracteata	--	--	Lab.	Liu et al. (1971), Liu & Davies (1972a, b, c, d)
"	Larvae	T. fibrata	--	Canada	Field	Steinhaus & Marsh (1962)
Simulium sp.	Larvae	Nosema stricklandi	--	--	Lab.	Briggs (1972)
"	Larvae	P. simulii	--	--	Lab.	"
"	Larvae	T. fibrata	--	--	Lab.	"
"	Larvae	T. fibrata	--	--	Lab.	Maurand & Boriux (1969)
"	Larvae	P. debaisieuxi	--	--	Lab.	Maurand & Manier (1968)

PATHOGENS OF SIMULIIDAE (BLACKFLIES) (continued)

Host	Host stage infected	Pathogen	% Incidence	Locality	Lab. or field study	Reference
Simulium sp.	Larvae	P. simulii	--	--	Lab.	Manier & Maurand (1966), Maurand (1966), Maurand & Manier (1968)
"	Larvae	Stempellia simulii	--	--	Lab.	Maurand & Manier (1968)
"	Larvae	T. bracteata	--	--	Lab.	Manier & Maurand (1966), Maurand (1966), Maurand & Manier (1968)
Tetanopteryx maculata	--	Microsporida	10, 22.2	USSR	Field	Shipitsina (1963)
Simuliidae	Larvae	Thelohania sp.	Freq. in 1 location	Scotland	Field	Maitland & Penney (1967)
"	Larvae, pupae	Microsporida	sporadic	USA	Field	Thomas & Poinar (1973)
"				USA	Field	Anderson & Dicke (1960)
"	Larvae	Caudospora simulii	3.5	USSR	Field	Bobrova (1971)
"	Larvae		--	USA	Field	Briggs (1966)
"	Larvae	Protozoan	--	Scotland	Field	Maitland & Penney (1967)
NEMATODES						
Boophthora erythrocephala	Larvae	Aproctonema sumuliophaga	--	USSR	Field	Rubtsov (1966c)
"	Pupae, adults	Atractonema sp.	--	USSR	Field	Rubtsov (1963)
"	Larvae	Gastromermis boophthorae	1-10	USSR	Field	Rubtsov (1967a)
"	Larvae	G. crassifrons	--	USSR	Field	Rubtsov (1967a)
"	Larvae	G. virescens actipenis	--	USSR	Field	Rubtsov (1967a)
"	Larvae	" virescens	--	USSR	Field	Rubtsov (1967a)
"	Larvae	Isomermis rossica	--	USSR	Field	Rubtsov (1967b)
"	Adults	Mermithonema acicularis	--	USSR	Field	Rubtsov (1966c)
"	Adults	Mermithonema brevis	--	USSR	Field	Rubtsov (1966c)
"	--	Tetradomermis angusta	--	USSR	Field	" "
"	Adults	T. decima	--	USSR	Field	" "
"	Larvae, pupae, adults	T. heterocella	--	USSR	Field	" "

PATHOGENS OF SIMULIIDAE (BLACKFLIES) (continued)

Host	Host stage infected	Pathogen	% Incidence	Locality	Lab. or field study	Reference
Boophthora erythrocephala	Larvae, pupae, adults	T. isocella	--	USSR	Field	Rubtsov (1966c)
"	Larvae	T. longistoma	--	USSR	Field	"
"	Larvae, adults	T. polycella	--	USSR	Field	"
"	Adults	T. varicella	--	USSR	Field	"
"	Pupae, adults	Tetradonema sp.	--	USSR	Field	"
Cnephia mutata	Larvae	Gastromermis sp.	--	USA	Field	Anderson & DeFoliart (1962)
"	Larvae	Isomermis sp.	--	USA	Field	"
"	Larvae	Neomesomermis flumenalis	--	Canada	Field	Ezenwa (1973)
"	Larvae	mermithid	--	USA	Field	Anderson & DeFoliart (1962)
Cnephia emergens	Larvae	Isomermis sp.	(14/16)	USA	Field	
Simulium (Eusimulium) aurem	Larvae	Gastromermis boophthorae	1-10	USSR	Field	Welch & Rubtsov (1965)
"	Larvae	G. clingogaster	--	USSR	Field	Rubtsov (1967a)
"	Larvae	mermithid	(1/1)	USA	Field	Anderson & DeFoliart (1962)
Eusimulium cryophilium	Larvae	G. boophthorae	1-10	USSR	Field	Welch & Rubtsov (1965)
"	Larvae	I. rossica	--	USSR	Field	Rubtsov (1968)
"	Larvae	Limnomermis cryophili	40-50	USSR	Field	Rubtsov (1967b)
"	Larvae	L. macronuclei	40-50	USSR	Field	"
Simulium (Eusimulium) latipes	Larvae	Isomermis sp.	(1/2)	USA	Field	Anderson & DeFoliart (1962)
"	Larvae	L. cryophili	40-50	USSR	Field	Rubtsov (1967b)
"	Larvae	L. macronuclei	40-50	USSR	Field	Jamnback (1970)
"	Larvae	Neomesomermis flumenalis	0.6	Canada	Field	Ezenwa (1974d)
"	Larvae	mermithid	6.5	USSR	Field	Rubtsov (1968)
"	Larvae	G. boophthorae	1-10	USSR	Field	Welch & Rubtsov (1965)
"	Larvae	I. rossica	--	USSR	Field	Rubtsov (1968)

PATHOGENS OF SIMULIIDAE (BLACKFLIES) (continued)

Host	Host stage infected	Pathogen	% Incidence	Locality	Lab. or field study	Reference
Eusimulium perteszi	Larvae	I. rossica	--	USSR	Field	Rubtsov (1968)
Eusimulium securiforme	Larvae	G. rosalbus	50-80	USSR	Field	Rubtsov (1967a)
Odagmia ornata	Larvae	G. boophthorae	1	USSR	Field	Welch & Rubtsov (1965)
"	Larvae	G. odagmiae	--	USSR	Field	Rubtsov (1967a)
"	Adults	nematode?	1.1	USSR	Field	Shipitsina (1963)
Prosimulium alpestre	Larvae	mermithid	4.6	USSR	Field	Bobrova (1971)
Prosimulium demarticulata	Larvae	mermithid	(1/3)	USA	Field	Anderson & DeFoliart (1962)
Prosimulium mixtum/fuscum	Larvae	Neomesomermis flumenalis	--	Canada	Field	Ezenwa & Carter (1975)
"	Larvae	Isomermis wisconsinensis	--	Canada	Field	" "
Simulium arcticum	Larvae	mermithid	--	USA	Field	Steinhaus & March (1962)
Simulium argyreatum	Larvae	G. boophthorae	100 in 1 location	USSR	Field	Welch & Rubtsov (1965)
Simulium corbis	Larvae	G. viridis	0.5	Canada	Field	Ezenwa (1974d)
"	Larvae	Gastromermis sp.	(1/13)	USA	Field	Anderson & DeFoliart (1962)
"	Larvae	N. flumenalis	4.8	Canada	Field	Ezenwa (1974d)
			--	Canada	Field	Ezenwa (1973)
Simulium damnosum	Larvae, adults	Hydromermis sp.	--	Guinea	Field	Briggs (1968)
"	Larvae, adult	Isomermis tansanieisis	--	Tanzania	Field	"
"	Larvae, adult	Mesomermis ethiopica	--	Tanzania	Field	"
"	Larvae	mermithid	1.0	Cameroons	Field	Lewis (1965)
"	Adult	nematode	rare	Ghana	Field	Marr & Lewis (1964)

PATHOGENS OF SIMULIIDAE (BLACKFLIES) (continued)

Host	Host stage infected	Pathogen	% Incidence	Locality	Lab. or field study	Reference
Simulium (nolleri) decorum	Larvae	Gastromermis sp.	0.9	USA	Field	Anderson & DeFoliart (1962)
"	Larvae	Isomermis sp.	0.9	USA	Field	" " "
"	Larvae	mermithid?	0.9	USA	Field	" " "
"	--	Tetradomermis longicorpis	--	USSR	Field	Rubtsov (1966c)
"	Larvae	Neomesomermis flumenalis	--	Canada	Field	Ezenwa (1973)
Simulium jenningsi	Larvae	Gastromermis sp.	5.7	USA	Field	Anderson & DeFoliart (1962)
Simulium luggeri	Larvae	Hydromermis sp.	14	USA	Field	" " "
"	Larvae	Isomermis sp.	14.8	USA	Field	Anderson & Dicke (1960)
Simulium morsitans	Larvae	G. boophthorae	1-10	USSR	Field	Welch & Rubtsov (1965)
"	Larvae	G. crassicauda	--	USSR	Field	Rubtsov (1967a)
"	Larvae	G. longispicula	--	USSR	Field	" "
"	Larvae	I. rossica	--	USSR	Field	Rubtsov (1968)
"	Larvae	Limnomermis aculeata	--	USSR	Field	Rubtsov (1967b)
"	Larvae	L. lanceicapta	--	USSR	Field	Rubtsov (1967b)
"	Larvae	L. teniucauda	--	USSR	Field	" "
"	Larvae, pupae, adults	T. heterocella	--	USSR	Field	Rubtsov (1966c)
"	Larvae, pupae, adults	T. angusta	--	USSR	Field	" "
"	Larvae, pupae, adults	T. isocella	--	USSR	Field	" "
Simulium ornatum nitidfrons	Pupae	nematode	--	Morocco	Field	Briggs (1967)
Simulium reptans	Adults	nematode	1.3, 1.9	USSR	Field	Shipitsina (1963)
Simulium rostratum	Larvae	mermithid	5.4	USSR	Field	Bobrova (1971)

PATHOGENS OF SIMULIIDAE (BLACKFLIES) (continued)

Host	Host stage infected	Pathogen	% Incidence	Locality	Lab. or field study	Reference
Simulium tuberosum	Larvae	Isomermis sp.	9.4	USA	Field	Anderson & DeFoliart (1962)
"	Larvae	mermithid	9.4	USA	Field	"
Simulium venustum	Larvae	Gastromermis viridis	0.1	Canada	Field	Ezenwa (1974d)
"	Larvae	Isomermis wisconsinensis	0.4	Canada	Field	"
"	Larvae	Neomesomermis flumenalis	--	USA	Field	Phelps & DeFoliart (1964)
			--	Canada	Field	Welch (1962)
			4.5 (7/127)	Canada	Lab.	Bailey et al. (1974)
				Canada	Field	Ezenwa (1974d)
				Canada	Field	Ezenwa (1973)
"	Larvae	Gastromermis sp.	17.3	USA	Field	Anderson & DeFoliart (1962)
"	Larvae	Isomermis sp.	17.3	USA	Field	"
"	Larvae	mermithid?	17.3	USA	Field	"
Simulium verecundum	Larvae	G. boophthorae	1	USSR	Field	Welch & Rubtsov (1965)
"	Larvae	G. rosalbus	50-80	USSR	Field	Rubtsov (1967a)
"	Larvae	I. rossica	--	USSR	Field	Rubtsov (1968)
Simulium vittatum	Larvae	G. viridis	5-53	USA	Field	Anderson & DeFoliart (1962)
"	Pupae		0-44	USA	Field	Phelps & DeFoliart (1964)
"	Adults	G. viridis	5	USA	Field	Welch (1962)
"	Larvae		37-63	USA	Field	Phelps & DeFoliart (1964)
"	Larvae	I. wisconsinensis	0-52	USA	Field	"
"	Pupae		--	USA	Field	Anderson & DeFoliart (1962), Welch (1962)
"	Adults	Gastromermis sp.	5	USA	Field	Phelps & DeFoliart (1964)
"	Larvae	Isomermis	37-63	USA	Field	"
"	Larvae	mermithid?	--	USA	Field	Anderson & DeFoliart (1962)
"	Larvae		--	USA	Field	"
"	Larvae	Neoaplectana carpocapsae	--	USA	Field	"
"	Larvae			USA	Lab.	Webster (1973)
Simulium vulgare	Larvae	mermithid?	41.1	USSR	Field	Bobrova (1971)
Tetanopteryx maculata	--	nematodes	2.8	USSR	Field	Shipitsina (1963)

PATHOGENS OF SIMULIIDAE (BLACKFLIES) (continued)

Host	Host stage infected	Pathogen	% Incidence	Locality	Lab. or field study	Reference
Simuliidae	Larvae	Gastromermis ambianeniss	--	France	Field	Rubtsov & Doby (1970)
"	Larvae	Hydromermis angusta	--	France	Field	" " "
"	Larvae	Isomermis rossica	--	France	Field	" " "
"	Larvae	Mesomermis simuliae	--	France	Field	" " "
"	Larvae	Hydromermis sp.	0-90	USA	Field	Anderson & Dicke (1960)
"	Larvae	mermithid	50-60	USSR	Field	Rubtsov (1963)
"	Larvae	"	--	USA	Field	Briggs (1966)
"	Larvae	nematodes	--	Scotland	Field	Maitland & Penney (1967)
OTHERS						
Simulium exigum	Larvae	Lecithodendrudae	35	Venezuela	Field	Lewis & Wright (1962)
Simulium venustum	Adults	Leucocytozoon simondi	--	--	Lab.	Desser & Yang (1973)

ABSTRACTS

Mary Ann Strand

Anderson, J. R. & DeFoliart, G. R. (1962).   Nematode parasitism of blackfly (Diptera:
Simuliidae) larvae in Wisconsin.   Ann. Entomol. Soc. Am., 55: 542-546.

    The occurrence of mermithid nematodes was observed in 16 species of field-collected
    simuliid larvae from several locations in Wisconsin.   Adult nematodes were observed
    only from S. vittatum, so the species of nematodes could not be determined in most
    cases.   However, larval parasites found in the univoltine species probably differ from
    those found in the multivoltine flies.   In most cases the incidence of mermithid
    parasitism did not exceed 25%.   Laboratory-reared, infected S. vittatum larvae did not
    pupate and eventually died.

Anderson, J. R. & Dicke, R. J. (1960).   Ecology of the immature stages of some Wisconsin
blackflies (Simuliidae: Diptera).   Ann. Entomol. Soc. Am., 53: 386-404.

    Immature stages of 23 species of blackflies were examined.   Microsporida were common
    parasites of larvae, as was a nematode, Hydromermis sp. which parasitized up to 90% of
    the larvae in some streams.   Infected larvae failed to pupate and eventually died.

Bailey, C. H. et al. (1974).   Procedure for mass collection of mermithid postparasites
(Nematoda: Mermithidae) from larval blackflies (Diptera: Simuliidae).   Can. J. Zool., 52(5):
660-661.

    Field collected larval blackflies were kept in the laboratory until their nematode
    parasites (Neomesomermis flumenalis, emerged.   Methods of collection, transport, and
    maintenance of the blackfly larvae are described.

Bailey, C. H. et al. (1975).   A cytoplasmic polyhedrosis virus from the larval blackflies
Cnephia mutata and Prosimulium mixtum (Diptera: Simuliidae).   J. Invertebr. Pathol., 25:
273-274.

    C. mutata and P. mixtum larvae infected with CPV were collected in Newfoundland.
    Incidence in C. mutata reached 22% in one locality, but the virus was only rarely found
    in P. mixtum.   Patently infected larvae were recognized by the opaque chalky appearance
    of the midgut.   Inclusion bodies were seen within the epithelial cells of the midgut.

Beaudoin, R. & Wills, W. (1965).   A description of Caudospora pennsylvanica sp. n.
(Caudosporidae, Microsporidia), a parasite of the larvae of the blackfly (Prosimulium magnum
Dyar and Shannon.   J. Invertebr. Pathol., 7: 152-155.

    A new microsporidan parasite of larval blackflies was found in Pennsylvania.   It was
    compared to Caudospora simulii.

Beaudoin, R. L. & Wills, W. (1968).   Haplosporidium simulii sp. n. (Haplosporida: Haplo-
sporidiidae) parasitic in larvae of Simulium venestum Say.   J. Invertebr. Pathol., 10: 374-
378.

    H. simulii was found infecting larval blackflies in Pennsylvania.   Several diseased
    larvae were collected from a stream.   The pathogen is compared to H. tipulae from
    craneflies.   The spores of H. simulii were over 2x the size of those of H. tipulae.

Bobrova, S. I. (1971).   Concerning the parasites and predators of blackflies.   Izv. Sib. Otd.
Akad. Nauk. SSSR Ser. Biol-Med Nauk., 10(2): 172-173.

The incidence of microsporida and mermithid parasitism of simuliids is reported from Krasnoyarsk and Gorno-Altai areas in the USSR. In the Krasnoyarsk area, 3.5% of the blackflies were infected by microsporidia and in Altai 41% of <u>Simulium vulgare</u> larvae were parasitized by mermithids. Other species of blackflies were infected to a lesser extent. /from the abstract in R. A. E. (b) (1972) <u>60</u>: 2272./

Briand, L. J. & Welch, H. E. (1963). Use of entomophilic nematodes for insect pest control. <u>Phytoprotection</u>, <u>44</u>(1): 37-41.

Present evidence indicates that entomophilic nematodes can be manipulated for insect control and that mermithids offer much promise for controlling blackflies.

Briggs, J. D. (1966). 1965 activities of the WHO International Reference Centre for diagnosis of diseases of vectors. WHO/EBL/66.73. 8pp.

Briggs, J. D. (1967). 1966 activities of the WHO International Reference Centre for diagnosis of diseases of vectors. WHO/VBC/67.8. 8pp.

Briggs, J. D. (1968). 1967 activities of the WHO International Reference Centre for diagnosis of diseases of vectors. WHO/VBC/68.97. 7pp.

Briggs, J. D. (1969). 1968 activities of the WHO International Reference Centre for diagnosis of diseases of vectors. WHO/VBC/69.171. 6pp.

Briggs, J. D. (1970). 1969 activities of the WHO International Reference Centre for diagnosis of diseases of vectors. WHO/VBC/70.250. 13pp.

Briggs, J. D. (1973). 1970 activities of the WHO International Reference Centre for diagnosis of diseases of vectors. WHO/VBC/73.422. 7pp.

Briggs, J. D. (1972). 1971 activities of the WHO International Reference Centre for diagnosis of diseases of vectors. WHO/VBC/72.408. 6pp.

Tentative and final determinations of pathogens of disease vectors including blackflies are given.

Burton, G. J. et al. (1973). Aerobic bacteria in the midgut of <u>Simulium damnosum</u> larvae. <u>Mosq. News</u>, <u>33</u>(1): 115-117.

Azotobacteraceae, Bacillaceae, Micrococcaceae, and Pseudomondaceae were cultured from midgut contents of 20 <u>Simulium damnosum</u> collected from a spillway in Ghana. Aerobic bacterial colonies ranged from 89×10^3 to 4×10^6 per midgut and fungal colonies from 1×10^3 to 7×10^3.

Chapman, H. C. (1973). Assessment of the potential of some pathogens and parasites of biting flies. D.71-77 In Biting Fly Control and Environmental Quality. Canada Defence Research Board, DR 217, Ottawa.

A summary of published incidence of insect pathogens in Canadian species of blackflies is given.

Crosby, T. K. (1974). Trichomycetes (Harpellales) of New Zealand <u>Austrosimulium</u> larvae (Diptera: Simuliidae). <u>J. Nat. Hist.</u>, <u>8</u>: 187-192.

<u>Harpella melusinae</u>, <u>Smittium</u> sp. and an unidentified genistellacean were found for the first time in the Southern Hemisphere in <u>Austrosimulium</u>. <u>H. melusinae</u> is abundant in

228

A. tillyardianum larvae and, where larval densities are high, 90-100% of the larvae may be infected. Infections have no apparent effect on the host larvae.

Desser, S. S. & Yang, Y. J. (1973). Sporogony of Leucocytozoon spp. in mammalophilic simuliids. Can J. Zool., 51(7): 793.

Simulium venustum became infected with Leucocytozoon simondi when force-fed on infected ducklings. Most died within 24 hours. Normal sporogonic stages were observed in the infected flies.

Doby, J. M. & Saguez, F. (1964). Weiseria, genre nouveau de Microsporidies et Weiseria laurenti n. sp., parasite de larves de Prosimulium inflatum Davies, 1957 (Diptères: Paranématocères). C. R. Hebd. Seances Acad. Sci. Ser. D. Sci. Nat. (Paris) 259: 3614-3617.

A new genus and species of microsporidan parasitizing P. inflatum are proposed. The infection is localized in the fat body of the larval blackflies. The new genus is characterized by the presence of external ornamentation on the posterior portion of the mature spore.

Doby, J. M. et al. (1965). Complément a l'étude de la morphologie et du cycle évolutif de Caudospora simulii Weiser 1947. Bull. Soc. Zool. Fr., 90: 393-399.

Observations of pansporoblasts suggest a closer relationship of Caudospora with Thelohania than with Pleistophora, contrary to the belief at the time of the original description.

Ebsary, B. A. & Bennett, G. F. (1973). Molting and oviposition of Neomesomermis flumenalis (Welch, 1962) Nickle, 1972, a mermithid parasite of blackflies. Can. J. Zool., 51: 637-639.

Postparasitic juveniles showed optimum development when reared at 12°C and 24°C was lethal. Mating began from a few hours to 14 days after molting to the adult stage. Most individuals mated only once. Oviposition occurred 36-59 days post mating. Most eggs, held at 12°C, hatched 39-46 days after they were laid.

Ebsary, B. A. & Bennett, G. F. (1974). Redescription of Neomesomermis flumenalis (Nematoda) from blackflies in Newfoundland. Can. J. Zool., 52(1): 65-68.

The egg and preparasitic stage of this nematode had not been previously encountered, so an amended description is given. The measurements of the adult stage given by various authors are compared.

Ezenwa, A. O. (1973). Mermithid and microsporidan parasitism of blackflies (Diptera: Simuliidae) in the vicinity of Churchill Falls, Labrador. Can. J. Zool., 51(10): 1109-1111.

Larvae from seven species of blackflies were collected from five streams. Four of these species were infected by one or more parasite species. Neomesomermis flumenalis infected 11.2%, Pleistophora multispora, Thelohania bracteata, and P. simulii infected 4.9%.

Ezenwa, A. O. (1974a). Ecology of free-living stages of mermithid parasites of blackflies. (Abstr.) Proc. 3rd Int. Congr. Parasitol. (Munich) 2: 862.

see Ezenwa, 1974c.

Ezenwa, A. O. (1974b).   Field observations on trends of mermithid and microsporidan parasitism of Simuliidae.   (Abstr.) Proc. 3rd Int. Congr. Parasitol. (Munich), 2: 927-928.

   see Ezenwa, 1974c.

Ezenwa, A. O. (1974c).   Ecology of Simuliidae, Mermithidae, and Microsporida in Newfoundland freshwaters.   Can. J. Zool., 52(5): 557-565.

   Seasonal succession of simuliid species and their parasites is gradual.   Simulium appears at the decline of Prosimulium and Cnephia.   The eggs of Prosimulium and Cnephia hatched synchronously with the eggs of Neomesomermis flumenalis.   Eggs of the first generation of Simulium species hatched at the same time as Gastromermis viridis and Isomermis wisconsinensis and the over-wintering eggs of N. flumenalis.   Caudospora simulii and C. brevicauda occur in Prosimulium and Cnephia during cold seasons. Thelohania bracteata, Pleistophora simulii, P. multispora, and Coelomycidium simulii occur during warm seasons.   The influence of stream temperature, oxygen concentration, and stream velocity on seasonal occurrence of the blackflies and their parasites is also discussed.

Ezenwa, A. O. (1974d).   Studies on host-parasite relationships of Simuliidae with mermithids and microsporidans.   J. Parasitol., 60(5): 809-813.

   Blackfly larvae from a number of streams in Newfoundland were sampled at regular intervals and their internal parasites were identified.   Nine pathogens distributed among 11 species of blackflies were found.   A table shows the incidence and host range of the pathogens.   Overall 9.42% of the flies were infected, 5.02% by mermithids, and the remaining percentage by microsporidans and fungi (Coelomycidium simulii).   The nematode Neomesomermis flumenalis was the most common pathogen and it occurred in four species of flies.

Ezenwa, A. O. & Carter, N. E. (1975).   Influence of multiple infections on sex ratios of mermithid parasites of blackflies.   Environmental Entomol., 4: 142-144.

   The ratio of males to females of Neomesomermis flumenalis and Isomermis wisconsinensis is positively correlated to the number of nematodes per host.   When one nematode per host was observed, 11.9% were male and 93.8% were male where there were four nematodes per host.   No specific external factor could be linked with observed rise in the proportion of males.

Frost, S. (1970).   Microsporidia (Protozoa: Microsporidia) in Newfoundland blackfly larvae (Diptera: Simuliidae).   Can. J. Zool., 48(4): 890-891.

   Of 125 larvae collected from 6 locations, 18 were infected with Pleistophora simulii and 2 with Caudospora simulii.

Frost, S. & Manier, J. F. (1971).   Notes on Trichomycetes (Harpellales: Harpellaceae and Genistellaceae) in larval blackflies (Diptera: Simuliidae) from Newfoundland.   Can. J. Zool., 49(5): 776-778.

   Penella hovassi and Harpella melusinae were found in the gut of blackfly larvae in Newfoundland.   Their morphology is described.

Frost, S. & Nolan, R. A. (1972).   The occurrence and morphology of Caudospora spp. (Protozoa: Microsporida) in Newfoundland and Labrador blackfly larvae (Diptera: Simuliidae). Can. J. Zool., 50(11): 1363-1366.

Blackfly larvae with conspicuously white abdomens were dissected. _Caudospora simulii_ was frequently found in _Prosimulium fuscum_ and _P. mixtum_ collected from streams in Newfoundland and Labrador. _C. brevicauda_ was observed in Newfoundland for the first time and a new spore form from _Cnephia mutata_ was found in Labrador. Collection data and spore measurements are given.

Gassouma, M. S. S. (1972). Microsporidan parasites of _Simulium ornatum_ Mg. in South England. _Parasitology, 65_(1): 27-45.

During 1967-9, _S. ornatum_ larvae were regularly collected from two rivers (Lea and Tilling Bourne) in South England. Five new species of _Thelohania_ and two _Pleistophora_ were named: _T. minuta, T. bertrami, T. canningi, T. simulii, T. avacuolata, P. tillingbournei,_ and _P. leasei._ Infection rates never exceeded 9% of the larvae.

Gassouma, M. S. S. & Ellis, D. S. (1973). The ultrastructure of sporogonic stages and spores of _Thelohania_ and _Plistophora_ (Microsporida, Nosematidae) from _Simulium ornatum_ larvae. _J. Gen. Microbiol., 74_: 33-43.

S. ornatum larvae parasitized by _P. tillingbournei_ and _T. minuta_ from the River Tilling Bourne in Surrey, England and other _S. ornatum_ larvae parasitized by _T. canningi_ from River Lea in Hertfordshire were used in this study. Sporonts were observed in the disintegrating cells of the host's fat body. _T. canningi_ spores are larger and have more polar filament coils than _T. minuta._ _P. tillingbournei_ spores were mid-way in size and had a more complex wall structure than the smooth walled _Thelohania_ species.

Gordon, R. et al. (1972). The potentialities of mermithid nematodes for the biocontrol of blackflies (Diptera: Simuliidae): a review. (WHO/VBC/72.396. 14pp.) _Exp. Parasitol. 33_(2): 226-238.

All three common genera (_Simulium, Prosimulium,_ and _Cnephia_) are susceptible to mermithid parasitism. Host specificity, mermithid distribution, and pathology of mermithid parasitism are reviewed. 88 references.

International Development Research Center (1972). Preventing onchocerciasis through blackfly control. A proposal for joint Afro-Canadian research into the feasibility of using mermithid parasites as biological control agents in the control of disease-transmitting blackflies. _Int. Dev. Res. Centre._ IDRC-006e. 11pp.

A collaborative project for investigating the feasibility of using mermithids as blackfly control agents in Africa is described.

Issi, I. V. (1966). _Stempellia rubtsovi_ sp. n. (Microsporidia, Nosematidae) a microsporidian parasite of _Odagmia caucasica_ larvae (Diptera, Simuliidae). _Acta Protozool., 6_: 345-352. (R, e)

A new species of microsporidan was found in larvae collected at Aksu Falls in Caucasus. The parasite attacks the salivary glands and causes the destruction of their cells. The new species is compared to other _Stempellia_ species and is the only one of those compared that is located in the salivary gland.

Jamnback, H. A. (1970). _Caudospora_ and _Weiseria,_ two genera of Microsporida parasitic in blackflies. _J. Invertebr. Pathol., 16_: 3-13.

Four new species of Microsporida parasitizing blackfly larvae in Alaska, New York, and Ghana are described.

Jamnback, H. A. (1973).   Recent developments in control of blackflies.   Ann. Rev. Entomol., 18: 281-304.

This is a review article which contains 144 references.

Kelly, D. C. & Robertson, J. S. (1973).   Icosahedral cytoplasmic deoxyriboviruses. J. Gen. Virol., 20(Suppl.): 17-41.

A comprehensive catalogue of ICDVs isolated from animals and plants is given.   The list includes one from blackflies.

Laird, M. et al. (1973).   Invertebrate pathology and the integrated control of blackflies. Abstr. Pap. 5th Int. Colloq. Insect Pathol. Microb. Control. (Oxford), p. 87.

This is a report of intended research by Memorial University of Newfoundland's Research Unit on Vector Pathology.

Lewis, D. J. (1965).   Features of Simulium damnosum population of the Kumba area in West Cameroon.   Ann. Trop. Med. Parasitol., 59: 365-374.

Fungal hyphae and cyst-like bodies probably of fungal origin were observed infecting nulliparous blackflies.   Some could not be identified, others appeared to be related to Entomophthora and Aspergillus.   Parasites resembling Haplosporida were seen in the ovaries of 0.3% of nullipars and ciliates were seen in the abdominal cavity of 1%. Mermithids were found in 1% of nullipars.   Principal parasites of the nullipars were not found in parous flies, suggesting that the parasites killed their hosts.

Lewis, D. J. & Wright, C. A. (1962).   A trematode parasite of Simulium.   Nature, 193: 1311-1312.

Encysted, digenetic-trematode metacercariae were found in S. exiguum in Venezuela.   Adult flies were reared and examined for worms;   35% were found to contain the trematode. The worms were observed in the heads and abdomens of infected flies.   The parasite may be a member of the Lecithodendriidae family.

Lichwardt, R. W. (1964).   Axenic culture of two new species of branched Trichomycetes. Am. J. Bot., 51: 836-842.

Smittium simulii was isolated from the hind-gut of Simulium argus larvae.   It was grown on a potato dextrose-yeast extract medium.

Liu, T. P. et al. (1971).   Preliminary observations on the fine structure of the pansporoblast of Thelohania bracteata (Strickland, 1913) (Microsporidia, Nosematidae) as revealed by freeze-etching electron microscopy.   J. Protozool., 18(4): 592-596.

The only cytoplasmic structures seen in the pansporoblasts were fluid filled vesicles and elevations which were evenly distributed.

Liu, T. P. & Davies, D. M. (1972a).   Fine structure of frozen-etched spores of Thelohania bracteata emphasizing the formation of the polarplast.   Tissue and Cell, 4: 1-10.

Pansporoblast formation is traced during spore development.

Liu, T. P. & Davis, D. M. (1972b).   Organization of frozen-etched Thelohania bracteata (Strickland, 1913) (Microsporida, Nosematidae) emphasizing the fine structure of the posterior vacuole.   Parasitology, 64: 341-345.

A double membrane surrounding the posterior vacuole was observed. This organelle is considered important in providing pressure for sporoplasm extrusion.

Liu, T. P. & Davies, D. M. (1972c). Ultrastructure of the cytoplasm in fat-body cells of the blackfly, Simulium vittatum, with microsporidian infection; a freeze-etching study. J. Invertebr. Pathol., 19(2): 208-214.

The cytoplasm of cells infected with Thelohania bracteata and uninfected cells in the same fat body were compared. In infected cells, protein granules were concentrated around the parasites or within the pansporoblast wall.

Liu, T. P. & Davies, D. M. (1972d). Ultrastructure of the nuclear envelope from blackfly, fat-body cells with and without microsporidan infection. J. Invertebr. Pathol., 20(2): 176-182.

The frozen-etched envelope of the nucleus in Thelohania bracteata-infected, fat-body cells exhibited distinct features which differed greatly from those in uninfected cells. The diameter of nuclear pores was larger and their numbers were fewer in the infected cells.

Loubes, C. & Manier, J. F. (1970). Sur Coelomycidium simulii Debaisieux, phycomycete pathogene mortel pour les larves de simulies. (Abstr.) Proc. 3rd Int. Congr. Parasitol. (Munich), 2: 926-927.

Electronmicrography was used to observe the development of C. simulii. Zoospores were characterized by a nuclear cap.

Maitland, P. S. & Penney, M. M. (1967). The ecology of the Simuliidae in a Scottish river. J. Anim. Ecol., 36: 179-206.

Only a few parasites of the blackflies were noted (mainly Protozoa and Nematoda), but one of these, Thelohania sp. was very common in larvae from one location.

Manier, J. F. (1969). Trichomycetes de France. Ann. Sci. Nat. Bot. Biol. Veg. Ser. 12, 10(4): 565-672.

A list of insect hosts and species of Trichomycetes is given. Species of Simuliidae are infected by a wide range of these fungi.

Mainer, J. F. & Maurand, J. (1966). Sporogonie de deux microsporidies de larves de Simulium: Thelohania bracteata (Strickland 1913) et Plistophora simulii (Lutz et Splendore 1904). J. Protozool., 13(Suppl.): 39. (Abstr.)

A brief description of sporogony on the two species of Microsporida is given. P. simulii generally has 1-15 sporonts per pansporoblast and at least 16 sporoblasts per sporont. T. bracteata characteristically has one sporont per pansporoblast and eight sporoblasts per sporont.

Marr, J. D. M. & Lewis, D. J. (1964). Observations on the dry-season survival of Simulium damnosum Theo. in Ghana. Bull. Entomol. Res., 55: 547-564.

A survey of parasites of S. damnosum in Ghana was made. Spherical ciliate protozoa were found in the abdomen of a fly with small ovaries. A nematode was found in the Malpighian tubes of parous flies. The worms were immature and could not be identified with certainty.

Maurand, J. (1966). <u>Plistophora simulii</u> (Lutz et Splendore 1904). Microsporidie parasite des larves de <u>Simulium</u>; cycle, ultrastructure, ses rapports avec <u>Thelohania bracteata</u> (Strickland, 1913). <u>Bull. Soc. Zool. Fr.</u>, <u>91</u>: 621-630.

Light and electron microscopy were used to observe the development stages of <u>P. simulii</u>. Sporogony of <u>P. simulii</u> is compared with that of <u>T. bracteata</u>.

Maurand, J. (1967). Relations microsporidies-larves de <u>Simulium</u>: spécificité parasitaire d'hôte; spécificité parasitaire tissulaire. <u>Ann. Parasitol. Hum. Comp.</u>, <u>42</u>: 285-290.

Although the two microsporidans, <u>Thelohania bracteata</u> and <u>Plistophora simulii</u>, are not host specific, they do seem to be specific in the tissue they attack. They are most frequently found in the fat tissue, hypodermic cells and hemocytes in all hosts examined. Also, this paper reports the first finding of <u>T. bracteata</u> in <u>S. monticola</u>, <u>S. ornatum</u>, and <u>S. variegatum</u>. New hosts, <u>S. monticola</u> and <u>S. ornatum</u>, were observed for <u>P. simulii</u>.

Maurand, J. (1974). Influence des microsporidies sur le development des larves de simulies. (Abstr.) Proc. 3rd Int. Congr. Parasitol. (Munich), <u>2</u>: 925-926.

A method for assessing the influence of Microsporida on the growth-habit of simuliids has been devised. The technique involves an analysis of the mathematical relationship between the width of the fronto clypeus and the length of the larvae. This method has provided evidence that parasitized larvae undergo more molts than do unparasitized ones.

Maurand, J. & Bouix, G. (1969). Mise en évidence d'un phénomène sécrétoire dans le cycle de <u>Thelohania fibrata</u> (Strickland 1913), Microsporidie parasite des larves de Simulium. <u>C. R. Hebd. Seances Acad. Sci. (Paris)</u>, <u>269</u>: 2216-2218.

Sulfomucopolysaccharides are secreted during formation of the sporoblast by <u>T. fibrata</u>.

Maurand, J. & Manier, J. F. (1967). Une microsporidie nouvelle pour les larves de simulies. <u>J. Protozool.</u>, <u>14</u>(Suppl.): 47 (Abstr.).

A new species of microsporidan, <u>Stempellia simulii</u>, was described from <u>Simulium bezzii</u> collected near Montpellier, France. Adipose tissue is invaded and the infection is fatal for the host.

Maurand, J. & Manier, J. F. (1968). Actions histopathologiques comparées de parasites coelomiques des larves de simulies (Chytridiales, Microsporidies). <u>Ann. Parasitol. Hum. Comp.</u>, <u>43</u>: 79-85.

The coelomic parasites in <u>Simulium</u> larvae can be classed into three types: <u>Coelomycidium simulii</u>, which attacks the adipose tissue and chromatocytes. The Microsporida, <u>Thelohania bracteata</u>, <u>Pleistophora simulii</u>, and <u>P. debaisieuxi</u> which attack the adipose tissue, and produce hypertrophy in the nuclei, blood cells and intestinal cells. And <u>Stempellia simulii</u> which causes nuclear and cytoplasmic hypertrophy in the parasitized fat body.

Moss, S. T. (1970). Trichomycetes inhabiting the digestive tract of <u>Simulium equinum</u> larvae. <u>Trans. Br. Mycol. Soc.</u>, <u>54</u>(1): 1-13.

<u>S. equinum</u> larvae were collected at monthly intervals from a stream in Dorset. Both mid- and hind-guts of the larvae were found to contain several species of Trichomycetes. Thalli attached to the peritrophic membrane lining the mid-gut belonged to <u>Harpella</u>

234

melusinae, while those attached to the hind-gut were Paramoebidium chattonii, Smittium sp. and two unidentified species of Stipella.

Phelps, R. J. & DeFoliart, G. R. (1964).   Nematode parasitism of Simuliidae.   Res. Agric. Exp. Stn. Univ. Wis., 245, 78pp.

This study of nematode parasitism of blackflies concentrated on populations of Simulium vittatum from streams in Wisconsin.   Mesomermis flumenalis was found in larvae appearing early in the spring with S. venustum being most heavily infected.   Gastromermis viridis and Isomermis wisconsinensis were recorded only from S. vittatum.   In the streams, 37-63% of the adult flies were infected and larval mortality was estimated at 50%.

Rubtsov, I. A. (1963).   /On mermithids parasitizing simuliids./   Zool. Zh., 42: 1768-1784. (R) (WHO/EBL/19) (E).

Many species of mermithids and other nematodes were found in blackflies near Leningrad. The incidence of parasitism reaches 100%, resulting in the elimination of the blackflies in some localities.   Parasitized larvae most commonly harbour two or three parasites, only one of which will emerge.

Rubtsov, I. A. (1966a).   /A new parasitic species from blackflies and errors in the instinct of the host./   Dokl. Akad. Nauk SSSR, 169: 1236-1238. (R).

A new species, Mesomermis brevis, is described from the blackflies Prosimulium isos and P. hirtipes.   Eleven female nematode specimens were examined.

Rubtsov, I. A. (1966b).   /Host-parasite interrelationships.   Responses of simuliids to microsporidia taken as an example./   Zh. Obshch. Biol., 27: 647-661.   (R, e)

Responses of simuliid larvae parasitized by Microsporida may involve the whole body or just the infected parts.   The reactions include: increased blood cell production, delay in pupal or adult organ development, disintegration of the fat body, alterations in the gut epithelial cells, degeneration of other organs, and death.   The relationship of these responses to natural selection and the evolution of the host-parasite relation are discussed.

Rubtsov, I. A. (1966c).   /Nematodes of the family Tetradonematidae, parasitizing in members of the family Simuliidae (Diptera)./   Helminthologica, 7: 165-198.   (R, e. g)

Blackflies collected near Leningrad were examined for nematode infections.   From these observations, a table was prepared for differentiating the genera Tetradomermis and Isomermis based on diagnostic characters.   Also, several new species and a new genus, Tetradomermis, were named.

Rubtsov, I. A. (1967a).   /Mermithidae (Nematoda), endoparasites of blackflies (Diptera, Simuliidae).   II. New species of the genus Gastromermis Micoletzky, 1923./   Tr. Zool. Inst. Akad. Nauk SSSR., 43: 59-92.   (R)

Eight new species of Gastromermis are named.   These species were described from limited material, in some cases only adults or larvae were examined or only specimens of one sex. All collections were from the Leningrad region.

Rubtsov, I. A. (1967b).   Mermithidae parasitizing simuliids, IV. New species of the genus Limnomermis, Dad. Zool. Zh., 46: 24-34.   (R, e)

Mermithids were reared from larvae of <u>Eusimulium latipes</u> and <u>E. cryophilum</u>. Two new species were named, <u>L. macronuclei</u> and <u>L. cryophili</u>. Forty to fifty percent of the blackflies were infected by these parasites in some years. Three species were described from other simuliids but their hosts could not be determined. These latter species were described from limited material.

Rubtsov, I. A. (1968). /A new species of <u>Isomermis</u> (Nematoda, Mermithidae) - parasite of blackflies - and its variability./ Zool. Zh., <u>47</u>: 510-524. (R, e)

Isomermis rossica is described from blackfly larvae. The anatomical dimensions and other features distinguish it from other <u>Isomermis</u> species; however, intermediate forms do exist. Several blackfly species were found as hosts.

Rubtsov, I. A. (1969). /Variability and relationships of coelomycidians with the host./ Zh. Obshch. Biol., <u>30</u>(2): 165-173. (R, e)

Coelomycidium simulii is a widespread pathogen of blackflies. Characteristic responses of the hosts to this parasite include: arrest of the development of pupal organs and imaginal discs, reduction of gonadal primordia, disintegration of fat body cells growth of infected cells, exhaustion of other tissues, and weakening of the organisms resistance.

Rubtsov, I. A. (1973). /Mechanisms of population density regulation in mermithids./ Zh. Obshch. Biol., <u>34</u>(1): 81-89. (R, e)

Sharp fluctuations of the mermithid populations are associated with their high reproductive capacity. The rapid increase in their numbers can result in nearly exterminating their hosts and consequently, themselves. The population is controlled when food supplies are inadequate, by sex redetermination (females to males), and by mass death.

Rubtsov, I. A. (1974). Natural enemies of the blackflies. (Abstr.) Proc. 3rd Int. Congr. Parasitol. (Munich), <u>2</u>: 924-925.

Coelomycidium simulii, microsporidans, and mermithids are proposed as the most promising agents for biological control of blackflies.

Rubtsov, I. A. & Doby, J. M. (1970). Mermithides parasites de simulies (Diptères) en provenance du nord et de l'ouest de la France. Bull. Soc. Zool. Fr., <u>95</u>(4): 803-836. (F, e)

Several new species and subspecies of mermithid nematodes are described. They were found parasitizing several species of blackflies in N. and W. France.

Shipitsina, N. K. (1963). /Infection of blackflies (Simuliidae: Diptera) with parasites and its effect on the function of ovaries./ Zool. Zh., <u>42</u>: 291-294. (R, e)

Of 2415 female blackflies dissected, 7% were infected by mermithid nematodes, 2.2% by Microsporida, and 0.6% by other organisms. <u>Simulium reptans</u> had the highest incidence of nematode parasitism, 11% in 1959; <u>Titanopteryx maculata</u> was most frequently infected by Microsporida, 31.5%. The presence of parasites, especially mermithids, interferes with the normal development of the ovaries.

Steinhaus, E. A. & Marsh, G. A. (1962). Report of diagnoses of diseased insects 1951-1961. Hilgardia, <u>33</u>: 349-490.

236

The location, host, submission date, and individual requesting diagnosis are given for each tentative diagnosis.   For details see the host-parasite table.

Thomas, G. M. & Poinar, G. O., jr. (1973).   Report of diagnoses of diseased insects 1962-1972. Hilgardia, 42: 261-350.

The location, host, submission date, and individual requesting diagnosis are given for each tentative diagnosis.   For details see the host-parasite table.

Thomson, H. M. (1960).   A list and brief description of the microsporidia infecting insects. J. Insect Pathol., 2: 346-385.

Descriptions of 139 species of Microsporida are given.   Known hosts are given for each taxon.

Vávra, J. (1965).   Étude au microscope électronique de la morphologie et du développement de quelques Microsporidies.   C. R. Hebd. Seances Acad. Sci. Ser. D Sci. Nat. (Paris), 261: 3467-3470.

The development of the microsporidians, Thelohania bracteata, T. fibrata, T. varians, Plistophora debaisieuxi, and P. simulii (parasites of simuliids), was observed using the electron microscope.

Vávra, J. (1968).   Ultrastructural features of Caudospora simulii Weiser (Protozoa, Microsporidia).   Folia Parasitol., 15(1): 1-9.

Infected Simulium latipes larvae were collected from a brook near Guichen, France. From these specimens, the fine structure of the plasmodia, sporoblasts, and spores of C. simulii are described.   The study was based on limited material.

Webster, J. M. (1973).   Manipulation of environment to facilitate use of nematodes in bio-control of insects.   Exper. Parasitol., 33(2): 197-206.

Aquatic larvae of Simulium vittatum feed fastest at 20°C, so engulf more infective larvae of Neoaplectana carpocapsae (DD-136) at that temperature.

Weiser, J. A. (1961).   Die Mikrosporidien als Parasiten des Insekten.   Monogr. Angew. Entomol. No. 17, 149 pp.

This monograph contains an extensive review which includes pathogens of blackflies.

Weiser, J. (1964).   Parasitology of blackflies.   Bull. WHO, 31: 483-485.

This is a review article with 26 references.   A list of pathogens current to 1964 is given.

Weiser, J. (1968).   Iridescent virus from the blackfly Simulium ornatum Meigen in Czechoslovakia.   J. Invertebr. Pathol., 12(1): 36-39.

Virogenic stromata were observed in fat body cells, hypodermis, connective tissue, tracheal matrix, and some muscle cells of Simulium ornatum larvae collected near Chotěboř, Czechoslovakia.   The virus has a particle size of 1400-1600 Å.   The disease was rare.

Welch, H. E. (1962).   New species of Gastromermis, Isomermis, and Mesomermis (Nematoda: Mermithidae) from blackfly larvae.   Ann. Entomol. Soc. Am., 55: 535-542.

Gastromermis viridis and Isomermis wisconsinensis are described from larvae of Simulium vittatum in Wisconsin and Mesomermis flumenalis from larvae of S. venustum in Ontario. All were found in the hemocoele of their host.

Welch, H. E. (1964). Mermithid parasites of blackflies. (WHO/EBL/16, WHO/VBC/57, 17 pp.) Bull. WHO, 31: 857-863.

This is a review article with 43 references. Past records of pathogenicity of the nematodes to various species of simuliids are given.

Welch, H. E. & Rubtsov, I. A. (1965). Mermithids (Nematoda: Mermithidae) parasitic in blackflies (Insecta: Simuliidae). I. Taxonomy and bionomics of Gastromermis boophthorae sp. n. Can. Entomol., 97: 581-596.

A new species, G. boophthorae, is described from simuliids collected near Leningrad. Six species of blackflies are found to host this nematode. Infected larvae usually ranged from 1 to 10% of the population but reached 100% in one sample.

XI. PATHOGENS OF TABANIDAE (HORSEFLIES)

Darrell W. Anthony

USDA, Agricultural Research Service
Insects Affecting Man Research Laboratory
Gainesville, FL 32611, USA

PATHOGENS OF TABANIDAE (HORSE FLIES)

Host	Host stage infected	Pathogen	% incidence	Locality	Lab. or field study	Reference
Chrysops furcata	Larvae	Bathymermis sp.	16-37	Canada	Field, lab.	Shamsuddin (1966)
Chrysops italicus	Larvae	Mucor sp.	Unknown	USSR	Unknown	Koval & Andreeva (1971)
Chrysops mitis	Larvae	Bathymermis sp.	to 70	"	Lab.	Shamsuddin (1966)
Chrysops relictus	Eggs	Aspergillus flavus	Unknown	"	Field, lab.	Koval & Andreeva (1971)
		A. niger	"	"	"	"
		A. fumigatus	"	"	"	"
		Beauveria densa	"	"	"	"
		Fusarium avenaceum	"	"	"	"
		F. solani agrillacea	"	"	"	"
Chrysops relictus	Larvae	Coelomomyces milkoi	18-30	"	"	Andreeva (1972)
		Aspergillus flavus	Unknown	"	"	"
		A. niger	"	"	"	"
		A. fumigatus	"	"	"	"
		Beauveria densa	"	"	"	"
		Fusarium avenaceum	"	"	"	"
		F. solani agrillacea	"	"	"	"
		Metarrhizium anisopliae	"	"	"	"
Tabanus atratus	Larvae	Thelohania tabani	20 from one area in Mississippi	USA (Mississippi)	Field, lab.	Gingrich (1965)
Tabanus autumnalis	Larvae	Coelomomyces milkoi	Up to 95	USSR	"	Andreeva (1972)

PATHOGENS OF TABANIDAE (HORSE FLIES) (continued)

Host	Host stage infected	Pathogen	% incidence	Locality	Lab. or field study	Reference
Tabanus autumnalis	Larvae, pupae, adults	Aspergillus flavus	Unknown	USSR	Field (?)	Andreeva (1972)
		A. niger	"	"	"	"
		A. fumigatus	"	"	"	"
		Beauveria densa	"	"	"	"
		Fusarium avenaceum	"	"	"	"
		F. solani agrillacea	"	"	"	"
		Metarrhizium anisopliae	"	"	"	"
Tabanus lineola	Larvae	Microsporida (to be described)	Unknown	USA (Florida)	Field	Hazard & Knell (1973)
Tabanus subsimilis	Larvae	Microsporida (to be described)	"	USA (Mississippi)	Field, lab.	Harlan (1973)

ABSTRACTS

Mary Ann Strand

Andreeva, R. V. (1972). /Some data on pathogenic fungi of larvae of horse flies in the vicinity of Kiev./ Probl. Parasitol. Trans. Sci. Conf. Parasitol. USSR, Part 1, 33 pp.

Coelomomyces milkoi was found infecting 95% of the larvae of Tabanus autumnalis in the spring of 1971. Previously 18-30% of the larvae were infected. The greater incidence in 1971 is possibly related to the humid, warm conditions of the summer and autumn of 1970. A description of the disease is given. Larvae of T. autumnalis and Chrysops relictus were found to be susceptible to Metarrhizium anisopliae, Beauveria densa and Aspergillus flavus. Adults were also susceptible to M. anisopliae and B. bassiana.

Gingrich, R. E. (1965). Thelohania tabani sp. n., a microsporidian from the larvae of the black horse fly, Tabanus atratus Fabricius. J. Invertebr. Pathol., 7: 236-240.

A new species of microsporidia is described from T. atratus collected in Mississippi. The pathogen develops in the muscle tissue and infection can be detected by the whitish, opaque areas visible through the transparent integument. Because of the predaceous feeding habits of the larvae, ingested spores from infected prey are the most probable source of infection.

Goodwin, J. T. (1968). Notes on parasites of immature Tabanidae (Diptera) and descriptions of the larvae and puparium of Carinosillus pravus (Diptera, Tachinidae). J. Tenn. Acad. Sci., 43: 107-108.

The dipterous parasites of Tabanidae are discussed.

Harlan, D. P. (1973). Personal communication, November 1973.

Hazard, E. I. & Knell, J. D. (1973). Personal communication, April 1973.

James, H. G. (1963). Larval habits, development and parasites of some Tabanidae (Diptera) in southern Ontario. Can. Entomol., 95: 1223-1232.

Immatures of 29 species of Tabanidae were collected and reared to obtain information on their parasites and other biotic control agents. Most numerous natural control agents appeared to be insect parasites. The nematode, Bathymermis sp., was found infecting 11/143 larvae, especially Chrysops.

Jackson, J. O. & Wilson, B. H. (1965). Observations on some predators of horse fly and deer fly (Diptera, Tabanidae) eggs in Louisiana. Ann. Entomol. Soc. Am., 58: 934-935.

Arthropod predators of Tabanidae eggs captured in the field included Coleomegilla maculata, Collops bipunctatus and Orchelimum sp. Laboratory studies on the acceptability of the eggs to other predators were also conducted.

Jones, C. M. & Anthony, D. W. (1964). The Tabanidae (Diptera) of Florida. USDA, ARS, Tech. Bull., No. 1295, 85 pp.

An extensive review of the biology and ecology of the Tabanidae is given. Control measures including parasites and predators are presented. No microbial pathogens are mentioned.

Koval', E. Z. & Andreeva, R. V. (1971). ⎣On studying pathogenic mycoflora of horse-flies (Diptera, Tabanidae) in the Ukraine.⎦ Dopov. Akad. Nauk Ukr., Series B, 11: 1042-1044. (Ukrainian, English summary).

A list of fungal diseases found in the field and under laboratory conditions on the larvae of Tabanus autumnalis and Chrysops relictus is given. Coelomomyces sp. and Metarrhizium anisopliae were found most frequently on the larvae. A number of other fungi were also found including Beauveria densa.

Shamsuddin, M. (1966). A Bathymermis species (Mermithidae, Nematoda) parasitic on larvae Tabanids. Quaest. Entomol., 2: 253-256.

A species of Bathymermis was found parasitizing Chrysops furcata in Alberta, Canada. Larvae are yellowish-green, during parasitism this changes to pale-yellow and later to black. Parasites could be readily seen in the mature larvae which become transparent. Infected larvae failed to pupate, usually did not feed, and were more sluggish in their movements.

Tesky, H. J. (1969). Larvae and pupae of some eastern North American Tabanidae (Diptera). Mem. Entomol. Soc. Can., No. 63, 147 pp.

Key based on larvae or pupal characteristics and diagnostic descriptions of these stages are given.

Tidwell, M. A. (1973). The Tabanidae (Diptera) of Louisiana. Tulane Stud. Zool. Bot., 18(1,2), 93 pp.

This volume is a good general reference on the biology of Tabanidae.

XII. PATHOGENS OF *MUSCA DOMESTICA* AND *M. AUTUMNALIS* (HOUSEFLIES AND FACE FLIES)

J. D. Briggs & Sheila E. Milligan

Department of Entomology
Ohio State University
Columbus, OH 43201, USA

PATHOGENS OF <u>MUSCA DOMESTICA</u> AND <u>M. AUTUMNALIS</u> (HOUSEFLIES AND FACE FLIES)

Host	Host stage infected	Pathogen	% incidence	Locality	Lab. or field study	Reference
<u>Musca domestica</u>	Larvae	<u>Bacillus cereus</u>[a]	-	-	Lab.	Briggs (1960)
"	Larvae, adults	"	-	USA (California)	Field	Thomas & Poinar (1973)
"	Larvae	<u>B. thuringiensis</u>[a]	99 mortality	-	Lab.	Briggs (1960)
"	Larvae	"	-	-	Lab.	Burgerjon & Galichet (1965), Cantwell et al. (1964), Connor & Hansen (1967), Feigin (1963), Galichet (1966), Gingrich (1965), Greenwood (1964), Harvey & Howell (1965), Millar (1965)
"	Larvae	<u>Bacillus</u> spp.	-	-	Lab.	Rogoff et al. (1969)
"	Adults	A non-spore forming bacterium[a] described in error as <u>Bacillus lutzae</u>	-	-	Lab.	Brown (1927), Brown & Heffron (1929)
"	Larvae	<u>Bacterium delendae-muscae</u> (not a recognized taxon)	-	-	Lab.	Roubaud & Descazeaux (1923)
"	Adults	<u>Bacterium mathisi</u> (not a recognized taxon)	-	-	Lab.	Roubaud & Treillard (1935)
"	Adults	<u>Pseudomonas</u> (near <u>septica</u>)	-	-	Lab.	Amonkar et al. (1967)

PATHOGENS OF <u>MUSCA DOMESTICA</u> AND <u>M. AUTUMNALIS</u> (HOUSEFLIES AND FACE FLIES) (continued)

Host	Host stage infected	Pathogen	% incidence	Locality	Lab. or field study	Reference
<u>Musca domestica</u>	Adults	Serratia marcescens	-	USA (California)	Field	Steinhaus (1951)
"	Adults	Staphyloccus[a] muscae (not a recognized taxon)	-	-	Lab.	Glaser (1924)
		Fungi:				
"	Larvae, adults	<u>Aspergillus flavus</u>[a]	90-97 mortality	-	Lab.	Amonkar & Nair (1965)
"	Larvae	"	-	USA (lab. colony)	Lab.	Beard & Walton (1965)
"	Adults	"	-	USA (California)	Field	Thomas & Poinar (1973)
"	Adults	<u>A. parasiticus</u>	-	Brazil	Field	Marchionatto (1945)
"	Larvae, adults	<u>Beauveria bassiana</u>	-	-	Lab.	Dresner (1950)
"	Adults	<u>Entomophthora</u>[a] <u>kansana</u>	Epizootic	USA (Kansas)	Lab., field	Hutchison (1962)
"	Adults	<u>E. muscae</u>[a]	-	-	Lab.	Güssow (1913), Hesse (1913)
		"	-	USA (California)	Field	Steinhaus (1951), Steinhaus & Marsh (1962), Thomas & Poinar (1973)
		"	48	USA (Idaho)	Field	Yeager (1939)

PATHOGENS OF <u>MUSCA DOMESTICA</u> AND <u>M. AUTUMNALIS</u> (HOUSEFLIES AND FACE FLIES) (continued)

Host	Host stage infected	Pathogen	% incidence	Locality	Lab. or field study	Reference
<u>Musca domestica</u>	Adults	Macrosporium sp.	-	-	Lab.	Damodar et al. (1964)
"	Adults	Herpetomonas muscae-domesticae	8	Surinam	Lab., field	Flu (1911)
"	Adults	"	50	India	Field	Ross & Hussain (1924)
"	Larvae	H. muscarum	(4/61)	USA (Illinois)	Field	Kramer (1961)
"	Larvae		0.1	-	-	Laird (1959)
"	Adults	Leptomonas muscae-domesticae	-	Surinam	Lab., field	Flu (1911)
"	Adults	Nosema kingi [a]	-	-	Lab.	Kramer (1964a)
"	Adults	Octosporea muscae-domesticae	8-90	Surinam	Lab., field	Flu (1911)
"	Adults	Octosporea muscae-domesticae [a]	-	USA (Illinois)	Lab., field	Kramer (1964b)
"	Adults	Octosporea muscae-domesticae	-	-	-	Kramer (1965)
"	Adults	Octosporea muscae-domesticae [a]	-	-	Lab.	Kramer (1966)
"	Adults	Octosporea muscae-domesticae	4	-	Field	Laird (1959)
<u>Musca autumnalis</u>	Adults	Octosporea muscae-domesticae [a]	-	USA	Lab.	Kramer (1973)
"	Adults	Heterotylenchus [a] autumnalis	16	USA (Massachusetts)	Field	Chitwood & Stoffolano (1971)

PATHOGENS OF <u>MUSCA DOMESTICA</u> AND <u>M. AUTUMNALIS</u> (HOUSEFLIES AND FACE FLIES) (continued)

Host	Stage	Pathogen	% incidence	Locality	Lab. or field study	Reference
<u>Musca autumnalis</u>	Adults	<u>Heterotylenchus autumnalis</u>	23, 33	USA (Nebraska)	Field	Jones & Perdue (1967)
"	Adults	<u>Heterotylenchus autumnalis</u>	25	USA (New York)	Field	Stoffolano & Nickle (1966)
"	Adults	<u>Heterotylenchus autumnalis</u>	26.2	USA (Missouri)	Field	Thomas et al. (1972)
"	Adults	<u>Heterotylenchus autumnalis</u>	50	USA (lab. colony)	Lab.	Treece & Miller (1968)
"	Adults	<u>Heterotylenchus autumnalis</u>	-	USA	Lab.	Nappi (1973), Nappi & Stoffolano (1972), Nickle (1967), Stoffolano (1967, 1969, 1970a, b, 1971, 1973), Stoffolano & Streams (1971)
"	Adults	<u>Thelazia</u> sp.	(1/155)	USA (Massachusetts)	Field	Chitwood & Stoffolano (1971)

<u>a</u> Etiology confirmed.

ABSTRACTS

Mary Ann Strand

Amonkar, S. V. et al. (1967). Mechanism of pathogenicity of Pseudomonas in the house fly.
J. Invertebr. Pathol., 9: 235-240.

Pathogenicity of P. septica strain EF and its autolytic variant CL to Musca domestica
nebulo is associated with an exotoxin produced by the bacteria during growth. Mortality
of adults occurred earlier in experiments where the toxin rather than the bacteria was
fed to the flies. However when the toxin was injected, a lag of two days occurred
prior to onset of death. No lag occurred when the bacteria were injected.

Amonkar, S. V. & Nair, K. K. (1965). Pathogenicity of Aspergillus flavus Link to Musca
domestica Fabricius. J. Invertebr. Pathol., 7: 513-514.

Flies fed on a spore suspension (conc. 1×10^3 - 1×10^9 spores/ml) and dilute honey
became infected; by the fourteenth day, 90-97% mortality was observed at all concentrations. Flies were also dusted with spores. By the seventh day there was 100%
mortality. Following the death of a fly, numerous conidiophores emerged from its body.

Beard, R. L. & Walton, G. S. (1965). An Aspergillus toxin lethal to larvae of the house fly.
J. Invertebr. Pathol., 7: 522-523.

A. flavus var. columnaris was found in laboratory colonies of Musca domestica maggots.
No mycelial development could be seen in moribund larvae. Aqueous extracts of the
culture medium, from which mycelial mats had been removed, proved toxic. These observations point to the water soluble toxin elaborated by the growing fungus; potency varied
with the culture medium and strain of fungus.

Briggs, J. D. (1960). Reduction of adult house-fly emergence by the effects of Bacillus spp.
on the development of immature forms. J. Insect Pathol., 2: 418-432.

Exotoxins from Bacillus thuringiensis var. thuringiensis and B. cereus were found to
interfere with adult house fly emergence. The exotoxin produced by B. thuringiensis
was more potent. Spore preparations retained their effectiveness in the faeces of
chickens which were fed the preparations. When 3 gm per day were fed to hens, 99%
reduction of adult fly emergence resulted.

Brown, F. M. (1927). Descriptions of new bacteria found in insects. Am. Mus. Novit., 251:
1-11

Bacillus lutzae (not a recognized taxon) was found to be pathogenic to house flies.

Brown, F. M. & Heffron, H. M. (1929). Mendelism among bacteria? Science, 69: 198-200.

A strain of yellow pigmented bacterium was isolated from the wood fly Lucillia sericata.
Attempts to identify the bacterium and the changes in its properties which occurred
during subculturing are described.

Burgerjon, A. & Galichet, P. F. (1965). The effectiveness of the heat-stable toxin of Bacillus thuringiensis var. thuringiensis Berliner on larvae of Musca domestica Linnaeus. J. Invertebr. Pathol., 7: 263-264.

Three preparations from B. thuringiensis cultures were added to the nutritive medium of the fly larvae. When a medium rich in soluble matter (industrial filtrate) and poor in spores and crystals was added, no adult flies emerged. It appears that the heat-stable toxin is responsible for the mortality observed in the larval and pupal stage.

Cantwell, G. E. et al. (1964). The production of an exotoxin by various crystal-forming bacteria related to Bacillus thuringiensis var. thuringiensis Berliner. J. Insect. Pathol., 6: 466-480.

When grown on adequate medium, the bacterium produces an exotoxin. This toxic compound interferes with the pupation of the house fly. The toxin is produced in culture at the time of sporulation or in the presence of dipicolinic acid.

Chitwood, M. B. & Stoffolano, J. G., jr (1971). First report of Thelazia sp. (Nematoda) in the face fly, Musca autumnalis, in North America. J. Parasitol., 57: 1363-1364.

A female face fly parasitized by nine Thelazia sp. larvae was found in Massachusetts. This fly was one of 155 face flies examined for parasites. Twenty-five (16%) were infected with Heterotylenchus autumnalis.

Connor, R. M. & Hansen, P. A. (1967). Effect of valine, leucine and isoleucine on the production of fly toxin by Bacillus thuringiensis and related organisms. J. Invertebr. Pathol., 9: 114-125.

Cultures of B. thuringiensis var. thuringiensis, var. sotto, and B. entomocidus var. entomocidus were grown in a citrate-salts medium containing combinations of valine, leucine, and isoleucine. The potency of the heat-stable exotoxin produced by var. thuringiensis was estimated by bioassay against Musca domestica. Varieties sotto and entomocidus produced no detectable fly toxin.

Damodar, P. et al. (1964). The toxicity of solvent extract of the fungus, Macrosporium sp. to flies and mosquitoes. Indian J. Entomol., 26: 110-112.

A petroleum ether extract of the cultured fungal mat was used in the bioassay tests as a space spray and for topical applications. Adult Musca nebulo were tested. The extract was much less effective than DDT. Mixed with DDT, the space spray had no synergistic effect against houseflies.

Dresner, E. (1950). The toxic effect of Beauveria bassiana (Bals.) Vuill., on insects. J. N. Y. Entomol. Soc., 58: 269-279.

Experiments indicated that the houseflies were paralyzed on exposure to germinating spores in much less time than has been determined histologically for the penetration of hyphae into the body cavity. The germinating spores apparently produce a toxin which is a contact poison. The toxin also acted through the alimentary tract, resulting in 100% mortality in flies fed on milk used as a culture medium for the fungus.

Feigin, J. M. (1963). Exposure of the housefly to selection by Bacillus thuringiensis. Ann. Entomol. Soc. Am., 56: 878-879.

In spite of selection pressure, there was no noticeable decrease in mortality resulting from bacterial infection over 27 generations of flies. The mortality occurred primarily in the larval stage whether eggs or larvae were intially exposed to the bacteria.

252

Flu, P. C. (1911). Studien über die im Darm der Stubenfliege, Musca domestica, vorkommenden protozoären Gebilde. Zentralbl. Bakteriol. Parazitenkd. Infetionskr. Hyg. Abt. I. Orig., 57(6): 522-535.

Octosporea muscae-domesticae, Herpetomonas muscae-domesticae, and Leptomonas muscae-domesticae were found in the guts of houseflies in Surinam. The flagellates did not apparently affect the life cycles of the flies. About 8% of those examined were infected with Herpetomonas. Octosporea did produce pathological conditions resulting in mortality. At the beginning of observations about 8% were infected with Octosporea but 8-10 days later the incidence was 80-90%.

Galichet, P.-F. (1966). Administration aux animaux domestiques d'une toxine thermostable sécrétée par Bacillus thuringiensis Berliner, en vue d'empecher la multiplication de Musca domestica Linnaeus dans les fèces. Ann. Zootech., 15: 133-145.

The faeces of animals is a poorer medium for the reproduction of M. domestica, when the heat-stable toxin from B. thuringiensis is incorporated into their feed. The efficiency of the treatment was the same whether the filtrate was given in solid or liquid form

Gingrich, R. E. (1965). Bacillus thuringiensis as a feed additive to control dipterous pests of cattle. J. Econ. Entomol., 58: 363-364.

B. thuringiensis fed to cattle inhibited larval development of Musca domestica in the cattle faeces. Three commercial preparations were used in the tests.

Glaser, R. W. (1924). A bacterial disease of adult house flies. Am. J. Hyg., 4: 411-415.

Greenwood, E. S. (1964). Bacillus thuringiensis in the control of Lucilla sericata and Musca domestica. N. Z. J. Sci., 7: 221-226.

Suspensions of B. thuringiensis v. thuringiensis were incorporated into the larval rearing medium of flies. High dose rates killed the larvae and sub-lethal doses retarded development throughout the life cycle. Surviving adults produced fewer viable eggs.

Güssow, H. T. (1913). Empusa muscae and the extermination of the house-fly. Rept. Local Govt. Board Public Health Med. Subjects (London) (n.s.), 85: 10-14.

Harvey, T. L. & Howell, D. E. (1965). Resistance of the house fly to Bacillus thuringiensis Berliner. J. Invertebr. Pathol., 7: 92-100.

Resistance to B. thuringiensis was induced in Musca domestica by selecting survivors from treated larval medium during 50 generations of rearing.

Hesse, E. (1913). A parasitic mould of the house fly. Br. Med. J., 1: 41-42.

Hutchison, J. A. (1962). Studies on a new Entomophthora attacking calyptrate flies. Mycologia, 54: 258-271.

A new species, E. kansana, was found infecting flies in epizootic proportions in Kansas. The pathogen digested head, thorax and abdomen tissues of the host. It was highly virulent in laboratory tests for four months, but was abruptly lost.

Jones, C. M. & Perdue, J. M. (1967). Heterotylenchus autumnalis, a parasite of the face fly. J. Econ. Entomol., 60: 1393-1395.

H. autumnalis was found parasitizing face flies in Nebraska. The nematodes were present in the host population throughout their active season and in the flies entering hibernation in the fall. About 22% of the flies visiting manure were infected and 32% of the flies developing in manure samples were infected. Over winter, mortality of infected and uninfected flies was about the same.

Kramer, J. P. (1961). Herpetomonas muscarum (Leidy) in the haemocoele of larval Musca domestica L. Entomol. News, 72: 165-166.

Of 61 sluggish fly larvae collected in Illinois, four were infected with H. muscarum. The parasites could not be detected without dissection. Although considered to be a benign parasite which is restricted to the alimentary canal, it apparently does gain entry to the haemocoel in some cases.

Kramer, J. P. (1964a). Nosema kingi sp. n., a microsporidian from Drosophila willistoni Sturtevant, and its infectivity for other muscoids. J. Insect Pathol., 6: 491-499.

Musca domestica was found to be susceptible to N. kingi, a monosporoblastic microsporidian. Spores in a sucrose solution were fed to the flies and extensive infection developed. A significant rise in death rate of the population was observed.

Kramer, J. P. (1964b). The microsporidian Octosporea muscaedomesticae Flu, a parasite of calypterate muscoid flies in Illinois. J. Insect Pathol., 6: 331-342.

The microsporidian was found in adult Musca domestica and other flies. The parasite is always found in the columnar epithelium of the proximal intestine of infected flies. The results of cross infectivity tests show that spores are infective for the species from which they were isolated as well as other ecologically related muscoids.

Kramer, J. P. (1965). The microsporidian Octosporea muscaedomesticae Flu, a little-known pathogen of muscoid flies. WHO/EBL/53.65. 8 pp.

Flies parasitized by Octosporea muscaedomesticae have been found world-wide.

Kramer, J. P. (1966). On the octosporeosis of muscoid flies caused by Octosporea muscaedomesticae Flu (Microsporidia). Am. Midl. Nat., 75: 214-220.

The microsporidian had a significant lethal effect on Musca domestica in laboratory tests. The effects of the disease are partly masked by the short life span of the flies after eclosion.

Kramer, J. P. (1973). Susceptibility of sixteen species of muscoid flies to the microsporidian parasite Octosporea muscaedomesticae. J. N. Y. Entomol. Soc., 81(1): 50-53.

One-day-old adult flies from disease-free laboratory cultures were tested for susceptibility. Spores were fed to the flies and 10 days after ingestion the alimentary tract of each fly was examined for lesions. Eight species were susceptible to varying degrees. Musca autumnalis was slightly susceptible, 2 of 11 flies contained inapparent infections. The other nine were not infected. In general, susceptibility was not correlated to taxonomic affinities.

Laird, M. (1959).   Protozoa, including an entozoan of guinea pigs, from house flies.   Can. J. Zool., 37: 467-468.

Of 179 houseflies examined, 4% harboured Octosporea muscaedomesticae, and 0.1% harboured Herpetomonas muscarum.

Marchionatto, J. B. (1945).   Nota sobre algunos hongos entomógenos.   Publ. Inst. Sanid. veg. (Buenos Aires), (A), 1, No.9, 11 pp.

Millar, E. S. (1965).   Bacillus thuringiensis in the control of flies breeding in the droppings of caged hens.   N. Z. J. Agric. Res., 8: 721-722 (WHO/EBL/40.65, 4 pp.).

The bacterium was more effective in controlling Musca domestica when applied to the droppings than when administered in the food of the hens.

Nappi, A. J. (1973).   Effects of parasitization by the nematode, Heterotylenchus autumnalis, on mating and oviposition in the host, Musca autumnalis.   J. Parasitol., 59(6): 963-969.

Infected male face flies had no abnormalities of the testes.   They were capable of mating and insemination.   Non-infected males would mate with infected females, but nematodes were not passed to the males during copulation.   Mating is apparently required for the parasitized females to "mock oviposit" and deposit nematodes.   Infected females are castrated by the parasite.

Nappi, A. J. & Stoffolano, J. G., jr (1971).   Heterotylenchus autumnalis: Hemocytic reactions and capsule formation in the host, Musca domestica.   Exp. Parasitol., 29: 116-125.

The infective, gamogenetic stage of the nematode is encapsulated in M. domestica and therefore cannot become established in the host.   The capsule is haemocytic in origin. The number of haemocytes in parasitized larvae is apparently greater than in controls of the same age.

Nappi, A. J. & Stoffolano, J. G., jr (1972).   Distribution of haemocytes in larvae of Musca domestica and Musca autumnalis and possible chemotaxis during parasitization.   J. Insect Physiol., 18: 169-179.

In parasitized fly larvae, large numbers of haemocytes circulate throughout the haemocoel and encapsulate parasites in various regions of the body.   In uninfected larvae, most of the haemocytes accumulate in the haemocoel in the last two segments of the body. Evidence suggests that the haemocytes exhibit chemotaxis.   The stimulus may come from the parasites directly or from the haemocytes when they contact the parasites.

Nickle, W. R. (1967).   Heterotylenchus autumnalis sp. n. (Nematoda: Spaerulariidae), a parasite of the face fly, Musca autumnalis De Geer.   J. Parasitol., 53: 398-401.

Thousands of nematodes are found in the body cavity, ovaries, and thorax of the parasitized adult face fly.   The ovaries contain packets of males and unmated females where normally eggs are found.   The nematodes are deposited in manure when the fly attempts oviposition.   Impregnated female nematodes enter the body cavity of the fly maggot and lay eggs in the haemolymph.

Rogoff, M. H. et al. (1969).   Insecticidal activity of thirty-one strains of Bacillus against five insect species.   J. Invertebr. Pathol., 14: 122-129.

The insecticidal activity of various strains of Bacillus grown on δ- and endotoxin-promoting medium and on β-exotoxin-promoting medium was tested against Musca domestica by mixing the _Bacillus_ cultures with the insect rearing diet.   Thirteen strains from the first medium and the same strains from the second medium produced more than 50% mortality in the flies.

Ross, W. C. & Hussain, M. (1924).   On the life history of _Herpetomonas muscae-domesticae_. A preliminary note.   Indian Med. Gaz., 59: 614-615.

Fifty per cent. of 7000 flies examined were infected by the flagellate.   The life history of _H. muscae-domesticae_ was observed in living preparations of flies and confirmed by stained specimens.

Roubaud, E. & Descazeaux, J. (1923).   Sur un agent bactérien pathogène pour les mouches communes: _Bacterium delendae-muscae_ n. sp.   C. R. Hebd. Seances Acad. Sci. (Paris), 177: 716-717.

The cultural properties of the new species are described.   Infections contracted in the larval stage are fatal by the end of the nymphal stage.   Infections are also fatal to adults and virulence apparently increased with transmission.

Roubaud, E. & Treillard, M. (1935).   Un coccobacille pathogène pour les mouches Tsétsé. C. R. Hebd. Seances Acad. Sci. (Paris), 201: 304-306.

House flies were experimentally infected with a coccobacillus.   The bacterium produced appreciable mortality.

Saccà, G. (1964).   Comparative bionomics in the genus _Musca_.   Ann. Rev. Entomol., 9: 341-358.

This is a review article with 79 references.   It includes a brief section on pathogens.

Steinhaus, E. A. (1951).   Report on diagnoses of diseases insects 1944-1950.   Hilgardia, 20: 629-678.

House flies infected with _Entomophthora_ (_Empusa_) _muscae_ and _Serratia marcescens_ were collected near Berkeley, California.

Steinhaus, E. A. & Marsh, G. A. (1962).   Report of diagnoses of diseased insects 1951-1961. Hilgardia, 33: 349-490.

Numerous isolations of _Entomophthora muscae_ from _Musca domestica_ were reported.

Stoffolano, J. G., jr (1967).   The synchronization of the life cycle of diapausing face flies, _Musca autumnalis_, and of the nematode, _Heterotylenchus autumnalis_.   J. Invertebr. Pathol., 9: 395-397.

Nondiapausing flies contain immature nematodes whereas diapausing flies do not.   Both contain gamogenetic and parthenogenetic female nematodes, but no eggs are found in the parthenogenetic females.   There is apparently some block in reproduction of the nematodes.   Several mechanisms for producing this block are postulated.

Stoffolano, J. G., jr (1969).   Nematode parasites of the face fly and the onion maggot in France and Denmark.   J. Econ. Entomol., 62: 792-795.

Face flies from three locations in France and one in Denmark were examined for parasites. Of 844 females observed in France about 2.5% and about 13% of the 22 males were infected with _Heterotylenchus autumnalis_.   The nematode was not observed in Denmark.   Finding _H. autumnalis_ in France indicates a Paleartic origin for the species.

Stoffolano, J. G., jr. (1970a).   Nematodes associated with the genus _Musca_ (Diptera:
Muscidae).   Bull. Entomol. Soc. Am., 16: 194-203.

This is a review article with 130 references.   All reported nematode parasites are
listed and the major genera are described.

Stoffolano, J. G., jr (1970b).   Parasitism of _Heterotylenchus autumnalis_ Nickle (Nematoda:
Sphaerulariidae) to the face fly, _Musca autumnalis_ De Geer (Diptera: Muscidae).   J. Nematol.,
2: 324-329.

Laboratory experiments demonstrated that the nematode is capable of penetrating all stages
of fly larvae.   Infected larvae frequently had brown spots on their cuticles, however
uninfected larvae also occasionally had the brown spots.   Female flies are castrated by
the nematode, but infected males produced normal sperm.

Stoffolano, J. G., jr (1971).   Darkening of the anal organ of larvae of _Musca autumnalis_ and
Orthellia caesarion when parasitized by the nematode _Heterotylenchus autumnalis_.   J. Invertebr.
Pathol., 17: 3-8.

In preliminary tests, five parasitized face fly larvae were observed with dark patches
surrounding the anus.   These flies had drowned in the water surrounding the dish
containing manure and parasitized larvae.   Larvae removed from the manure did not have
the dark patches.   The anal organ of some uninfected larvae which had drowned, also,
darkened.   The pigment is probably melanin.

Stoffolano, J. G. (1973).   Maintenance of _Heterotylenchus autumnalis_, a nematode parasite of
the face fly, in the laboratory.   Ann. Entomol. Soc. Am., 66(2): 469-470.

Three methods of maintaining the nematode are described: natural seeding of manure with
infected hosts, artifically seeding on a small scale with dissected, infected ovaries,
and artificially seeding on a large scale with nematodes separated from the insect tissue.

Stoffolano, J. G., jr & Nickle, W. R. (1966).   Nematode parasite (_Heterotylenchus_ sp.) of
face fly in New York State.   J. Econ. Entomol., 59: 221-222.

The nematode parasite was first observed infecting female flies in a laboratory colony at
Cornell University.   Subsequently samples were collected from 14 counties in New York
State.   The nematode infections occurred in all locations and ranged from 8-44% with an
average of 25%.   Male and female face flies had about equal percentage infected.

Stoffolano, J. G., jr & Streams, F. A. (1971).   Host reactions of _Musca domestica_, _Orthellia
caesarion_, and _Ravinia l'herminieri_ to the nematode _Heterotylenchus autumnalis_.   Parasitology,
63: 195-211.

Teskey, H. J. (1969).   On the behavior and ecology of the face fly, _Musca autumnalis_
(Diptera: Muscidae).   Can. Entomol., 101: 561-576.

General account of the biology of face flies is given.   The field observations were made
in pastures near Guelph, Ontario.   Pathogens were not observed in the studied populations.

Thomas, G. D. et al. (1972).   Further studies of field parasitism of the face fly by the
nematode _Heterotylenchus autumnalis_ in central Missouri, with notes on the gonadotrophic cycles
of the face fly.   Environ. Entomol., 1(6): 759-763.

Mean seasonal parasitism at two farms in Missouri was 35.5 and 26.2%.   Parasitism was
higher in flies collected from the surface of manure pats than those collected from either

the sides of a truck or the faces of the cattle.   The frequency of nematode parasitism was not found to be related to fly density.

Thomas, G. M. & Poinar, G. O., jr (1973).   Report of diagnoses of diseased insects 1962-1972. Hilgardia, 42: 261-360.

Entomophthora muscae, Bacillus cereus, and Aspergillus flavus are reported to infect houseflies in California.

Treece, R. E. & Miller, T. A. (1968).   Observations on Heterotylenchus autumnalis in relation to the face fly.   J. Econ. Entomol., 61: 454-456.

About 50% of the flies in a laboratory culture were infected with the nematode.   The percentage infected did not decline with age of the flies.   Infected flies completed only one gonadotrophic cycle while uninfected flies completed three during the 13 day study period.   All uninfected females contained ova, but 94.7% of those infected contained ovaries with no eggs.   Apparently only flies more than eight days old can transmit gamogenetic nematodes.

Yeager, C. C. (1939).   Empusa infections of the house-fly in relation to moisture conditions of northern Idaho.   Mycologia, 31: 154-156.

During dry months few or no infected flies found and in humid or rainy months many infections were seen.   In June which had unusually high precipitation, 48% of the flies tested were infected with Entomophthora (Empusa) muscae.

XIII. PATHOGENS OF
STOMOXYS CALCITRANS
(STABLE FLIES) [a]

Bernard Greenberg

Professor of Biological Sciences
University of Illinois at Chicago Circle
Chicago, IL 60680, USA

[a] The author thanks Vijtas Bindokas for help in gathering the information for this table.

PATHOGENS OF STOMOXYS CALCITRANS (STABLE FLIES)

Host	Host stage infected	Pathogen	% Incidence	Locality	Lab. or field study	Reference
Stomoxys calcitrans	Larvae	Cillopasteurella delendae-muscae (Roubaud & Descazeaux) Prévot (=Bacterium delendae-muscae)	--	France	Lab., field	Roubaud & Descazeaux (1923)
"	Adults	Brucella (=Micrococcus) melitensis (Hughes) Meyer & Shaw	0	Malta	Lab.	Kennedy (1906)
"	Larvae	Bacillus thuringiensis var. thuringiensis Heimpel & Angus	Bioferm LD$_{50}$ ~150 mg/100 g faeces; Bakthane LD$_{50}$ ~200 mg/100 g faeces; Biotrol LD$_{50}$ ~1000 mg/100 g faeces	Texas	Lab.	Gingrich (1965)
"	Adults	? Entomophthora Fresenius	--	Dahomey	Lab., field	Roubaud (1911)
"	Larvae	Entomophthora (=Empusa) muscae Cohn	~25% of larvae infected died between 9 and 11 days	Sahara	Lab., field	Surcouf (1923)
"	Larvae	Rhabditis axei (Cobbold)	--	England	Lab.	Hague (1963)
"	Larvae, pupae, adults	Habronema muscae (Carter), H. megastoma (Rudolphi) Seurat, H. microstoma (Schneider)	100% induced infection with H. microstoma. H. muscae and H. megastoma do not develop in S. calcitrans	Australia	Lab.	Bull (1919)

PATHOGENS OF *STOMOXYS CALCITRANS* (STABLE FLIES) (continued)

Host	Host stage infected	Pathogen	% Incidence	Locality	Lab. or field study	Reference
Stomoxys calcitrans (continued)	Adults	Filaria stomoxeos (=? Habronema microstoma or ? Seteria cervi)	2 found infected out of 41 examined	--	Field	Von Linstow (1875)
"	Larvae, pupae, adults	Habronema microstoma	Among 63 flies, 10 pupae, and 12 larvae caught outdoors, only 1 adult was infected. In lab., a high % of adults could be infected in the larval stage	Australia	Lab., field	Hill (1919)
"	Adults	Habronema microstoma	Two infected out of 22 examined	Australia (Brisbane)	Lab., field	Johnston & Bancroft (1920)
"	Adults (?)	Habronema microstoma	--	Australia (Brisbane)	Lab., field	Johnston (1920)
"	Larvae, adults	Habronema microstoma	--	France	Lab.	Roubaud & Descazeaux (1922)
"	Adults (?)	Habronema muscae	--	Australia (Brisbane)	Field	Johnston (1913)
"	Adults	Microfilaria sanguinus equi africano	--	Philippines	Lab.	Mitzmain (1914)
"	Adults	Setaria cervi (Rudolphi) (=? Filaria labiato papillosa)	Only 3-4% of Stomoxys were infected in any locality despite high filarial rates in livestock	Italy	Lab., field	Noè (1913)

262

ABSTRACTS

Bernard Greenberg & Mary Ann Strand

Bull, L. B. (1919). A contribution to the study of habronemiasis: A clinical, pathological, and experimental investigation of a granulomatous condition of the horse - habronemic granuloma. Trans. R. Soc. S. Aust., 43: 85-141.

No mention of nematode killing fly. It is probable that heavily infected flies sicken and die sooner as noted in Musca fergusoni (Johnston and Bancroft, 1920) and Musca domestica (Patton and Cragg, 1913).

Gingrich, R. E. (1965). Bacillus thuringiensis as a feed additive to control dipterous pests of cattle. J. Econ. Entomol., 58: 363-364.

Each of three commercial preparations of B. thuringiensis was mixed with bovine faeces and also was fed directly to cattle: Haematobia irritans most susceptible; S. calcitrans most resistant; and Musca domestica, intermediate.

Hague, N. G. M. (1963). The influence of Rhabditis (Rabditella) axei (Rabditinae) on the development of Stomoxys calcitrans. Nematologica, 9: 181-184.

Nematode found in large numbers in routine cultures of the fly. Rate of emergence and number of flies emerging were decreased, thought due to competition for available food source between nematode and fly larvae, rather than pathogenicity.

Hill, G. F. (1919). Relationship of insects to parasitic diseases in stock. 1. The life history of Habronema muscae, Habronema microstoma, and Habronema megastoma. Proc. R. Soc. Victoria, n.s., 31: 11-107.

No mention of nematode killing fly.

Johnston, T. H. (1913). Notes on some Entozoa. Proc. R. Soc. Queensl., 24: 63-91.

First known isolation of larval nematode in Australia. Effect on flies not noted.

Johnston, T. H. (1920). Flies as transmitters of certain worm parasites of horses. Sci. Ind. (Melbourne), 2: 369-372.

No mention of nematode killing fly.

Johnston, T. H. & Bancroft, M. J. (1920). The life history of Habronema in relation to Musca domestica and native flies in Queensland. Proc. R. Soc. Queensl., 32: 61-88.

No mention of nematode killing fly.

Kennedy, J. C. (1906). Experiments on mosquitoes and flies. Rep. Comm. Medit. Fever. R. Soc. (London), 4: 83-84.

In a search for a vector of B. melitensis among goats, dissections of S. calcitrans which had been exposed to the pathogen and subsequent culturing of the organs revealed no Brucella.

Mitzmain, M. B. (1914).   An experiment with Stomoxys calcitrans in an attempt to transmit a filaria of horses in the Philippines.   Amer. J. trop. Med., 2: 759-763.

Parasitized flies suffered increased mortality during first 10 days of infection.

Noè, G. (1903).   Studî sul ciclo evolutivo della Filaria labiato papillosa Alessandrini. Atti R. Accad. Lincei, Ser. 5, 12: 387-393.

S. calcitrans appears refractory to infection; flies feeding on infected bovines never had more than 3 larvae in their gut.

Patton, W. S. & Cragg, F. W. (1913).   A textbook of medical entomology.   Christian Literature Society for India, London, 345 p.

Roubaud, E. (1911).   Etudes sur les Stomoxydes du Dahomey.   Bull. Soc. Pathol. Exot., 4: 122-132.

In a limited trial, 1/5 S. calcitrans died 24 hours after ingesting Entomophthora spores.

Roubaud, E. & Descazeaux, J. (1922).   Evolution de l'Habronema muscae Carter chez la mouche domestique et de l'H. microstoma Schneider chez le Stomoxe.   (Note prélim.).   Bull. Soc. Pathol. Exot., 15: 572-574.

No mention of nematode killing fly.

Roubaud, E. & Descazeaux, J. (1923).   Sur un agent bactérien pathogène pour les mouches communes: Bacterium delendae-muscae n. sp.   C. R. Hebd. Seances Acad. Sci., 177: 716-717.

Infection appeared spontaneously in laboratory rearings of the stable fly and in stable flies naturally breeding in rabbit droppings.   All specimens died between 2 and 30 days after eclosion when the larvae were infected; inoculated, the organism kills flies in 18 to 24 hours.   No deaths occurred in adults fed the organism, although adult house flies did die.

Surcouf, J. M. R. (1923).   Deuxième note sur les conditions biologiques du Stomoxys calcitrans L.   Bull. Mus. Natl. Hist. Nat. Paris, 29: 168-172.

A water solution of M. domestica which had been killed by E. muscae was sprayed on Stomoxys larvae at 23°C and high humidity.   Infection established in 9 days as 23°C and high humidity; one quarter of the larvae died between the ninth and the eleventh days.

Von Linstow (1875)   (See G. F. Hill)

First record in stable fly.

XIV. PATHOGENS OF *GLOSSINA* (TSETSE FLIES) [a]

Richard A. Nolan

Department of Biology
Memorial University of Newfoundland
St. John's, Newfoundland, Canada A1C 5S7

[a] This study was supported by National Research Council of Canada grant A-6665.

PATHOGENS OF GLOSSINA (TSETSE FLIES)

Host	Host stage infected	Pathogen	% Incidence	Locality	Lab. or field study	Reference
Glossina austeni	Adults	Bacteroids	100	Canada, England	Lab.	Huebner & Davey (1974)
G. brevipalpis	Adults	Rickettsia-like organism	-	Uganda, Tanzania, England	Field, lab.	Pinnock & Hess (1974)
"	Adults	Phycomycete	-	Somalia	Field	Moggridge (1936)
"	Adults	Mermithid larvae	0.02 (1/5000)	Uganda	Field	Moloo (1972)
G. fusca congolensis	Pupae	Absidia repens	45	Central African Republic	Field, lab.	Vey (1971)
"	Pupae	Penicillium lilacinum	50	"	Field	-
"	-	Penicillium janthinellum	-	"	-	Vago & Meynadier (1973)
"	-	Aspergillus flavipes	-	"	-	"
G. fuscipes	Adults	Rickettsia-like organism	-	Uganda, Tanzania, England	Field, lab.	Pinnock & Hess (1974)
"	Adults	Bacillus	~10	Uganda	Field	Carpenter (1912)
"	Adults	Bacillus	100	"	Field	Carpenter (1913)
"	Adults	Ascomycete	(1/455)	"	Field	Carpenter (1912)
"	Adults	Fungi Imperfecti	(3/455)	"	Field	"
"	Adults	Algal filaments	(2/455)	"	Field	"
"	Adults	Nematode larva	(1/205)	"	Lab.	"

PATHOGENS OF GLOSSINA (TSETSE FLIES) (continued)

Host	Host stage infected	Pathogen	% Incidence	Locality	Lab. or field study	Reference
G. fuscipes	Adults	Nematode	<0.4	Uganda	Field	Carpenter (1913)
G. f. fuscipes	Adults	Intranuclear virus like particles	-	"	Field, lab.	Jenni and Steiger (1974)
G. morsitans	Adults	Rickettsia like organism	-	Uganda, Tanzania, England	Field, lab.	Pinnock & Hess (1974)
"	Adults	Bacterium mathisi (unrecognized taxon)	-	Tanzania	Field	Roubaud & Treillard (1935)
"	Adults	Pseudomonas sp. / Aeromonas sp.	70-80 mortality	Austria	Lab.	Bauer (1974)
"	Adults	Leptomonas sp.	-	Nigeria	Field	Buxton (1955)
"	Adults	Cicadomyces sp.	~100	Federal Republic of Germany	Lab.	Nogge (1974)
"	Adults	Fungus	20	Tanzania	Field	Nash (1933)
"	Adults	Fungus	33 of old females	Nigeria	Field	Lester (1934)
"	Adults	Hexamermis sp.	-	Zambia	Field	Nickle (1973)
"	Adults	Mermis sp.	0.20	Tanzania	Field	Thomson (1947)
"	Adults	Nematode larvae (Mermis?)	-	Zaire	Field	Rodhain et al. (1913)
"	Adults	Immature nematode	1.0 (3/300)	Zambia	Field, lab.	Leiper (1912)
"	Adults	Herpetomonad	0.08 (3/3603)	Nigeria	Field	Lloyd (1924)

PATHOGENS OF GLOSSINA (TSETSE FLIES) (continued)

Host	Host stage infected	Pathogen	% Incidence	Locality	Lab. or field study	Reference
G. morsitans	Adults	Symbiont	~100	England	Lab.	Hill et al. (1973)
G. m. centralis	Adults	Virus like particles	-	Tanzania	Field	Jenni (1973)
G. m. morsitans	Adults	Bacteroids	100	Kenya	Lab.	Ma & Denlinger (1974)
G. pallicera	Adults	Flagellate	1/8	Liberia	Field	Foster (1964)
G. pallidipes	Adults	Rickettsia-like organism	-	Uganda, Tanzania England	Field, lab.	Pinnock & Hess (1974)
"	Adults	Bacteria-like organism	50	Uganda	Field, lab.	Rogers (1973)
G. palpalis	Adults	Borrelia glossinae	33 (2/6)	-	Field	Novy & Knapp (1906)
"	Adults	Bacterium mathisi	-	France	Lab.	Roubaud & Treillard (1936)
"	Adults	Entomophthora muscae	Does not infect	Angola	Lab.	Vanderyst (1923)
"	Adults	Phycomycete	-	Tanzania	Field	Swynnerton (1936)
"	Larvae, pupae, adults	Cicadomyces sp.	~100	Congo region (Zaire?)	Lab.	Roubaud (1919)
"	Adults	"	88 in lab., 50-91 in field	Uganda	Field, lab.	Wallace (1931)
"	Adults	Fungal mycelium	-	Congo region (Zaire?)	Lab.	Van Hoof & Henrard (1934)
"	Adults	Fungus with septate hyphae	1.3 (1/75)	Ghana	Field	Macfie (1915)
"	Adults	Haemogregarine (Hepatozoon sp.)	0.86 (4/465)	Dahomey	Field	Chatton & Roubaud (1913)

PATHOGENS OF GLOSSINA (TSETSE FLIES) (continued)

Host	Host stage infected	Pathogen	% Incidence	Locality	Lab. or field study	Reference
G. palpalis	Adults	Myxosporidium heibergi	∿50	Congo region (Zaire?)	Field	Dutton et al. (1907)
"	Adults	Hepatozoon pettiti	-	-	Field	Buxton (1955)
"	Adults	Haemogregarine	2.7 (2/75)	Ghana	Field	Macfie (1915)
"	Adults	Treponema macfiei	1.3 (1/75)	Ghana	Field	Macfie (1915)
"	Adults	Immature female Mermis	1 found	Uganda	Field	Leiper (1910)
"	Adults	Larval mermithid (Agamomermis)	0.37 (15/4001)	Liberia	Field	Foster (1963)
"	Adults	Larval mermithid	3 found	Nigeria	Field	Nickle (1974)
"	Adults	Flagellate	4.25 (170/4001)	Liberia	Field	Foster (1964)
G. tachinoides	Adults	Spirochaete	1 found	Nigeria	Field	Macfie (1914a)
"	Larvae, pupae, adults	Cicadomyces sp.	∿100	Congo region	Lab.	Roubaud (1919)
"	Adults	Fungus	11 of old females	Nigeria	Field	Lester (1934)
"	Adults	Reptilian haemogregarine	42, 45, 54	Nigeria	Field	Lloyd et al. (1924)
"	Adult females	Treponema macfiei	1 found	Nigeria	Field	Macfie (1914b)

PATHOGENS OF GLOSSINA (TSETSE FLIES) (continued)

Host	Host stage infected	Pathogen	% Incidence	Locality	Lab. or field study	Reference
Glossina sp.	Pupae	Coccoid bacterium	-	Chad	Field	Vey (1974)
"	Pupae	Aspergillus ochraceus	-	Chad	Field	Vey (1974)
"	Adults	Phycomycete	-	Tanzania	Field	Nash (1970)
"	Adults	Fusarium semitectum var. majus	1 found	South Africa	Field	Doidge (1950)
"	Adults	Entomophthora muscae	Does not infect	Dahomey	Lab.	Roubaud (1911)

ABSTRACTS

Richard A. Nolan

Bauer, B. (1974).   Bacterial infections in Glossina morsitans Westwood fed through membranes. (Abstr.) Proc. 3rd Int. Cong. Parasitol. (Munich), 2: 943.

   Pseudomonas and Aeromonas were isolated from dead and dying flies with black abdomens. Transmission of the bacteria by saliva of artificially infected flies was demonstrated. Most died within 8 days.

Buxton, P. A. (1955).   The natural history of tsetse flies.   An account of the biology of the genus Glossina (Diptera).   Lond. Sch. Hyg. Trop. Med. Mem., No. 10, 816 pp.

   Buxton reports a Leptomonas sp. infecting G. morsitans in Nigeria.   Also reference is made to an observation by Hoare in 1932 of G. palpalis infected by Hepatozoon pettiti. The tsetse had acquired the haemogregarine by feeding upon Crocodilus niloticus which was infected with Haemogregarina pettiti Thiroux.   It was assumed that transmission of the haemogregarine occurred when a fly was crushed in the crocodile's mouth.   This was the first report definitely establishing a Dipteran as an intermediate host for a haemogregarine.

Carpenter, G. D. H. (1912).   Progress report on investigations into the bionomics of Glossina palpalis, July 27, 1910 to August 5, 1911.   Rep. Sleeping Sickness Bur. Lond., 12: 79-111.

   The fly was originally reported as being G. palpalis, but this error was corrected in Carpenter (1913), and the identification changed to G. fuscipes.   Gut contents of 455 flies were examined.   An unidentified ascomycete with hyaline ascospores, a member of the Fungi Imperfecti, and filaments of an unidentified alga were infrequently observed in the gut contents.   Also, a larval nematode was found coiled in the gut of a single fly just after emergence.   Bacilli were also observed and Carpenter felt that there was a marked incompatibility between bacilli and trypanosomes in the gut of the fly.

Carpenter, G. D. H. (1913).   Second report on the bionomics of Glossina fuscipes (palpalis) of Uganda.   Rep. Sleeping Sickness Bur. Lond., 14: 1-37.

   When large numbers of bacilli were present, the fore-gut wall seemed thin and friable, and Carpenter felt it possible that the flies were "sickly".   Also, 4 nematodes were found in the coelomic cavity of separate flies of over a thousand flies dissected.

Chatton, E. & Roubaud, E. (1913).   Sporogonie d'une hémogrégarine chez une tsétsé (Glossina palpalis R. Desv.).   Bull. Soc. Pathol. Exot., 6: 226-233.

   The first report of a haemogregarine from Glossina.   It was found in the coelomic cavity.

Doidge, E. M. (1950).   The South African fungi and lichens to the end of 1945.   Bothalia, 5: 1-1094.

   Recognized as a valid fungal taxon, Fusarium semitectum Berk. and Rav. is considered to be extremely common in tropical and subtropical countries and to be a secondary invader of plant tissue and seldom found to be pathogenic.   However, some diseases of animals are produced by toxin-producing species of Fusarium (fusariotoxicoses).

Dutton, J. E. et al. (1907).   Concerning certain parasitic protozoa observed in Africa.
Part II.   Ann. Trop. Med. Parasitol., 1: 287-370.

   Myxosporidium heibergi should be considered of uncertain affinities due to the incomplete
   nature of the study.

Foster, R. (1963).   Infestation of Glossina palpalis R.-D. 1830 (Diptera) by larval
Mermithidae Braun 1883 (Nematoda) in West Africa, with some comments on the parasitization of
man by the worms.   Ann. Trop. Med. Parasitol., 57: 347-358.

   Of 4001 G. palpalis collected in Liberia, 15 were infested with larval mermithid.   The
   nematodes were found in the abdominal cavity lying loosely intertwined with the gut and
   Malpighian tubules.   This article is a good review of nematode records through
   Thomson (1947).   Adults of G. fusca Walker, G. pallicera Bigot and G. nigrofusca
   Newstead were also examined and found to be free of nematode infections.   Foster felt
   that the tsetse became infected while resting on the ground.   The 15 nematodes infecting
   G. palpalis were found in the coelomic cavity.

Foster, R. (1964).   An unusual protozoal infection of tsetse flies (Glossina Weidemann 1830
spp.) in West Africa.   J. Protozool. 11(1): 100-106.

   The flagellate was similar to, but not synonymous with Trypanosoma grayi Novy (1906).
   This unidentified flagellate accounted for approximately 90% of the total flagellate
   infections.   The infection was usually confined to the midgut with hind-gut infections
   being rare.   In about 8.5% of the infections, the flagellates occurred in the coelomic
   cavity only;   and in approximately 8.5% of the infections, they occurred in both the
   coelomic cavity and the midgut.   This is the first report of a trypanosome developing
   in the coelomic cavity of tsetse;   the mouthparts were never invaded.

Hill, P. et al. (1973).   The production of "symbiont-free" Glossina morsitans and an
associated loss of female fertility.   Trans. R. Soc. Trop. Med. Hyg., 67: 727-728.

   This reference is included in the coverage of the literature because it has implications
   for the use of microbial control of tsetse.   The microbial control envisaged here
   involves the elimination or reduction of a microorganism by chemical means in order to
   reduce the level of tsetse instead of the normal concept of introducing and/or increasing
   the level of a pathogenic microorganism.

Huebner, E. & Davey, K. G. (1974).   Bacteroids in the ovaries of a tsetse fly.   Nature,
249 (5454): 260-261.

   Bacteroids were present in the cytoplasm of oocytes and nurse cells of the ovary and
   occasionally in follicle cells and sheath cells.   The authors suggest that transovarian
   transmission of the symbiont may occur.

Jenni, L. (1973).   Virus-like particles in a strain of G. morsitans centralis, Machado (1970).
Trans. R. Soc. Trop. Med. Hyg., 67(2): 295.

   The VLP were found in cytoplasmic vesicles of salivary glands and were first seen in
   flies four days after emergence.   The VLP were composed of an electron-dense core
   approximately 240Å across and a surrounding envelope measuring 500-550Å in diameter.   The
   VLP were also associated with special rod-like structures.   This study and the discovery
   of a VLP in G. fuscipes fuscipes /Jenni, L. & Steiger, R. (1974)/ point out the need for
   established, susceptible tsetse cell lines to allow viral replication in vitro in order
   that a proper evaluation of viruses from Glossina can be made.

Jenni, L. & Steiger, R. (1974).   Viruslike particles in <u>Glossina fuscipes fuscipes</u> Newst.
1910.   <u>Acta Trop.</u>, <u>31</u>: 177-180.

The viruslike particles measure about 350-390Å in diameter and form intranuclear pseudo-
crystalline arrays in midgut epithelial cells.   Young flies were found to have disrupted
midguts, and the occurrence of VLP aggregates in older flies could be regarded as a
latent infection.   There was a high mortality rate in young flies (within 10 days after
emergence) and a low hatching rate.

Leiper, R. T. (1910).   /Exhibition of a series of Entozoa/.   <u>Proc. Zool. Soc. Lond.</u>, <u>79</u>:
147.

A single immature female <u>Mermis</u> was observed infecting <u>G. palpalis</u>.   The specimen was
three inches in length and from the coelomic cavity of the tsetse.

Leiper, R. T. (1912).   Mr Lloyd records a "New Nematode parasite of <u>Glossina morsitans</u>".
<u>J. Lond. Sch. Hyg. Trop. Med.</u>, <u>2</u>: 41-42.

Four immature nematodes were found in the tsetse coelomic cavity.   Two were in one fly.
Leiper refers to a similar immature <u>Mermis</u> found previously by a Professor Minchin a few
years prior to this report in <u>Glossina palpalis</u> in Uganda.

Lester, H. M. O. (1934).   Report of the tsetse investigation.   <u>Rep. Med. Health Serv.</u>
<u>Nigeria 1933</u>, pp. 74-83.

Flies infected with 2 unidentified fungi were found at the end of the wet season.

Lloyd, L. (1924).   Note on the occurrence of a herpetomonad in <u>Glossina morsitans</u>.   <u>Bull.</u>
<u>Ent. Res.</u>, <u>15</u>: 185-186.

The flagellate was found in the proboscis and in the midgut.   This organism was not
encountered in 10 000 <u>G. tachinoides</u> examined by Lloyd's group.   Wenyon, C. M. (1926,
p. 434) speculated that this organism might be a <u>Leishmania</u> as human leishmaniasis occurs
in Nigeria.

Lloyd, L. et al. (1924).   Second report of the tsetse-fly investigation in the Northern
Provinces of Nigeria.   <u>Bull. Ent. Res.</u>, <u>15</u>: 1-27.

An unidentified reptilian haemogregarine was found in <u>G. tachinoides</u>,   The incidence in
the fly depended upon the reptilian host: crocodile, frogs, and colubrine snake, 42%;
monitors and lizards (<u>Varanus</u>), 45%; and chameleon, 54%.   No evidence of pathogenicity
to <u>G. tachinoides</u> was found.

Ma, W.-C. & Denlinger, D. L. (1974).   Secretory discharge and microflora of milk gland in
tsetse flies.   <u>Nature</u>, <u>247</u>(5439): 301-303.

The results in <u>G. morsitans morsitans</u> which indicated the presence of the possibly gram-
negative bacteria in the lumen of the milk gland were also confirmed by comparative
observations of <u>G. austeni</u> Newstead and <u>G. longipalpis pallidipes</u> Austen.   This study
indicates some ultrastructural differences between the bacteroids of the mycetome and the
bacteria in the milk gland lumen.

Macfie, J. W. S. (1914a).   Notes on some blood parasites collected in Nigeria.   Ann. Trop. Med. Parasitol., 8: 439-465.

One tsetse fly "out of a number examined" was infected by an unidentified spirochaete. Macfie felt that this spirochaete was distinct from Spirochaeta glossinae described from G. palpalis by Novy & Knapp (1906).   Macfie kept the tsetse in captivity for 22 days before it was killed for examination, and the fly did not appear to be affected by the presence of the large number of active spirochaetes.

Macfie, J. W. S. (1914b).   The occurrence of a spirochaete in the gut of Glossina tachinoides. Ann. Trop. Med. Parasitol., 8: 464-465.   The spirochaete was apparently non-pathogenic.

Macfie, J. W. S. (1915).   The results of dissections of tsetse flies at Accra.   Rep. Accra Lab. Lond., pp. 49-54, 98-99.

Seventy-five flies were dissected and 3 possibly parasitic organisms were found.   The abdomen of one infected fly was swollen; and it was also hard and pale-coloured posteriorly.   Death was felt to be due, without doubt, to a fungal infection.   Reproduction of the fungus appeared to occur via arthrospore formation.   The hyphae formed a dense network in which the lower half of the gut and the testicles were firmly embedded. Spirochaetes were found in the gut of another fly and appeared to be morphologically identical to those found by Macfie in the gut of a specimen of G. tachinoides in Nigeria. Also, sporocysts were found in the abdominal cavities of 2 flies and were similar to those found by Chatton and Roubaud (1913) in the abdominal cavity of the same species.

Moggridge, J. Y. (1936).   Some observations on the seasonal spread of Glossina pallidipes in Italian Somaliland with notes on G. brevipalpis and G. austeni.   Bull. Ent. Res., 27: 449-466.

An unidentified Phycomycete was found in G. brevipalpis.   Moggridge noted that the fungus was similar to a pathogenic fungus known in certain areas of Tanzania and that the fungus was present on the abdomen of the flies.

Moloo, S. K. (1972).   Mermithid parasite of Glossina brevipalpis Newstead.   Ann. Trop. Med. Parasitol., 66: 159.

Two mermithid larvae were found in the coelomic cavity of the one infected adult tsetse. The 5000 flies examined included G. fuscipes, G. brevipalpis and G. pallidipes.   Moloo felt that the infection of Glossina was accidental.

Nash, T. A. M. (1933).   The ecology of Glossina morsitans, Westw., and two possible methods for its destruction.   Part I.   Bull. Ent. Res., 24: 107-157.

The article includes the statement that a Mr Scott did some work with a fungus and proved that it was definitely pathogenic to G. morsitans.   Infected larvae had black spots on their abdomens.

Nash, T. A. M. (1970).   Control by parasites and predators of Glossina, pp. 521-532. In Mulligan, H. W. (ed.), The African Trypanosomiases, Allen & Unwin Ltd, London.

A personal communication from W. H. Potts reports that infection by an unidentified Phycomycete neither prevented completion of development of tsetse larvae nor their deposition.

Nickle, W. (1973).   Personal communication to R. F. Myers.   A nematode, tentatively identified as Hexamermis sp., was found in the coelomic cavity of G. morsitans.   The collection was made by S. N. Okiwelu.

Nickle, W. (1974).   Personal communication to R. F. Myers.   Three unidentified larval mermithid were found in the coelomic cavities of G. palpalis adults.   The collection was made by K. Riordan.

Nogge, C. (1974).   Investigations on the role of symbionts in tsetse flies (Glossina morsitans).   Proc. 3rd Int. Cong. Parasitol. (Munich) 2: 947.

  Symbiont-free flies were obtained by oral treatment with lysozyme.   Attempts were made to compensate for the loss of the symbionts by feeding defibrinated blood supplemented by B vitamins.   Other techniques such as the raising and lowering of the temperature, the feeding of blood with added sodium oxalate, 7-chlortetracycline, and lysozyme in different concentrations as well as the injection of lysozyme into the coelom resulted in destruction of the symbionts as well as a decrease in the longevity of the flies.   The blood meals of these latter flies remained undigested.

Novy, F. G. & Knapp, R. E. (1906).   Studies on Spirillum obermeieri and related organisms. J. Infect. Dis., 3: 291-393.

  A bacterium, Borrelia glossinae, was described as Spirillum glossinae in 1906 from smears of flies' stomach contents.   It is listed in Buchanan and Gibbons (Buchanan, R. E. & Gibbons, N. E.   1974.   Bergey's Manual of Determinative Bacteriology.   8th Edition. The Williams and Wilkins Company.   Baltimore. p. 190) as a Species incertae sedis because of inadequate description.

Pinnock, D. E. & Hess, R. T. (1974).   The occurrence of intracellular rickettsia-like organisms in the tsetse flies, Glossina morsitans, G. fuscipes, G. brevipalpis and G. pallidipes.   Acta Trop., 31: 70-79.

  Rickettsia-like organisms were observed in electron microscopy preparations of cells of the midgut epithelium, including the mycetome, in cells associated with the fat body and in developing oocytes.   The authors concluded that the presence of a lytic zone around the organisms and that the disruption of cells at the ultrastructural level suggested that the rickettsia-like organisms were parasitic.

Rodhain, J. et al. (1913).   Rapport sur les Travaux de la Mission Scientifique du Katanga (Octobre 1910 à Septembre 1912).   Hayez. Bruxelles. 258 p.

  Single nematode larvae (possibly Mermis sp.) were found in 4 flies.   The total number examined was not given.

Rogers, A. (1973).   Microorganisms in spermathecae of wild Glossina pallidipes.   Trans. R. Soc. Trop. Med. Hyg., 67: 299.

  These microorganisms occurred in the lumen of the spermathecae in both inseminated and uninseminated females in wild populations and also occurred at high concentrations in the spermathecae of laboratory-reared females.

Roubaud, E. (1911).   Etudes sur les Stomoxydes du Dahomey.   Bull. Soc. Pathol. Exot., 4, 122-132.

  Roubaud attempted to infect by means of direct contact with Entomophthora muscae: 8 G. palpalis, 4 G. tachinoides, and one G. longipalpis and by means of ingestion of the spores: 10 Glossina palpalis. The Glossinas were not infected by either means. The sporangiospores used had been attached to potato leaves and had originated from an infected Stomóxys cálcitrans L. However, E. Müller-Kogler (Pilzkrankheiten bei Insekten. Parey. Berlin. 1965, pp. 160-161) has previously pointed out the lack of controls in this study.

Roubaud, E. (1919). Les particularités de la nutrition et la vie symbiotique chez les mouches tsétsés. Ann. Inst. Pasteur, 33: 489-536.

Cicadomyces sp. is a symbiote of G. palpalis and G. tachinoides. The mycetome is located in the middle of the anterior segment of the mid-gut, where absorption but no digestion occurs (Wigglesworth, V. B. (1972). The Principles of Insect Physiology. Chapman & Hall. London, p. 523).

Roubaud, E. & Treillard, M. (1935). Un coccobacille pathogène pour les mouches Tsétsés. C. R. Hebd. Séances Acad. Sci. (Paris) 201: 304-306.

This microorganism, Bacterium mathisi, which initially showed promise as a biocontrol agent has neither been maintained in culture nor is it a recognized valid taxon. Roubaud & Treillard (1935) found that the bacterium was taken in via the proboscis during feeding on guinea pigs whose hair and skin had been coated with the bacterium. The flies are apparently killed by a septicemic infection within the coelom within 1-8 days. Several insects were shown to be susceptible when inoculated with bacterial suspensions; however, Culex pipiens, Stegomyia sp. and Aedes aegypti survived ingestion of the microbe.

Roubaud, E. & Treillard, M. (1936). Infection expérimentale de Glossina palpalis par un coccobacille pathogène pour les muscides. Bull. Soc. Pathol. Exot., 29: 145-147.

No naturally occurring infection by Bacterium mathisi was found in the several hundred Glossina palpalis shipped from Uganda in the pupal stage. The low number of flies used in the experiments makes interpretation difficult; however, it should be noted that a septicemia was not indicated in all cases of adult mortality. When G. morsitans from Tanzania were used in experiments, they died of a septicemia in 1 to 3 days after feeding on contaminated guinea pigs.

Swynnerton, C. F. M. (1936). The tsetse flies of East Africa. A first study of their ecology, with a view to their control. Trans. R. Entomol. Soc. Lond., 84: 1-580.

An unidentified Phycomycete was found infecting G. palpalis. Material for the cultivation of the fungus was sent to Johannesburg; however, the study does not appear to have been continued. Swynnerton felt that the hope of success using this particular fungus was small because it seemed to be fatal only during periods of wet weather.

Thomsón, W. E. F. (1947). Nematodes in tsetse. Ann. Trop. Med. Parasitol., 41: 164.

The nematodes were located in the coelomic cavity and had an average length of 79 mm.

Vago, C. & Meynadier, S. pp. 24, 27, In Strains of Entomophagous Microorganisms. International Organization for Biological Control of Noxious Animals and Plants. WPRS Bulletin 1973/3.

Aspergillus flavipes and Penicillium janthinellum were isolated from G. fusca congolensis in 1965. Both fungi have been previously isolated from soil in Africa.

Vanderyst, P. H. (1923).   La prophylaxie contre la trypanose humaine.   Rev. Méd. Angola, 4: 32-42.

A local fly "Nkulu nyansi" was very rapidly infected with Entomophthora muscae when placed in a tube with one of the infected M. domestica L. received from Europe; however, G. palpalis was not infected under the same conditions.   The experiment was later supplemented by adding locally obtained M. domestica, and these were also not infected. One of the tsetses used did die during the course of the experiment but did not show any signs of a disease.   The infection of the fly "Nkulu nyansi" was felt to be enhanced by injury during insect procurement.   Other experiments with this local fly gave negative results.   This is the first attempt at assessing the virulence of a microbial pathogen derived from outside Africa.

van Hoof, L. & Henrard, C. (1934).   La transmission cyclique de races résistantes de Trypanosoma gambiènse par Glossina palpalis.   Ann. Soc. Belge Med. Trop., 14: 109-144.

Hyphae of an unidentified fungus occurred in the coelomic cavity, in the gut, in the musculature of the thorax and in the proventriculus of G. palpalis.   These authors cite a reference (Lloyd & Johnson, 1922), in the text to a similar infection but do not include it in the bibliography.

Vey, A. (1971).   Recherches sur les champignons pathogènes pour les glossines.   Etudes sur Glossina fusca congolensis Newst. et Evans. en République Centrafricaine.   Rev. Elev. Méd. Vét. Pays Trop. (N.S.) 24: 577-579.

Two fungal species were isolated from tsetse pupae, Absidia repens and Penicillium lilacinum.   Both are common soil inhabitants.   Experimental infections of tsetse larvae are reported.   Larvae were wounded prior to incubation with the spore suspensions.

Vey, A. (1974).   Personal communication to R. Nolan.   An unidentified coccoid bacterium was isolated from Glossina pupae at the Laboratory I.E.M.U.T. de Farcha (Chad).   Aspergillus ochraceus was isolated from Glossina pupae into pure culture in the Laboratory I.E.M.U.T. de Farcha (Chad).   A. ochraceus has also been found in material of Theobroma cacao from the Gold Coast (Dade, H. A. 1940.   A revised list of Gold Coast Fungi and Plant Diseases. Royal Botanic Gardens.   Kew Bulletin.   Bulletin of Miscellaneous Information No. 6, pp. 205-247).

Wallace, J. M. (1931).   Micro-organisms in the gut of Glossina palpalis.   Ann. Trop. Med. Parasitol., 25: 1-19.

A fairly thorough study which would indicate that G. palpalis does not readily pick up bacteria from contaminated surroundings, such as fly boxes; however, during feeding experiments organisms can be introduced into the gut.   Wallace felt that some earlier workers had introduced bacteria which were surface contaminants into gut preparations during dissection.   He also observed that in every instance where Trypanosoma grayi was present the bacteroids were also present.   This latter finding is contrary to those of Carpenter (1912). Wallace's comments with regard to the general inability of G. palpalis to acquire bacteria within the gut by feeding through contaminated skin or hair because of a possible downward flow of saliva or because saliva is bactericidal is contrary to the findings of Roubaud & Treillard (1935) using G. morsitans.

XV. PATHOGENS OF
SIPHONAPTERA (FLEAS) [a]

Mary Ann Strand

Boyce Thompson Institute
Yonkers, NY 10701, USA

---

[a] The literature for this table and bibliography covers the period 1962–1975.

PATHOGENS OF SIPHONAPTERA (FLEAS)

Host	Host stage infected	Pathogen	% Incidence	Locality	Lab. or field study	Reference
Catallagia sculleni rutherfordi	-	Psyllotylenchus viviparus	2-5	USA (California)	Field	Poinar & Nelson (1973)
Catallagia sp.	-	Psyllotylenchus viviparus	"	"	"	"
Ceratophyllus gallinae	-	Yersinia (Pasteurella) pestis	-	-	-	Jenkins (1964g,v)
"	-	Legerella parva	20-40	-	-	Jenkins (1964p)
"	-	Actinocephalus parvus	-	-	-	Jenkins (1964p,y)
"	-	Steinina rotundata	-	-	-	Jenkins (1964a)
Ceratophyllus laeviceps	-	Yersinia (Pasteurella) pestis	-	USSR	Lab.	Kondrashkina et al. (1968)
"	-	Psyllotylenchus (Heterotylenchus) pavlovskii	-	"	Field	Kurochkin (1961)
"	-	Mastophorus muris	15.5	"	"	Akopyan (1968)
Ceratophyllus tesquorum	-	Yersinia (Pasteurella) pestis	-	"	Lab.	Kondrashkina et al. (1968)
Cotptopsylla lamellifera	-	Psyllotylenchus (Heterotylenchus) pavlovskii	-	"	Field	Kurochkin (1961)
Ctenocephalides canis	-	Filamentous rickettsiae	-	-	-	Jenkins (1964e)
"	-	Yersinia (Pasteurella) pestis	-	-	-	Jenkins (1964g,v)

PATHOGENS OF SIPHONAPTERA (FLEAS) (continued)

Host	Host stage infected	Pathogen	% Incidence	Locality	Lab. or field study	Reference
Ctenocephalides canis	-	Crithidia ctenocephali	-	-	-	Jenkins (1964r)
"	-	Herpetomonas ctenocephali	3	-	-	Jenkins (1964i,1)
"	-	Leptomonas ctenocephali	-	-	-	Jenkins (1964b,x)
"	-	Nosema cteno-cephali/pulicis	17	-	-	Jenkins (1964j)
"	-	Dirofilaria immitis	-	-	-	Jenkins (1964c)
Ctenocephalides felis	-	Yersinia (Pasteurella) pestis	-	-	-	Jenkins (1964g,v)
"	-	Spirochaeta ctenocephali	-	-	-	Jenkins (1964q)
"	-	Crithidia sp.	-	-	-	Jenkins (1964s,t)
"	-	Herpetomonas ctenocephali	3	-	-	Jenkins (1964i)
"	-	Nosema cteno-cephali	-	-	-	Jenkins (1964j,k)
"	-	Dipylidium caninum	35 mortality	-	-	Jenkins (1964d)
"	-	Dirofilaria immitis	-	-	-	Jenkins (1964c)
Ctenophthalmus avernus	-	Mastophorus muris	Rare	France	Field	Beaucournu & Chabaud (1963)
"	-	Mermithid	Rare	"	Field	Rothschild (1969)
Diamanus montanus	-	Yersinia (Pasteurella) pestis	-	-	-	Jenkins (1964g,v)

PATHOGENS OF SIPHONAPTERA (FLEAS) (continued)

Host	Host stage infected	Pathogen	% Incidence	Locality	Lab. or field study	Reference
Diamanus montanus	-	Psyllotylenchus viviparus	2-5	USA (California)	Field	Poinar & Nelson (1973)
Leptopsylla segnis	-	Pasteurella pestis	-	-	-	Jenkins (1964g,v)
"	-	Herpetomonas ctenopsyllae	-	-	-	Jenkins (1964m)
Monopsyllus ciliatus protinus	-	Psyllotylenchus viviparus	2-5	USA (California)	Field	Poinar & Nelson (1973)
Monopsyllus wagneri	-	Psyllotylenchus viviparus	"	USA (California)	Field	" "
Myoxopsylla laverani	-	Mermithid	Rare	France	Field	Rothschild (1969)
Neopsylla setosa	-	Pasteurella pestis	-	USSR	Lab.	Konkrashkina et al. (1968)
Nosopsyllus fasciatus	-	Filamentous rickettsiae	-	-	-	Jenkins (1964u)
"	-	Yersinia (Pasteurella) pestis	-	-	-	Jenkins (1964g,v)
"	-	Malpighiella refringens	-	-	-	Jenkins (1964n)
"	-	Herpetomonas pattoni	85	-	-	Jenkins (1964m)
"	-	Trypanosoma lewisi	-	-	-	Jenkins (1964f,o)
"	-	Legerella fasciatus	-	-	-	Jenkins (1964p)
"	-	Agrippina bona	-	-	-	Jenkins (1964w)
"	-	Spiroptera obtusa?	-	-	-	Jenkins (1964h)
"	-	Hymenolepis diminuta	-	-	-	" "
"	-	H. murina	-	-	-	" "
Pulex irritans	-	Filamentous rickettsiae	-	-	-	Jenkins (1964e)

PATHOGENS OF SIPHONAPTERA (FLEAS) (continued)

Host	Host stage infected	Pathogen	% Incidence	Locality	Lab. or field study	Reference
Pulex irritans	-	Yersinia (Pasteurella) pestis	-	-	-	Jenkins (1964g,v)
"	-	Crithidia pulicis	-	-	-	Jenkins (1964s,t)
"	-	Herpetomonas ctenophali	3	-	-	Jenkins (1964i)
"	-	Leptomonas pulicis	-	-	-	Jenkins (1964)
Spilopsyllus cuniculi	-	Mermithid	Rare	-	Lab.	Rothschild (1969)
"	-	Leptomonas sp.	50	-	Field	"
Xenopsylla cheopis	-	Yersinia (Pasteurella) pestis	80.4% on A1122, 5.8 on TRU	USA	Lab.	Kartman & Quan (1964)
"	-	"	-	-	-	Jenkins (1964g,v)
"	-	"	-	-	-	Kondrashkina et al. (1968)
"	-	Salmonella enteritidis	-	USSR	Lab.	Vaschenok et al. (1971)
"	-	Leptomonas ctenocephali	-	India	Lab.	Amonkar & Mushi (1965)

284

Akopyan, M. M. (1968).   Parasitic castration of the flea.   In: N. D. Levine, ed. &
F. K. Plous, jr, tr.   Natural nidality of diseases and questions of parasitology.   Urbana,
Univ. Illinois Press, pp. 399-401.

The nematode Mastophorus muris caused castration of male Ceratophyllus laeviceps.   The
castrates made up an average of 15.5% of the collections.   In other reports from the
USSR, more than 50% were castrates.   Helminths including roundworms and tapeworms are
also reported to cause castration in fleas.

Amonkar, S. V. & Munshi, D. M. (1965).   On the presence of leptomonads in the digestive
tract of the Indian rat flea Xenopsylla cheopis (Roths).   J. Anim. Morphol. Physiol., 12(1):
32-41.

The appearance of various forms of leptomonads in the rat flea are described.   The
leptomonads are morphologically indistinguishable from Leptomonas ctenocephali described
from dog fleas.   The dialaled pyloric region of the ileum of infected fleas is found to
be practically blocked by a large number of these flagellates.

Beaucournu, J.-C. & Chabaud, A.-G. (1963).   Infestation spontanée de puces par le spiruride
Mastophorus muris (Gmelin).   Ann. Parasitol. Hum. Comp., 38: 931-934.

The flea Ctenophthalmus avernus, collected in France was found to be infected by
M. muris.   The nematode was also found in field mice which were hosts to the fleas.
Although this nematode has been experimentally transferred to insect hosts, it has been
rarely found infecting insects in nature.

Jenkins, D. W. (1964).   Pathogens, parasites, and predators of medically important arthropods.
Bull. WHO, 30 (Suppl.), 150 pp.

The following papers were cited by Jenkins.   Information for the flea pathogen table
was derived from Jenkins' listings.

(a)   Ashworth, J. H. & Rettie, T. (1912).   On a gregarine: Steinina rotundata, nov.
      sp., present in the mid-gut of bird-fleas of the genus Ceratophyllus.   Proc. R. Soc.
      Lond. Ser. B. Biol. Sci., 86: 31-38.

(b)   Basile, C. (1920).   Leishmania, Herpetomonas and Crithidia in fleas.
      Parasitology, 12: 366.

(c)   Breinl, A. (1921).   Preliminary note on the development of the larvae of
      Dirofilaria immitis in dog-fleas Ctenocephalus felis and canis.   Ann. Trop. Med.
      Parasitol., 14: 389-392.

(d)   Chen, H. T. (1934).   Reactions of Ctenocephalides felis to Dipyliduum canimum.
      Z. Parasitenkd., 6: 603-637.

(e)   Cowdry, E. V. (1923).   The distribution of Rickettsia in the tissues of insects
      and arachnids.   J. Exp. Med., 37: 431-456.

(f)   Garnham, P. C. C. (1959).   Some natural protozoal parasites of mosquitos with special reference to Crithidia.   Trans. 1st Inst. Conf. Insect Pathol. Biol. Control (Prague), pp. 287-294.

(g)   Holdenried, R. (1952).   Sylvatic plague studies.   VIII. Notes on the alimentary and reproductive tracts of fleas, made during experimental studies of plague. J. Parasitol., 38: 289-292.

(h)   Johnson, T. H. (1913).   Notes on some entozoa.   Proc. R. Soc. Queensl., 24: 63-91.

(i)   Khodukin, N. I. (1927).   /On the protozoa of the intestins of canine fleas in Tashkent and their role in the epidemiology of canine Leishmaniasis./   Med. Mysl' Uzes., 2: 69-73.

(j)   Korke, V. T. (1916).   On a Nosema (Nosema pulicis n. sp.) parasitic on the dog flea (Ctenocephalus felis).   Indian J. Med. Res., 3: 725-730.

(k)   Kudo, R. R. (1924).   A biologic and taxonomic study of the Microspiridia. Ill. Biol. Monogr., 9: 76-344.

(l)   Laveran, A. & Franchini, G. (1913).   Infections expérimentales de mammifères par des flagelles du tube digestif de Ctenocephalus canis et d'Anopheles maculipennis. C. R. Hebd. Seances Acad. Sci., 157: 744-747.

(m)   Laveran, A. & Franchini, G. (1914).   Infection naturelle du rat et de la souris au moyen de puces de rat parasitées par Herpetomonas pattoni.   C. R. Hebd. Seances Acad. Sci., 158: 450-453.

(n)   Minchin, E. A. (1910).   On some parasites observed in the rat flea (Ceratophyllus fasciatus).   In: Festschrift zum 60.   Geburtstage Richard Hertwigs, Jena, Fischer, Vol. 1, pp. 289-302.

(o)   Minchin, E. A. & Thomson, J. D. (1915).   The rat-trypanosome, Trypanosoma lewisi, in its relation to the rat-flea, Ceratophyllus fasciatus.   Q. J. Microsc. Sci., 60: 463-692.

(p)   Nöller, W. (1914).   Die Übertragunsweise der Rattentrypanosomen.   II.   Arch. Protistenkd., 34: 295-335.

(q)   Patton, W. S. (1912).   Spirochaeta ctenocephali, sp. nov., parasitic in the alimentary tract of the Indian dog-flea Ctenocephalus felis.   Ann. Trop. Med. Parasitol., 6: 357.

(r)   Patton, W. S. & Rao, S. (1921).   Studies on the flagellates of the genera Herpetomonas, Crithidia, and Rhynchoidomonas.   2.   The morphology and life-history of Crithidia ctenocephali, sp. nov., parasitic in the alimentary tract of Ctenocephalus canis Curtice.   Indian J. Med. Res., 8: 593-612.

(s)   Patton, W. S. & Rao, S. (1921).   Studies on the flagellates of the genera Herpetomonas, Crithidia, and Rhynchoidomonas.   5.   The morphology and life-history of Herpetomonas pulicis, sp. nov., parasitic in the alimentary tract and malpighian tubes of Pulex irritans, L.   Indian J. Med. Res., 8: 621-628.

(t)   Peus, F. (1938).   Die Flöhe.   Leipzig, Schöps, 106 pp.

(u) Porter, A. (1911).   The structure and life-history of _Crithidia pulicis_, n. sp., parasitic in the alimentary tract of the human flea _Pulex irritans_.   _Parasitology_, _4_: 237-254.

(v) Steinhaus, E. A. (1946).   Insect microbiology; an account of the microbes associated with insects and ticks, with special reference to the biologic relationships involved, _Ithaca, N. Y., Comstock_, 763 pp.

(w) Strickland, C. (1912).   _Agrippina bona_ nov. gen. et nov. sp. representing a new family of gregarines.   _Parasitology_, _5_: 97-108.

(x) Tyzzer, E. E. & Walker, E. L. (1919).   A comparative study of _Leishmania infantum_ of infantile kala azar and _Leptomonas (Herpetomonas) ctenocephali_ parasitic in the gut of the dog flea.   _J. Med. Res._, _40_: 129-176.

(y) Wellmer, L. (1910).   Sporozoen Ostpreussischer Arthropoden.   _Schr. Phys.-Ökon. Ges. Königsb. i. Pr._, _52_: 102-164.

Kartman, L. & Quan, S. F. (1964).   Notes on the fate of avirulent _Pasteurella pestis_ in fleas.   _Trans. R. Soc. Trop. Med. Hyg._, _58_(4): 363-365.

Xenopsylla cheopis fleas were fed six avirulent strains of _Yersinia (P.) pestis_ in suspensions of rat blood.   Three of the strains multiplied in the fleas and caused blockage of the proventricular valve.   The blocking rate was in direct relation to the number of bacilli in the blood meal.

Kondrashkina, K. I. et al. (1968).   /Some questions of the mutual adaptation of _Pasteurella pestis_ and fleas./ _Parazitologiya_, _2_(6): 543-547).   (R, e).

Uninfected fleas used twice as much oxygen as ones infected with _Yersinia (P.) pestis_. Uninfected fleas were slightly larger.   _Yersinia (P.) pestis_ apparently exerts a toxic effect on the fleas; however, lysogenous substances are formed in the fleas in response to the infection and fleas may rid themselves of the microbes.

Kurochkin, Y. V. (1961).   _Heterotylenchus pavlovskii_ sp. n., a nematode castrating plague-carrying fleas.   _Dokl. Akad. Nauk SSSR Biol. Sci. Sect._ (Transl.), _135_: 952-954.

A new nematode species was described from _Coptopsylla lamellifer_ and _Ceratophyllus laeviceps_ collected in Astrakhan region.   The parasite is present in all life stages of the fleas and invariably causes castration.   It is widely distributed and causes a high rate of infection.

Poinar, G. O., jr & Nelson, B. C. (1973).   _Psyllotylenchus viviparus_ n. gen., n. sp. (Nematodea: Tylenchida: Allantonematidae) parasitizing fleas (Siphonaptera) in California. _J. Med. Entomology_, _10_(4): 349-354.

A new species of nematode was found in the body cavities of four species of fleas. Under favourable moist conditions, nematode parasites may depress certain flea populations significantly.   Infection resulted in partial castration in one male of five examined.

Rothschild, M. (1969).   Notes on fleas with the first record of a mermithid nematode from the order.   _Proc. Brit. Entomol. Nat. Hist. Soc._, _1_(1): 1-8.

A mermithid nematode has been observed infecting _Spilopsyllus cuniculi_, _Myoxopsylla laverani_, and _Ctenophthalmus avernus_ in France.   In all cases the infected individuals were females.   The infection is very rare.   Flagellate parasites, _Leptomonas_ spp., are more common.   They infect a large number of fleas including dog, cat, and human fleas. They are primarily parasites of the hindgut, but no pathological effects are reported.

Vashchenok, V. S. et al. (1971).   /The ability of fleas <u>Xenopsylla cheopis</u> Roths. to preserve and transmit <u>Salmonella enteritidis</u> (Gärtner)./   Parazitologiya, 5: 15-19.   (R, e)

The flea can become infected with <u>S. enteritidis</u> as a consequence of feeding on infected white mice.   The bacterium was pathogenic to the fleas.   Uninfected fleas had an average life span twice as long as infected ones.   However, when given access to healthy hosts, the fleas could rid themselves of the infection.

XVI. PATHOGENS OF BLATTIDAE (COCKROACHES) [a]

Mary Ann Strand

Boyce Thompson Institute
Yonkers, NY 10701, USA

and

Marion A. Brooks

Department of Entomology, Fisheries, and Wildlife
University of Minnesota
St. Paul, MN 55101, USA

---

[a] The literature for this table and bibliography covers the period 1962–1975.

PATHOGENS OF BLATTIDAE (COCKROACHES)

Host	Host stage infected	Pathogen	% Incidence	Locality	Lab. or field study	Reference
BACTERIA						
Blaberus craniifer	Adults	Pseudomonas aeruginosa	-	USA	Lab.	Anderson et al. (1973)
"	Adults	Escherichia coli	-	USA	Lab.	"
"	Adults	Escherichia freundii	-	USA	Lab.	Briscoe et al. (1961, 1963)
"	Adults	Salmonella typhosa	-	USA	Lab.	Anderson et al. (1973)
"	Adults	Salmonella choleraesuis	-	USA	Lab.	Briscoe et al. (1961, 1963)
"	Adults	Diplococcus pneumoniae	-	USA	Lab.	Anderson et al. (1973)
"	Adults	Aerobacter aerogenes	-	USA	Lab.	Briscoe et al. (1961, 1963)
"	Adults	Aerobacter cloacae	-	USA	Lab.	"
"	Adults	Paracolobactrum aerugenoides	-	USA	Lab.	Briscoe et al. (1961)
"	Adults	Paracolobactrum intermedium	-	USA	Lab.	Briscoe et al. (1961, 1963)
"	Adults	Proteus morganii	-	USA	Lab.	Briscoe et al. (1961)
"	Adults	Klebsiella sp.	-	USA	Lab.	Briscoe et al. (1963)
Blaptica dubia	Larvae	Pseudomonas aeruginosa	-	Federal Republic of Germany	Lab.	Sauerlander & Ehrhardt (1961), Sauerlander & Kohler (1961)
"	Larvae	Serratia marcescens	-	Federal Republic of Germany	Lab.	Sauerlander & Ehrhardt (1961)
Blatta orientalis	-	Bacillus thuringiensis	-	USSR	Lab.	Yarnykh & Tonkozhenko (1969)
Blattella germanica	-	Streptococcus	30	Poland	Lab.	Ulewicz & Zawistowski (1973)
"	-	Escherichia coli	70	Poland	Lab.	"
"	-	Serratia marcescens	-	USA	Lab.	Henry (1965)
"	-	Pseudomonas aeruginosa	-	USA	Lab.	"
"	-	Aerobacter aerogenes	-	USA	Lab.	"
"	-	Bacillus lentimoribus	-	USA	Lab.	"

PATHOGENS OF BLATTIDAE (COCKROACHES) (continued)

Host	Host stage infected	Pathogen	% Incidence	Locality	Lab. or field study	Reference
Blattella germanica	-	B. cereus	-	USA	Lab.	Henry (1965)
"	-	B. alvei	-	USA	Lab.	"
"	-	B. thuringiensis	-	USA	Lab.	"
"	-	"	-	USSR	Lab.	Yarnykh & Tonkozhenko (1969)
Leucophaea maderae	-	Citrobacter sp.	-	USA (Oregon)	Field	Thomas & Poinar (1973)
"	-	Haffnia (Erwinia?)	-	USA (Oregon)	Field	"
"	-	Streptococcus faecalis	-	USA (oregon)	Field	"
Periplaneta americana	-	Bacillus thuringiensis	-	Australia	Lab.	Ryan & Nicholas (1972)
"	-	Escherichia coli	-	Australia	Lab.	"
"	-	Corynebacterium sp.	-	Australia	Lab.	"
"	-	Pseudomonas aeruginosa	-	Federal Republic of Germany	Lab.	Sauerlander & Ehrhardt (1961), Sauerlander & Kohler (1961)
"	-	"	-	Pakistan	Field	Zuberi et al. (1969)
"	-	Serratia marcescens	-	Federal Republic of Germany	Lab.	Sauerlander & Ehrhardt (1961)
"	-	"	-	USA	Lab.	Steinhaus & Marsh (1962)
"	-	"	-	USA	Lab.	"
Shelfordella tartara	-	Bacillus thuringiensis	-	USSR	Lab.	Yarnykh & Tonkozhenko (1969)
Supella supellectilium	-	Staphylococcus aureus	48	USA (Kansas)	Field	Lord et al. (1964)
FUNGI						
Blabera fusca	Larvae	Mucor hiemalis	-	France	Lab.	Heitor (1962)
Blatta orientalis	Larvae, adults	Herpomyces stylopygae	-	USA	Lab.	Tavares (1966)
Blattella germanica	-	Aspergillus flavus	-	USA	Lab.	Henry (1965)

PATHOGENS OF BLATTIDAE (COCKROACHES) (continued)

Host	Host stage infected	Pathogen	% Incidence	Locality	Lab. or field study	Reference
PROTOZOA						
Blabera fusca	Adults	Gregarina blaberae	-	France	Lab.	Tuzet et al. (1965)
"	Adults	Trichomonas vaginalis	-	France	Lab.	Gorbert et al. (1974)
Blaberus craniifer	-	Lophomonas blattarum	-	USA	Lab.	Briscoe (1971)
"	-	Retortomonas blattae	-	USA	Lab.	"
"	-	Acanthamoeba hyalina	-	USA	Lab.	"
"	-	Endamoeba thomsoni	-	USA	Lab.	"
"	-	Endolimax blattae	-	USA	Lab.	"
"	-	Gregarina blattarum	-	USA	Lab.	"
"	-	Balantidium blattarum	-	USA	Lab.	"
"	-	Nyctotherus ovalis	-	USA	Lab.	Briscoe (1971)
"	Larvae	Nosema sp.	-	USA	Lab.	Fisher & Sanborn (1964)
Blaberus giganteus	-	Nyctotherus ovalis	-	England	Lab.	Hoyte (1961 a, b, c)
"	-	Lophomonas striata	-	England	Lab.	Hoyte (1961a)
"	-	L. blattarum	-	England	Lab.	"
"	-	Monocercomonoides opthopterorum	-	England	Lab.	"
"	-	Endamoeba blattae	-	England	Lab.	"
Blaberus sp.	-	Monocercomonas sp.	-	Mexico	Field	Perez-Reyes (1966)
Blatta orientalis	-	Endolimax blattae	-	England	Lab.	Warhurst (1963, 1967)
"	-	Nyctotherus ovalis	-	England	Lab.	Hoyte (1961 a, b, c)
"	-	Lophomonas striata	-	England	Lab.	Hoyte (1961a)
"	-	L. blattarum	-	England	Lab.	"
"	-	Monocercomonoides opthopterorum	-	England	Lab.	"

PATHOGENS OF BLATTIDAE (COCKROACHES) (continued)

Host	Host stage infected	Pathogen	% Incidence	Locality	Lab. or field study	Reference
Blatta orientalis	-	Endamoeba blattae	-	England	Lab.	Hoyte (1961a)
Blattella germanica	-	Nyctotherus ovalis	-	England	Lab.	Hoyte (1961 a, b, c)
"	-	Lophomonas striata	-	England	Lab.	Hoyte (1961a)
"	-	L. blattarum	-	England	Lab.	"
"	-	Monocercomonoides opthopterorum	-	England	Lab.	"
"	-	Endamoeba blattae	-	England	Lab.	"
"	-	Nosema periplanetae	-	Federal Republic of Germany	Lab.	Selmair (1962)
"	-	Nephridiophaga blattellae	-	USA	Lab.	Woolever (1966)
Gromphadorhina portentosa	Larvae, adults	Gregarine	-	USA	Lab.	Bhatnagar & Edwards (1970)
Periplaneta americana	-	Endolimax blattae	-	England	Lab.	Warhurst (1963, 1967)
"	-	Gregarina blattarum	-	France	Lab.	Desportes (1966)
"	Larvae	Nosema sp.	-	USA	Lab.	Fisher & Sanborn (1964)
"	-	Nyctotherus ovalis	-	England	Lab.	Hoyte (1961 a, b, c)
"	-	Lophomonas striata	-	England	Lab.	Hoyte (1961a)
"	-	L. blattarum	-	England	Lab.	"
"	-	Monocercomonoides opthopterorum	-	England	Lab.	"
"	-	Endamoeba blattae	-	England	Lab.	"
"	-	Polymastix periplanetae	100	India	Field	Qadri & Rao (1963)
"	-	Polymastix sp.	6	England	Lab.	Warhurst (1966)
"	-	Rhizomastix periplanetae	90	India	Field	Rao (1963)
"	-	Tetrahymena pyriformis	-	USA	Lab.	Seaman & Clement (1970) Seaman & Robert (1968) Seaman & Tosney (1967) Seaman et al. (1972)

PATHOGENS OF BLATTIDAE (COCKROACHES) (continued)

Host	Host stage infected	Pathogen	% Incidence	Locality	Lab. or field study	Reference
Periplaneta australasiae	-	Lophomonas striata	-	England	Lab.	Hoyte (1961a)
"	-	L. blattarum	-	England	Lab.	"
"	-	Monocercomonoides opthopterorum	-	England	Lab.	"
"	-	Endamoeba blattae	-	England	Lab.	"
"	-	Nyctotherus ovalis	-	England	Lab.	"
Pycnoscelus surinamensis	-	Fusiona geusi	-	Venezuela	Field	Stejskal (1965)
NEMATODES						
Blaberus sp.	-	Thelastoma magalhaesi	(1/1)	Brazil	Field	Kloss (1966)
Blatta orientalis	Adults	Hammerschmidtiella diesingi	-	Tunisia (Tunis)	Field	Jarry & Jarry (1963)
"	-	"	-	Czechoslovakia	Lab.	Groschaft (1965)
"	Larvae, adults	"	-	Malaya	Field	Leong & Paran (1966)
"	Adults	Leidynema appendiculata	-	Tunisia (Tunis)	Field	Jarry & Jarry (1963)
"	-	"	-	Czechoslovakia	Lab.	Groschaft (1965)
"	Larvae, adults	"	-	Malaya	Field	Leong & Paran (1966)
"	Adults	Thelastoma icemi	-	Tunisia (Tunis)	Field	Jarry & Jarry (1963)
"	Larvae, adults	Schwenkiella icemi	-	Malaya	Field	Leong & Paran (1966)
"	-	Thelastoma bulhoesi	-	Czechoslovakia	Lab.	Groschaft (1965)
"	Larvae, adults	Thelastoma singaporensis	-	Malaya	Field	Leong & Paran (1966)
Blattella fulginosa	-	Leidynema appendiculata	-	USA (La)	Field, Lab.	Feldman (1972)

PATHOGENS OF BLATTIDAE (COCKROACHES) (continued)

Host	Host stage infected	Pathogen	% Incidence	Locality	Lab. or field study	Reference
Blattella germanica	Adult	Blatticola blattae	-	USA (NY)	Field	Cali & Mai (1965)
Gromphadorhina portentosa	Larvae, adults	Leidynema appendiculata	-	USA	Lab.	Bhatnagar & Edwards (1970)
Leucophaea maderae	Larvae, adults	Leidynema sp.	-	USA (Oregon)	Field	Thomas & Poinar (1973)
Periplaneta americana	Larvae, adults	Leidynema appendiculata	-	USA (Florida)	Field, Lab.	Feldman (1972)
"	Larvae	"	27	Canada	Lab.	Hominck & Davey (1972)
"	Adults	"	80	Canada	Lab.	"
"		"	-	Tunisia (Tunis)	Field	Jarry & Jarry (1963)
"		"	-	Brazil	Field	Kloss (1966)
"		"	-	Canada	Lab.	Pawlik (1966)
"		"	-	Czechoslovakia	Lab.	Groschaft (1965)
"	Larvae, adults	"	-	Malaya	Field	Leong & Paran (1966)
"	Larvae	Hammerschmidiella diesingi	90	Canada	Lab.	Hominck & Davey (1972)
"	Adults	"	70	Canada	Lab.	"
"		"	-	Tunis	Field	Jarry & Jarry (1963)
"		"	-	Brazil	Field	Kloss (1966)
"		"	-	Czechoslovakia	Lab.	Groschaft (1965)
"	Larvae, adults	"	-	Malaya	Field	Leong & Paran (1966)
"	Larvae, adults	Thelastoma icemi	-	Tunisia (Tunis)	Field	Jarry & Jarry (1963)
"	Larvae, adults	T. singaporensis	-	Malaya	Field	Leong & Paran (1966)

PATHOGENS OF BLATTIDAE (COCKROACHES) (continued)

Host	Host stage infected	Pathogen	% Incidence	Locality	Lab. or field study	Reference
Periplaneta americana	-	T. bulhoesi	-	Czechoslovakia	Lab.	Groschaft (1965)
"	Adults	Thelastoma sp.	-	Tunisia (Tunis)	Field	Jarry & Jarry (1963)
"	Larvae, adults	Protrellina aurifluus	-	Malaya	Field	Leong & Paran (1966)
"	Larvae, adults	Severianoia severianoi	-	Malaya	Field	"
"	Larvae, adults	Schwenkiella icemi	-	Malaya	Field	"
Periplaneta australasiae	-	Leidynema appendiculata	(1/1)	Brazil	Field	Kloss (1966)
	-	Hammerschmidtiella diesingi	"	"	"	"
Platyzosteria novaeseelandiae	Larvae, adults	Blatticola tuapakae	86	New Zealand	"	Dale (1966)
"	Larvae, adults	Protrellina gurri	"	"	"	"
Polyphaga aegypticaca	-	Leidynema appendiculata	-	Egypt	Field	Steinhaus & Marsh (1962)
RICKETTSIAE						
Blatta orientalis	Larvae, adults	Rickettsiella blattae	1	Germany	Lab.	Huger (1964)
OTHER						
Periplaneta americana	-	Moniliformis dubius	-	Australia	Lab.	Mercer & Nicholas (1967)

ABSTRACTS

Mary Ann Strand

Anderson, R. S. et al. (1973). In vitro bactericidal capacity of Blaberus craniifer hemocytes. J. Invertebr. Pathol., 22: 127-135.

The ability of cultured cockroach hemocytes to phagocytose various bacteria was tested. Pseudomonas aeruginosa, Escherichia coli, Salmonella typhosa, and Diplococcus pneumoniae were phagocytosed but not killed. However some known insect pathogens, Staphylococcus albus, Serratia marcescens, and Proteus mirabilis, were killed by the hemocytes.

Bhatnagar, K. N. & Edwards, L. J. (1970). Parasites of the Madagascar cockroach, Gromphadorhina portentosa. Ann. Entomol. Soc. Am., 63: 620-621.

No external symptoms indicated the presence of the parasites in cockroaches from a stock culture. The nematode Leidynema appendiculata occurred in the gut of all stages and both sexes. A gregarine occurred within the lumen of the mid- and hind-gut. At times the cysts were so dense that they appeared to block the passage of other material through the gut.

Briscoe, M. S. (1971). A survey of the protozoan fauna of the cockroach Blaberus craniifer. J. Invertebr. Pathol., 17: 291.

Eight species of protosoans were found in the gut of the cockroach. The protozoa exhibited some degree of specificity for locations in the gut.

Briscoe, M. S., Moore, R. E. & Puckett, D. E. (1961). Microbial isolations from the gut of Blaberus craniifer Burmeister. J. Insect. Pathol., 3: 254-258.

Seven species of bacteria from the family Enterobacteriaceae were isolated from sections of gut removed from adult roaches. No pathological conditions were observed.

Briscoe, M. S., Moore, R. E. & Puckett, D. E. (1963). The sensitivity to antibiotics of microorganisms isolated from the gut of Blaberus craniifer Burmeister. AIBS Bull., 13: 27.

Six species of bacteria were isolated from the gut of Blaberus craniifer. Their sensitivity to low and high concentrations of 16 antibiotics was tested.

Brooks, M. A. & Richards, K. (1966). On the in vitro culture of intracellular symbiotes of cockroaches. J. Invertebr. Pathol., 8: 150-157.

Cockroach symbiotes, known as bacterioids because they resemble bacteria, are always found in the ovaries and in the mycetocytes. In vitro culture of the bacteroids was unsuccessful and this study discounts the claims of in vitro culture by other authors.

Cali, C. T. & Mai, W. F. (1965). Studies on the development of Blatticola blattae (Graeffe, 1860) Chitwood, 1932 within its host, Blattella germanica L. Proc. Helminthol. Soc. Wash., 32: 164-169.

Parasite-free roaches were fed eggs of the nematode parasite. The developmental stages of the nematode are described. Hatching occurred in the mid- or hind-gut 2-1/2 to 4 hours after ingestion. Adult female nematodes were recovered from the infected roaches on the 14th day after infection.

Dale, P. S. (1966).   Blatticola tuapakae and Protrellina gurri n. spp., nematode parasites of the black roach.   N. Z. J. Sci., 9: 538-544.

Two new species of thelastomatid nematodes are described from Platyzosteria novaeseelandiae collected in New Zealand.   The parasites were common and were found in the hindgut of nymphs and adults.

Desportes, I. (1966).   L'ultrastructure de la jonction entre le primite et le satellite des associations de Gregarina blattarum Sieb.   (Eugrégarines, Gregarinidae).   C. R. Hebd. Seances Acad.Sci. Ser. D Sci. Nat. (Paris), 262: 1869-1870.

A new structure was found at the point of connexion between associated male and female pairs.   An electron micrograph of a longitudinal section revealed the new structure.

Feldman, M. R. (1972).   Fine structural studies of the intestinal system of the nematode Leidynema appendiculata (Leidy, 1850).   Trans. Am. Microsc. Soc., 91(3): 337-347.

Adult female L. appendiculata were obtained from the hind-gut of naturally infected Periplaneta americana and Blattella fuliginosa collected in the New Orleans area. Electron microscopic techniques were employed to investigate the fine structure and function of the intestinal system of the nematode.

Fisher, F. M. jr & Sanborn, R. C. (1964).   Nosema as a source of juvenile hormone in parasitized insects.   Biol. Bull., 126: 235-252.

Surgical implants of Nosema in allatectomized cockroaches prevented adult development. Observations indicate that Nosema is capable of replacing the hormone normally produced by the host's corpora allata.

Gabriel, B. P. (1968).   Entomogenous microorganisms in the Philippines: new and past records. Philipp. Entomol., 1(2): 97-130.

From new surveys and past records, five species of nematodes, and three species of bacteria are reported to infect cockroaches in the Philippines.

Gobert, J. G. et al. (1974).   Etude de l'endoparasitisme expérimental de Trichomonas vaginalis chez Blabera fusca.   Ann. Parasitol. Hum. Comp., 49(2), 159-174.   (French, English.)

T. vaginalis was inoculated into the body cavity of B. fusca.   At least 6×10^6 flagellates per insect were required to produce infection.   The lesions induced by the parasite and the haemolymph in infected insects are described.

Groschaft, J. (1965).   /Incidence of nematodes (Oxyuroidea) in laboratory reared cockroaches (Blattodea)./   Cesk. Parasitol., 3, 67-74.   (CZ)

Hammerschmidtiella diesingi, Leidynema appendiculata and Thelatoma bulhoesi were found infecting Periplaneta americana and Blatta orientalis.

Heitor, F. (1962).   Parasitisme de blessure par le champignon Mucor hiemalis Wehmer chez les insectes.   Ann. Epiphyt. (Paris), 13: 179-203.   (French, English.)

Attempts were made to infect cockroach nymphs with Mucor hiemalis.   No pathological responses were seen when the fungus was ingested or when it was in contact with the insect's epidermis.   Infection only occurred when the fungus was injected into the haemolymph.   Mortality occurred in 20% of the injected nymphs.

Henry, S. M. (1965).   Intestinal microorganisms of the cockroach with and without intracellular symbiotes.   Proc. 12th Inst. Congr. Entomol. (London), p. 748.

Massive quantitites of 6 species of bacteria were fed to the Blattella germanica. Neither aposymbiotic nor normal insects were visibly affected by the bacteria during the 6 weeks following the initiation of feeding.

Hominick, W. M. & Davey, K. G. (1972).   The influence of host stage and sex upon the size and composition of the population of two species of thelastomatids parasitic in the hindgut of Periplaneta americana.   Can. J. Zool., 50(7):   947-954.

The interrelationship of the parasites Hammerschmidtiella diesingi and Leidynema appendiculata and their host is examined.   H. diesingi increased in numbers from the 7th instar then decreased in adults.   L. appendiculata juveniles could not survive moulting of the host and were uncommon in the nymphs.   Their numbers increase in the adults.   Competition may occur between the 2 species of nematodes in the hind-gut of the adult.   Male and female cockroaches differ in their suitability as hosts.   A greater incidence of L. appendiculata was noted in adult females and 8th instar males harboured more H. diesingi.

Hoyte, H. M. D. (1961a).   The protozoa occurring in the hind-gut of cockroaches. I.   Response to changes in environment.   Parasitology, 51:   415-436.

The well-being of protozoan parasites was closely linked to that of the host.   The effects of extreme humidity and O_2 poisoning on the protozoans Lophomonas striata, L. blattarum, Monocercomonoides orthopterorum, Endamoeba blattae, and Nyctotherus ovalis in Periplaneta americana, Blatta orientalis, Blattella germanica, and Blaberus giganteus were tested.

Hoyte, H. M. D. (1961b).   The protozoa occurring in the hind-gut of cockroaches. II.   Morphology of Nyctotherus ovalis.   Parasitology, 51:   437-463.

The size and shape of the trophozoite of N. ovalis in different species and cockroaches was found to vary considerably, and this was attributed to the presence or absence of large paraglycogen granules within the ciliate.   The constant size and shape of the cysts were not altered when the host moulted, but there was always a decrease in size and sometimes a change in shape when an infection was established in a new host.

Hoyte, H. M. D. (1961c).   The protozoa occurring in the hind-gut of cockroaches. III.   Factors affecting the dispersion of Nyctotherus ovalis.   Parasitology, 51:   465-495.

The only mode of transmission of N. ovalis is by way of ingestion of the encysted stage of the ciliate.   It could encyst in any of the 4 species of cockroach, Periplaneta americana, Blatta orientalis, Blattella germanica, and Blaberus giganteus, irrespective of the species of origin.   In vivo, the process of excystation took as little as 3 hours for completion.

Huger, A. (1964).   Eine Rickettsiose der Orientalischen Schabe, Blatta orientalis L., Verursacht durch Rickettsiella blattae nov. spec.   Naturwissenschaften, 51:   22.

Laboratory colony of cockroaches (Blatta orientalis) was infected by a new species of rickettsia, Rickettsia blattae.   The number of patently infected individuals remained on the average below 1%.   Heavily infected cockroaches appeared swollen; they passed from lethargy to paralysis often with brief periods of symptom remission.   The rickettsia appeared in nearly all tissues and organs of the host.   They were spindle form, 1-3 μ x 0.8-1.7 μ.

Jarry, D. M. & Jarry, D. T. (1963). Quelques Thelastomatidae (Nematoda: Oxyuroidea) parasites des Blattides à l'Institut Pasteur de Tunis. Arch. Inst. Pasteur Tunis, 40: 229-234.

Four species of Thelastomatidae were found in the hind-gut of Blatta orientalis and Periplaneta americana. In Tunis only P. americana was found to be infested with the nematodes of the genus Thelastoma, although B. orientalis occurs in the same niche.

Kloss, G. R. (1966). /Review of the nematodes from the cockroaches in Brazil./ Pap. Avulsos Dep. Zool. (São Paulo), 18 (16): 147-188. (Portuguese.)

Descriptions and notes on habitat and distribution are given for the nematodes infecting cockroaches in Brazil. A table lists host-parasite incidence.

Ladd, T. L. jr and Travis, B. V. (1965). Elimination of gregarines from the cockroach, Leucophaea maderae (Fab.). J. Parasitol., 51: 569.

Adult roaches were examined immediately after the last nymphal moult and were found to be free of gregarines. The midgut epithelium is destroyed by phagocytosis and a new epithelium is regenerated during the moulting process. Most gregarines are eliminated with the epithelium lining.

Leong, L. & Paran, T. P. (1966). A study of the nematode parasites of cockroaches in Singapore. Med. J. Malaya, 20: 349.

Protreffina aurifluus and Severianoia severianoi appear to be specific for Periplaneta americana. Leidynema appendiculata, Hammerschmidtiella diesigni, Schwenkiella icemi, and Thelastoma singaporensis were found parasitizing both Blatta orientalis and P. americana.

Lord, T. H., Foltz, V. D. & Van Sickle, R. (1964). Natural incidence of Staphylococcus aureus Rosenbach in the brown-banded cockroach. J. Insect Pathol., 6: 21-25.

The natural incidence of S. aureus in the cockroach, Supella supellectilium was determined by a trapping experiment. Forty-eight per cent. of the traps caught roaches carrying S. aureus.

Mercer, E. H. & Nicholas, W. L. (1967). The ultrastructure of the capsule of the larval stages of Moniliformis dubius (Acanthocephala) in the cockroach Periplaneta americana. Parasitology, 57: 169-174.

Electron and light microscopy were used to observe the capsule which surrounds the parasite M. dubius in its intermediate host. Encapsulation of invaders is associated with host resistance, however, it does not prevent growth, development, or the uptake of nutrients by M. dubius.

Norris, T. R. (1971). The physical and antigenic structure of the crystalline protein toxin of Bacillus thuringiensis. Proc. 13th Int. Cong. Entomol. (Moscow), 2: 87.

Intact protein had no effect on nervous transmission of cockroaches. Protein digested with Pieris gut juice yielded a fraction that effectively inhibited their nervous transmission.

Pawlik, J. (1966).   Control of the nematode Leidynema appendiculata (Leidy) (Nemata:Rhabditida: Thelastomatidae) in laboratory cultures of the American cockroach.   J. Econ. Entomol., 59: 468-469.

The nematode L. appendiculata is an intestinal commensal of the cockroach Periplaneta americana.   In laboratory culture, their numbers can be great enough to interfere with research on the digestive tract of the roaches.   Antihelminths used on human pinworms were found to be effective controls of the nematode.

Perez-Reyes, R. (1966).   Insect protozoa.   I.   Monocercomonas and other small flagellates of Mexican insects.   Proc. 1st Int. Cong. Parasitol. (Rome), pp. 600-601.

Monocercomonas, an entozoic flagellate, was found in the gut of Blabera sp.

Qadri, S. S. & Rao, T. B. (1963).   On a new flagellate Polymastix periplanetae from the common cockroach Periplaneta americana.   Riv. Parassitol., 24(3): 153-158.

Nearly all of 500 cockroaches collected near Hyderabad, India were heavily infected with a new species of flagellate, P. periplanetae.   The parasites were found in the lumen and hind-gut.

Rao, T. B. (1963).   On Rhizomastix periplanetae n. sp., from Periplaneta americana of Hyderabad A. P., India.   Riv. Parassitol., 24 (13): 159-162.

A new monoflagellate species was found in the common cockroach near Hyderabad. Almost 90% of the 200 cockroaches examined were found to harbour the parasite in the rectum.

Ryan, M. & Nicholas, W. L. (1972).   The reaction of the cockroach Periplaneta americana to the injection of foreign particulate material.   J. Invertebr. Pathol., 19, 299-307.

Bacillus thuringiensis, Escherichia coli, and Corynebacterium were injected into the hemocoel of P. americana.   The material injected was about 1% bacteria by volume. B. thuringiensis was lyzed in the hemolymph.   All cockroaches injected with one of the 3 bacteria died.

Sauerländer, S. & Ehrhardt, P. (1961).   Reduzierung der Proteinfraktionen in der Hämolymphe bakterienkranker Blattiden.   Naturwissenschaften, 48: 674-675.

The electrophoresis-diagram of the hemolymph of healthy cockroaches, Periplaneta americana and Blaptica dubia, exhibited 3 protein fractions.   Only two were found in similar preparations from late instar cockroaches infected with Pseudomonas aeruginosa or Serratia marcescens.   The reduction in protein fractions is not dependent upon the time since infection, but rather on the intensity of the disease.

Sauerländer, S. & Köhler, F. (1961) Erhöhung der Körpertemperatur von Periplaneta americana L. im Verlauf zweier Bakteriosen.   Experientia, 17, 397-398.

Two bacterial diseases (Serratia marcescens and Pseudomonas aeruginosa) were accompanied by increase in body temperature of the cockroaches.   Male adults were inoculated with the bacteria and the temperature increase began about 1 hour after inoculation and continued for at most 10 days.

Seaman, G. R. & Clement, J. J. (1970).   Tricarboxylic acid cycle enzymes in Tetrahymena pyriformis after infection of the cockroach Periplaneta americana.   J. Protozool., 17: 287-290.

Activities of enzymes of the TCA cycle in extracts of T. pyriformis recovered from the hemocoel of cockroaches were compared with activities in ciliates not exposed to the cockroach.   Several alterations in activity levels were noted and these alterations persisted for several hundred generations after recovery from the insect.

Seaman, G. R. & Robert, N. L. (1968).   Immunological response of male cockroaches to injection of Tetrahymena pyriformis.   Science, 161, 1359-1361.

At dosage that kills 50% of the roaches over a 3 day period after injection, ciliates cannot be recovered from the survivors.   This suggests that male cockroaches produce an immunological substance against the ciliates.   A separated fraction of the hemolymph from the inoculated roaches immobilized S strain ciliates in culture.

Seaman, G. R. & Tosney, T. J. (1967).   Alterations in morphology of Tetrahymena pyriformis S after facultative parasitism of the cockroach, Periplaneta americana.   J. Protozool., 14 (Supplement):   23-24.

After recovering the T. pyriformis from the roaches, the ciliates were almost completely rounded and contained large numbers of osmophilic granules.   As the ciliates became readapted to a free-living existence, the usual pyriform shape was reassumed.

Seaman, G. R. et al. (1972).   Infectivity and recovery of Tetrahymena pyriformis strain S from adult female cockroaches (Periplaneta americana).   J. Protozool., 19(4), 644-647.

Female cockroaches died within 72 hours after injection into the hemocoel of 5800-8000 cells of T. pyriformis/individual.   Injection with smaller numbers resulted in no apparent ill effects and ciliates could not be recovered from the roaches.

Selmair, E. (1962).   Über das Verhalten von Symbionten nach dem Tode des Wirtstieres und bei gleichzeitiger Infektion mit Mikrosporidien.   Arch. Mikrobiol., 43:   290-293.

After the death of the host Blattella germanica, the symbiotic bacteria begin to degenerate.   Also when the host is infected with the microsporidian, Nosema periplanetae, the symbionts perish.

Steinhaus, E. A. & Marsh, G. A. (1962).   Report of diagnoses of diseased insects 1951-1961. Hilgardia, 33:   349-490.

The location, host, submission date, and individual requesting diagnosis are given for each tentative diagnosis.   For details see the host-parasite table.

Stejskal, M. (1965).   Eine neue Gregarine, Fusiona geusi, n. sp., n. gen., n. fam., n. superfam., aus der Südamerikanischen Schabe Pycnoscelus surinamensis L. 1758. Z. Parasitenkd., 26:   215-220.   (German, English.)

The new superfamily, Fusionoidae, is placed in the suborder Cephalina.   The new family, Fusionidae, is characterized by the sexual processes beginning with a fusion of the entocyte.   The new species, F. geusi, was found in the gut and coeca of the cockroaches collected in Venezuela.

Tavares, I. I. (1966).   Structure and development of Herpomyces stylopygae (Laboulbeniales). Am. J. Bot., 53: 311-318.

    Characteristics which separate H. stylopygae from the other species of Herpomyces are given.   The fungus is parasitic on Blatta orientalis.

Thomas, G. M. & Poinar, G. O. jr (1973).   Report of diagnoses of diseased insects 1962-1972. Hilgardia, 42: 261-360.

    The location, host, submission date, and individual requesting diagnosis are given for each tentative diagnosis.   For details see the host-parasite table.

Tuzet, O., Ormieres, R. & Bergoin, M. (1965).   Cycle de Gregarina blaberae Frenzel, parasite de Blabera fusca Brünner (Dictyoptères).   Arch. Zool. Exp. Gen., 105: 299-303.

    The life cycle of a gregarine found in B. fusca was observed.   The authors contend it should be classed as Gregarina blaberae.   This contention is in agreement with Frenzel's original description and in opposition to other authors who put it in the genus Pileocephalus.

Ulewicz, K. & Zawistowski, S. (1973).   Experimental studies on the epidemiological importance of the cockroach, Blattella germanica (L.) in streptococcal infections on ships.   Zentralbl. Bakteriol. Parasitenke Infektionskr. Hyg. Abt. I. Orig. A., 223, 77-82.   (English, German.)

    Under experimental conditions, oral and anal introduction of the haemolytic streptococci lead to considerable infection in the cockroach.   However, the infection was not transmitted to the offspring and did not apparently affect the biology of the insects.

Warhurst, D. C. (1963).   The biology of Endolimax blattae Lucas, an endocommensal of cockroaches.   J. Protozool., 10 (Supp.): 28.

    Endolimax blattae is a small amoeba inhabiting the colon of Periplaneta americana and Blatta orientalis.   The colon of Periplaneta is an oxygen poor habitat and tests on agnotobiotically grown E. blatta show that oxygen is toxic to the cysts and its effect is reinforced by desiccation.

Warhurst, D. C. (1966).   A note on Polymastix Butschli, 1884, from the colon of Periplaneta americana (Insecta: Dictyoptera).   Parasitology, 56: 21-23.

    The flagellate was found in stained smears and live preparations from the colons of 3 out of over 200 nymphs.   A morphological description of the flagellate is given.

Warhurst, D. C. (1967).   Cultivation in vitro of Endolimax blattae Lucas, 1927 from the cockroach hind gut.   Parasitology, 57: 181-187.

    Samples of the colon contents from living cockroaches, Periplaneta americana and Blatta orientalis, were inoculated into a diphasic serum-saline culture medium.   Strains of the amoeba and its microbial associates were isolated.   E. blattae will grow in the absence of starch in the medium, but its growth is enhanced by the addition of starch.

Woolever, P. (1966).   Life history and electron microscopy of a haplosporidian, Nephridiophaga blattellae (Crawley) n. comb., in the Malpighian tubules of the German cockroach, Blattella germanica (L.).   J. Protozool., 13, 622-642.

    Cockroaches obtain N. blattellae by ingesting the spores rather than transovarially. All ages and both sexes are susceptible.   Progressive infection of the Malpighian tubules appears 15 days after ingestion of the spores.   The infection is spread mainly by schizogony.

Yarnykh, V. S. & Tonkozhenko, A. P. (1969). /Microbiological method in veterinary insect control./ Sel'skokhoz. Biol., 4(1): 98-103. (Russian, English.)

Results are given of a 3 year study on the use of bacterial preparations against various insects including 3 cockroaches, Blattella germanica, Blatta orientalis, and Shelfordella tartara.

Zuberi, R. I., Hafiz, S. & Ashrafi, S. H. (1969). Bacterial and fungal isolates from laboratory-reared Aedes aegypti (Linnaeus), Musca domestica (Linnaeus) and Periplaneta americana (Linnaeus). Pak. J. Sci. Ind. Res., 12: 77-82.

Pseudomonas aeruginosa was found to occur as part of the normal flora of the cockroach. Various species of Bacillus isolated from insects seemed to be the result of saprophytic contamination of their food. Adult cockroaches harboured Aspergillus niger and A. ustus.

XVII. PATHOGENS OF ANOPLURA AND MALLOPHAGA (LICE) [a]

Ronald A. Ward

Department of Entomology
Walter Reed Army Institute of Research
Walter Reed Army Medical Center
Washington, DC 20012, USA

[a] This work was supported by research contract DAMD-17-74-C-4086 from the US Army Medical Research and Development Command, Office of the Surgeon General and carried out at the Medical Entomology Project, Smithsonian Institution, Washington, DC 20560, USA.

PATHOGENS OF ANOPLURA AND MALLOPHAGA (LICE)

Host	Host stage infected	Pathogen	% incidence	Locality	Lab. or field study	Reference
ANOPLURA		RICKETTSIAE				
Pediculus humanus	-	Rickettsia akari	-	-	-	Jenkins (1973)
"	-	R. conorii	-	-	-	"
"	-	R. prowazeki	-	-	-	Jenkins (1973), Řeháček (1965)
"	-	R. typhi	(Maybe)	-	-	Jenkins (1973)
"	-	R. rickettsi	-	-	-	"
"	-	R. tsutsugamushi	-	-	-	"
"	-	Wolbachia persica	-	-	-	Jenkins (1973), Weyer (1973)
Polyplax spinulosa	-	Rickettsia prowazeki	Kills lice	Mexico	Lab.	Mooser et al. (1931)
Pediculus humanus	Adults	BACTERIA Bacillus pediculi	High	England	Lab.	Arkwright & Bacot (1921)
"	-	Staphylococcus aurens	-	-	-	Jenkins (1973)
"	-	Eberthella typhosa (not a recognized taxon)	-	-	-	"
"	-	Escherichia coli	Kills in 2 days	-	-	"
"	-	Yersinia pseudo-tuberculosis	60-90 mortality	-	-	"
"	-	Proteus vulgaris	Kills in 2 days	-	-	"

PATHOGENS OF ANOPLURA AND MALLOPHAGA (LICE) (continued)

Host	Host stage infected	Pathogen	% incidence	Locality	Lab. or field study	Reference
P. humanus (continued)	-	Francisella tularensis	Variable mortality kills in 4-7 days	USA	-	Price (1957)
"	-	Strickeria jurgensi	Sometimes very abundant	Germany	-	Stempell (1916)
"	-	Salmonella enteritidis	100 mortality in 2 days	-	-	Jenkins (1973)
"	-	Shigella dysenteriae	-	-	-	"
"	-	Yersinia enterocolitica	60-90 mortality	-	-	"
Linognathus sp.	-	Yersinia pestis	Kills lice	Trans-baikalia and Mongolia	Lab.	Jettmar (1925)
PROTOZOA						
Pediculus humanus	Adults	Cocconema pediculusvestimenti	-	USSR	Field	Popov & Manuilova (1926)
"	Nymphs, adults	Leptomonas pediculi	8	England	Field	Fantham (1912)
"	Adults	Toxoplasma gondii	71	Poland	Lab.	Dutkiewicz (1966)

PATHOGENS OF ANOPLURA AND MALLOPHAGA (LICE) (continued)

Host	Host stage infected	Pathogen	% incidence	Locality	Lab. or field study	Reference
MALLOPHAGA		BACTERIA				
Bovicola bovis	-	Bacillus thuringiensis (various preparations)	50 mortality in 23-172 hrs depending on preparation	USA (Texas)	Lab.	Gingrich et al. (1974)
Bovicola crassipes	-	"	"	"	"	"
Bovicola limbatus	-	"	"	"	"	"
Bovicola ovis	-	"	"	"	"	"
Lipeurus caponis	-	Bacillus thuringiensis	up to 100 mortality	USA (Texas)	Lab.	Hoffman & Gingrich (1968)
Menacanthus stramineus	-	"	"	"	"	"
Menopon gallinae	-	"	"	"	"	"
		FUNGI				
Bruelia cyclothorax	Adults	Trenomyces helveticus	-	USSR	Field	Lunkashu (1970)
Bruelia gracilis	Adults	"	-	"	"	"
Bruelia nebulosa	Adults	"	-	"	"	"
Bruelia uniinosa	Adults	"	-	"	"	"
Gonides gigas	Adults	Trenomyces histophtorus	rare	France	Field	Chatton & Picard (1909)

PATHOGENS OF ANOPLURA AND MALLOPHAGA (LICE) (continued)

Host	Host stage infected	Pathogen	% incidence	Locality	Lab. or field study	Reference
Menopon gallinae	Adults	Trenomyces histophtorus	10	France	Field	Chatton & Picard (1909)
Menopon gallinae	-	"	-	Italy	Field	Trinchieri (1910)
Menopon gallinae	-	"	-	Argentina	Field	Spegazzini (1917)
Picicola coutiguus	Adults	Trenomyces helveticus	-	USSR	Field	Lunkashu (1970)
Saemundssonia sp.	-	Trenomyces platensis	-	Argentina	Field	Spegazzini (1917)
Sturnidoecus ruficeps	Adults	Trenomyces helveticus	-	USSR	Field	Lunkashu (1970)
Many species of Mallophaga	-	Trenomyces (5 species)	-	Europe and North America	Field	Thaxter (1912)
NEMATODA						
Dennyus hirundis	-	Filaria cypseli	-	Gambia	Field	Dutton (1905)
Heterodoxus spiniger	Adult	Dipetalonema reconditum	17	Kenya	Field	Nelson (1962)

ABSTRACTS

Mary Ann Strand

Arkwright, J. A. & Bacot, A. (1921).   A bacillary infection of the copulatory apparatus of
Pediculus humanus.   Parasitology, 13: 25-26.

   A gram-negative cocco-bacillus was isolated from the excreta, guts, and copulatory
   apparati of the lice.   No pathogenic conditions were reported in the infected lice.

Chatton, E. & Picard, F. (1909).   Contribution à l'étude systématique et biologique des
Laboulbeniacees: Trenomyces histophtorus Chatton et Picard, endoparasite des poux de la poule
domestique.   Bull. Trimest. Soc. Mycol. Fr., 25: 147-170.

   The fungus, T. histophtorus, grows in the fat body of the lice Menopon gallinae
   (=M. pallidum) and Goniodes gigas (=Gonicotes abdominalis).   A detailed description of
   the growth habit of the fungus is given.   The taxonomic status of the genus Trenomyces
   in the Laboulbeniales is discussed.

Dutkiewicz, J. (1966).   Studies on the biological properties of Toxoplasma gondii in the body
of insects, using Pediculus humanus L. as test animals.   Acta Parasitol. Pol., 14: 187-199.

   T. gondii was introduced into the alimentary system of the lice by intrarectal injection.
   It was able to penetrate the cells of the intestinal epithelium and hemocytes and to
   multiply.   Infected lice died 120-144 hours after injection.   The mortality may be due
   to the damage to the tissues as a result of T. gondii multiplication.

Dutton, J. E. (1905).   The intermediate host of Filaria cypseli (Annett, Dutton, Elliot);
the filaria of the African swift, Apus affinis.   Thompson Yates (and Johnston).   Lab. Rep.,
6: 137-147.

   Reference not seen by M. A. Strand.

Fantham, H. B. (1912).   Herpetomonas pediculi nov. spec., parasitic in the alimentary tract
of Pediculus vestimenti, the human body louse.   Proc. R. Soc. Ser. B, 84: 505-517.

   H. pediculi was found in the alimentary tract and feces of lice collected from children
   in England.   It is transmitted through excretion and ingestion of the trypanosomes.
   Fantham does not mention the effect of infection on the lice.

Gingrich, R. E. et al. (1974).   Bacillus thuringiensis: Laboratory tests against four species
of biting lice (Mallophaga: Trichodectidae).   J. Invertebr. Pathol., 23: 232-236.

   The species of Bovicola were more susceptible to spore- -endotoxin complex than to
   beta-endoxin.

Hoffman, R. A. & Gingrich, R. E. (1968).   Dust containing Bacillus thuringiensis for control
of chicken body, shaft, and wing lice.   J. Econ. Entomol., 61: 85-88.

   One or more applications of a commercial preparation of B. thuringiensis gave complete
   control of the 3 species of lice on chickens for varying periods of time.

Jenkins, D. W. (1973).   Biologic control of human lice.   Pan. Am. Health Organ. Sci. Publ.,
263: 256-260.

    Body lice and head lice are normally free from pathogenic microorganisms.   Various
viruses pathogenic to humans have been introduced into lice, however, no pathogenicity
for the lice has been reported.   There are no known rickettsiae pathogenic for mice that
do not produce disease in humans, except possibly E strain of Rickettsia prowazeki.
Twenty species of bacteria have been tested for pathogenicity for lice, but all that were
pathogens of lice were also pathogenic for man.   Staphylococcus aureus passed many times
in lice did increase in virulence for lice in some cultures.   No spirochetes have been
confirmed to be pathogenic for lice.

Jettmar, H. M. (1925).   Beitrage zum Studium der Pest unter den Insekten. I. Mitteilung.
Die Tarbaganlaus.   Z. Hyg. Infektionskr., 104: 551-568.

    Plague bacilli (Yersinia pestis) multiply in the louse Linognathus sp. which dies
2-3 days after feeding.   Virulent bacilli are found in the gut and feces of infected
lice.

Lunkashu, M. I. (1970).   /Fungi of the genus Trenomyces from bird lice of birds from
Moldavia/.   Parazity Zhivotn. Rast., 5: 128-130 (Russian).

    T. helveticus was present on the thorax, abdomen, legs and head of several species of
avian Mallophaga.

Mooser, H. et al. (1931).   The transmission of the virus of Mexican typhus from rat to rat by
Polypax spinulosus.   J. Exp. Med., 54: 567-575.

    Rickettsia prowazeki was observed in lice after they had fed on an infected rat.
The rickettsiae multiply in the epithelial cells of the gut, destroying its lining, and
leading to the death of the louse.

Nelson, G. S. (1962).   Dipetalonema reconditum (Grassi, 1889) from the dog with a note on its
development in the flea, Ctenocephalides felis and the louse, Heterodoxus spiniger.
J. Helminthol., 36: 297-308.

    The larvae of D. reconditum developed in the fat body of H. spiniger taken from infected
dogs in Kenya.   Also filaria were seen in Dennyus hirundis.   No pathological conditions
were mentioned.

Popov, P. & Manuilova, M. (1926).   /On the discovery in the human body louse - Pediculus
vestimenti, Nitzsch a microsporidian parasite sp. nov./ Russ. Z. Trop. Med., 4(8): 43-49
(Russian, English summary).

    Cocconema pediculusvestimenti was observed in the cells of the intestine and fat body of
the louse.

Price, R. D. (1957).   A microscopic study of Pasteurella tularensis in the human body louse.
Parasitology, 47: 435-446.

    Multiplication of the bacteria is extracellular in the midgut lumen and the haemolymph
and is intracytoplasmic in the gut cells of the louse Pediculus humanus.   The gut cells
are disrupted and the bacteria enter the body cavity.   Their growth in the hemocoel is
fatal, death usually occurs in 4-7 days.

312

Rahǎček, J. (1965).   Development of animal viruses and rickettsiae in ticks and mites.
Ann. Rev. Entomol., 10: 1-24.

Among rickettsiae, only Rickettsia prowazeki is reported to kill its vector-lice.

Spegazzini, C. (1917).   Revision de las Laboulbeniales Argentinas.   An. Mus. Nac. B. Aires,
29: 445-688.

Spegazzini mentions two species of Trenomyces observed on all parts of the bodies of two
mallophagan species in Argentina.

Stempell, W. (1916).   Ueber einen als Erreger des Fleckfiebers verdächtigen Parasiten der
Kleiderlaus.   Dtsch. Med. Wochenschr., 42: 439-442.

A new genus and new species of flagellate, Strickeria jurgensi, is named.   It was found
in the intestine of Pediculus humanus.

Thaxter, R. (1912).   Preliminary descriptions of new species of Rickia and Trenomyces.
Proc. Am. Acad. Arts Sci., 48: 365-386.

T. histophtorus, T. lipeuri, T. laemobothrii, T. circinans, and T. gibbus are described;
the latter two as new species.   Thaxter examined thousands of lice and considered their
infestation with Trenomyces as rare.

Trinchieri, S. G. (1910).   Intorno a una Laboulbeniacea nouva per l'Italia (Trenomyces
histophtorus Chatton et Picard).   Boll. Soc. Nat. Napoli, 24: 18-22.

T. histophtorus was found on Menopon gallinae (=M. pallidum) in Italy.

Weidhaas, D. E. (1973) Biologic control of lice.   Pan. Am. Health Organ. Sci. Publ., 263: 256.

The use of biological agents to control human lice has received little attention by
researchers.   Effective agents which are safe for humans have not been identified and
other methods of control such as sanitation and chemical control are available.

Weyer, F. (1973).   Versuche zur Übertragung von Wolbachia persica auf Kleiderläuse.
Z. Angew. Zool., 60: 77-93.

W. persica was successfully transferred to the gut and hemolymph of body lice.   In all
cases, the organisms multiplied and had a damaging effect on the hosts which died within a
few days.   In the lumen of the louse gut, multiplication was predominantly extracellular,
although evidence of intracellular multiplication was seen in epithelial cells.
W. persica was found in the feces and in oviposited eggs but infection in the progeny was
not observed.

XVIII. PATHOGENS OF CIMICIDAE (BEDBUGS)

Mary Ann Strand

Boyce Thompson Institute
Yonkers, NY 10701, USA

PATHOGENS OF CIMICIDAE (BED BUGS)

Host	Host stage infected	Pathogen	% Incidence	Locality	Lab. or field study	Reference
Cimex lectularius	All	Symbiotes (Rickettsia) lectularius	100	England	Field, lab.	Arkwright et al. (1921)
"	All	"	100	France	Lab.	Louis et al. (1973)
"	All	"	100	Canada	Lab.	Chang & Musgrave (1973)
"	All	"	100	Canada	Lab.	Chang (1974)
"	All	Bacillus anthracis	-	USSR	Lab.	Rosenholz & Owsjannikowa (1921)
"	All	Francisella (Bacterium) tularensis	-	USA	Lab.	Francis (1927)
"	Adults	Brevibacterium (Bacterium) tegumenticola (classification doubtful)	-	USA	Lab.	Steinhaus (1941)
"	Adults	Corynebacterium paurometabolum	-	USA	Lab.	Steinhaus (1941)
"	Adults	Coccus (?)	-	USA	Lab.	"
"	Adults	Micrococcus luteus (=Sarcina flava)	-	USA	Lab.	"
"	-	Salmonella avium (not a recognized taxon)	-	USA	Lab.	Jenkins (1964)
"	All	Aspergillus flavus	100	USA	Lab. colony	Cockbain & Hastie (1961)
"	-	Leishmania (Herpetomonas) donovani	-	India	Lab.	Patton (1913)
"	-	Trypanosoma rangeli	-	India	Lab.	Grewal (1957, 1959)
C. hemipterus rotundatus	-	Leishmania (Herpetomonas) donovani	-	India	Lab.	Patton (1913)
"	-	Leishmania (Herpetomonas) tropica	-	India	Lab.	Patton et al. (1921)

PATHOGENS OF CIMICIDAE (BED BUGS) (continued)

Host	Host stage infected	Pathogen	% Incidence	Locality	Lab. or field study	Reference
C. hemipterus rotundatus	All	Wuchereria bancrofti	83 from bed of an infected patient	Guyana	Field	Burton (1963)
"	-	"	-	Ceylon	Field	Gunawardena (1972)
"	-	Nosema adie	-	-	-	Jenkins (1964)
Paracimex sp.	-	Aspergillus flavus	-	Malaya	Field	Usinger (1966)

ABSTRACTS

Mary Ann Strand

Arkwright, J. A. et al. (1921).   An hereditary Rickettsia-like parasite of the bed bug (Cimex lectularius).   Parasitology, 13: 27-36.

This paper contains the original observations of Symbiotes (Rickettsia) lectularius, a symbiote of the bed bug.

Burton, G. J. (1963).   Natural and experimental infection of bed bugs with Wuchereria bancrofti in British Guiana.   Am. J. Trop. Med. Hyg., 12: 541-547.

C. hemipterus adults and larvae were collected from a bed used by people infected with W. bancrofti.   Living or dead larvae of W. bancrofti were found in the abdomen, thorax, legs, and antennae of 83 of the bed bugs.   However in only 3 cases did the filariae develop into the infective stage.

Chang, K. P. (1974).   Effects of elevated temperature on the mycetome and symbiotes of the bed bug Cimex lectularius (Heteroptera).   J. Invertebr. Pathol., 23: 333-340.

The lower fecundity of C. lectularius after treatment at 36° may be due to a direct effect of the high temperature on the insects or to the loss of their symbiotes.

Chang, K. P. & Musgrave, A. J. (1973).   Morphology, histochemistry, and ultrastructure of mycetome and its rickettsial symbiotes in Cimex lectularius 1.   Can. J. Microbiol., 19: 1075-1081.

Two kinds of rickettsia-like microorganisms were observed in the mycetomes of C. lectularius.   One was a small rod and the other was pleomorphic.   The rod is probably Symbiotes (Rickettsia) lectularius, but the pleomorphic one is apparently different in that it has host-provided membranes without encapsulation.

Cockbain, A. J. & Hastie, A. C. (1961).   Susceptibility of the bed bug, Cimex lectularius Linnaeus, to Aspergillus flavus Link.   J. Insect Pathol., 3: 95-97.

A culture of bed bugs kept at 30°C and 90% R.H. became infected with A. flavus.   The culture was destroyed within 18 days after initial outbreak.   Infectivity tests showed that a combination of high humidity and temperature favoured the development of the disease.

Francis, E. (1927).   Microscopic changes of tularaemia in the tick, Dermacentor andersoni, and the bedbug, Cimex lectularius.   Publ. Health Rep., 42: 2763-2772.

In laboratory experiments Francisella (Bacterium) tularense multiplied in the fresh blood contents of the midgut in C. lectularius.   Heavy infections in epithelium cells of the midgut and occasional infections in the mapighian tubules were observed.

Grewal, M. S. (1957).   Pathogenicity of Trypanosoma rangeli Tejera, 1920 in the invertebrate host.   Exp. Parasitol., 6: 123-130.

Invasion of the haemolymph of Cimex lectularius by the trypanosome was rapid.   Mortality was higher in bugs with heavier infections in the haemolymph.

Grewal, M. S. (1969).   Studies on Trypanosoma rangeli, a South American human trypanosome. Res. Bull. Panjab Univ. Sci., 20: 449-486.

T. rangeli is not pathogenic to its vertebrate hosts but is pathogenic to its inverte-brate hosts including Cimex lectularius.   The bed bug cannot molt after a haemolymph infection is established.   Heavily infected individuals die.

Gunawardena, K. (1972).   A note on the susceptibility of the tropical bed bug Cimex hemipterus to Wuchereria bancrofti in Ceylon.   J. Med. Entomol., 9: 300.

Microfilaria ingested from infected patients survived in the bed bugs for up to 20 days. No infective forms were found in the bugs.

Hase, A. (1917).   Die Bettwanze (Cimex lectularius L.) ihr Leben und ihre Bekämpfung. Monogr. Angew. Entomol., 1: 1-144.

Although no bed bug pathogens are listed, this paper contains a fascinating account of control attempts before the use of DDT.

Jenkins, D. W. (1964).   Pathogens, parasites, and predators of medically important athropods. Bull. World Health Organ., 30 (Suppl.), 150 pp.

A review of the literature through 1963 is given.

Louis, C. et al. (1973).   Mobilité, ciliature et caractères ultrastructuraux des micro-organismes symbiotiques endo et exocellulaires des Cimex lectularius L. (Hemiptera Cimicidae). C. R. Hebd. Seances Acad. Sci., Ser. D, 277(6): 607-611.

The endo- and exocellular forms of Symbiotes (Wolbachia) lectularius were examined using the electron microscope.

Patton, W. S. (1913).   Further observations on the development of Herpetomonas donovani in Cimex rotundatus and Cimex lectularius.   Proc. 3rd Meet. Gen. Malaria Comm. (Madras), pp. 221-232.

Leishmania (H.) donovani is capable of carrying its development in the bed bug to the post-flagellate stage but does not multiply and eventually dies out.

Patton, W. S. et al. (1921).   Studies on the flagellates of the genera Herpetomonas, Crithidia, and Rhynchoidomonas.   No. 8.   Note on the behavior of Herpetomonas tropica Wright, the parasite of a cutaneous Herpetomonas (Oriental sore) in the bed bug, Cimex hemiptera, Fabr.   Indian J. Med. Res., 9: 240-251.

An intracellular stage of Leishmania (H.) tropica is described from the cells of the midgut of C. hemipterus.   In the alimentary canal of the bugs, L. tropica can live for months.

Rosenholz, H. P. & Owsjannikowa, O. W. (1929).   Über die rolle der Wanzen (Cimex lectularius) und Zecken (Ornithodorus moubata) bei Übertragung des Milzbrandes.   Zentralbl. Bacteriol. Parasitenkd. Infektionsdr. Hyg. Abt. I Orig., 110: 160-164.

The bugs became infected with Bacillus anthracis from feeding on infected guinea-pigs. The bacillus penetrated the gut wall and multiplied in the haemolymph.   Most bugs died within 2 days.   A few survived and carried the infection in the intestinal tract for over one month.

318

Steinhaus, E. A. (1941).   A study of the bacteria associated with thirty species of insects. J. Bacteriol., 42: 757-790.

The bacterial flora of various organs of Cimex lectularius was examined.   Bacteria were found on the integument and in the ovaries, mycetome and hind gut.

Usinger, R. L. (1966).   Monograph of Cimicidae (Hemiptera-Heteroptera).   Thomas Say Foundation (Entomol. Soc. Am.) Vol. 7, 585 pp.

Aspergillus flavus was found infected cultures of Paracimex received from Malaya.

XIX. PATHOGENS OF REDUVIIDAE (ASSASSIN BUGS)

Mary Ann Strand

Boyce Thompson Institute
Yonkers, NY 10701, USA

PATHOGENS OF REDUVIIDAE (ASSASSIN BUGS)

Host	Host stage infected	Pathogens	% Incidence	Locality	Lab. or field study	Reference
Eutriatoma (Triatoma) rubrovaria	-	Haemogregarina triatomae	-	South America	-	Steinhaus (1946)
Eutriatoma maculata	-	Machadoella triatomae	-	-	-	Mayer & Pifano (1949)
Panstrogylus megistus	All	Beauveria bassiana	-	Brazil	Lab.	Dias & Leão (1967)
"	-	Octosporea carloschagasi	-	-	Field	Kramer (1972)
"	-	O. muscae-domesticae	-	-	-	Chatton (1911)
"	Adults	Trypanosoma rangeli	-	Brazil	Field	Lucena & deVergetti (1973)
Rhodnius brethesi	-	T. rangeli	-	Colombia	Field	D'Alessandro et al. (1971)
Rhodnius ecuadoriensis	-	T. rangeli	7	Peru	Field	Cuba et al. (1972)
"	-	"	-	Peru	Lab.	Cuba (1972), Herrer et al. (1972)
Rhodnius pallescens	-	"	-	Colombia	Field	D'Alessandro et al. (1971)
Rhodnius prolixus	Nymphs, adults	Crystal inclusions	Rare	USA	Lab. colony	Steinhaus & Marsh (1962)
"	Adults	Bacillus thuringiensis	100 mortality	USA	Lab.	Nyirady (1973)
"	All	Diplococcus (possible symbiote)	100	-	Lab.	Bewig & Schwartz (1954)
"	-	Escherichia coli	-	USA	Lab.	Nyirady (1973)
"	-	Streptococcus faecalis	-	USA	Lab.	Nyirady (1973)
"	-	"	-	USA	Lab.	Brecher & Wigglesworth (1944)

PATHOGENS OF REDUVIIDAE (ASSASSIN BUGS) (continued)

Host	Host stage infected	Pathogens	% Incidence	Locality	Lab. or field study	Reference
Rhodnius prolixus (continued)	All	Nocardia rhodni	100	-	Lab.	Bewig & Schwartz (1954), Baines (1956), Lake & Friend (1968), Auden (1974). For more references see Jenkins (1964)
"	-	T. rangeli	-	Venezuela	Field	Tejara (1920)
"	-	"	-	Colombia	Field	D'Alessandro et al. (1971)
"	-	"	-	-	Lab.	Grewal (1957, 1969), Zeledon & de Monge (1966), Gomez (1967), Tobie (1965, 1968, 1970), Watkins (1971a, b)
		Blastocrithidia spp.	-	USA	Lab.	Hanson et al. (1968)
		Crithidia spp.	-	USA	Lab.	" " " "
Rhodnius robustus	-	T. rangeli	-	Colombia	Field	D'Alessandro et al. (1971)
Triatoma dimidiata	-	T. rangeli	-	Colombia	Field	" " " "
		Machadoella triatomae	-	-	-	Steinhaus (1946)
Triatoma infestans	All	Bacterium (symbiote)	100	-	Lab.	Halff (1956)
"	All	Diplococcus (possible symbiote)	100	-	Lab.	Bewig & Schwartz (1954)
"	All	Nocardia rhodnii (symbiote)	100	-	Lab.	" " " "
"	All	Beauveria bassiana	-	Brazil	Lab.	Dias & Leão (1967)
"	-	T. rangeli	-	Costa Rica	Lab.	Zeledon & de Monge (1966)
"	-	Blastocrithidia spp.	10	Brazil	Lab. colony	Rocha de Silva & Amaral (1971)
"	-	"	-	USA	Lab.	Hanson et al. (1968)
"	-	Crithidia spp.	-	USA	Lab.	" " " "
Triatoma protracta	Nymph	Bacillus thuringiensis	100 mortality	USA	Lab.	Nyirady (1973)

PATHOGENS OF REDUVIIDAE (ASSASSIN BUGS) (continued)

Host	Host stage infected	Pathogens	% Incidence	Locality	Lab. or field study	Reference
Triatoma protracta	-	Escherichia coli	-	USA	Lab.	Nyirady (1973)
"	-	Streptococcus faecalis	-	USA	Lab.	"
"	Nymph	Nocardia (Mycobacterium) sp.	100	USA	Lab.	Marchette & Hatie (1965) (gut flora)
"	Nymph	N. (M.) rhodochrous	100			"
"	Nymph	Corynebacterium sp.	100			"
"	Nymph	Yeast	100			"
Triatoma rubida	Nymph	Bacillus thuringiensis	-	USA	Lab.	Nyirady (1973)
"	-	Escherichia coli	-	USA	Lab.	"
"	-	Streptococcus faecalis	-	USA	Lab.	"
Triatoma vitticeps	All	Beauveria bassiana	-	Brazil	Lab.	Dias & Leão (1967)

ABSTRACTS

Mary Ann Strand

Auden, D. T. (1974). Studies on the development of Rhodnius prolixus and the effects of its symbiote Nocardia rhodnii. J. Med. Entomol., 11: 68-71.

N. rhodnii was shown not to be necessary for the development of the bugs when they are raised on mouse blood but it is essential when they are fed rabbit blood. The nutritional factors for proper molting are apparently lacking in the rabbit blood.

Baines, S. (1956). The role of the symbiotic bacteria in the nutrition of Rhodnius prolixus (Hemiptera). J. Exp. Biol., 33: 533-541.

The symbiote Nocardia rhodnii is described. The development of normal and symbiote-free nymphs is compared. The development of both groups was similar during the first and second instars. The symbiote-free nymphs were greatly retarded or failed to molt after the third instar.

Bewig, F. & Schwartz, W. (1954). Untersuchingen über die Symbiose der Triatomiden Rhodnius prolixus Stal und Triatoma infestans Klug. Naturwissenschaften, 41(18): 435.

The authors cultured the microbial population of the gut and faeces of both bugs. Their observations confirm the presence of Nocardia rhodnii in both species. Also, a diplococcus was isolated which grew weakly in culture and was frequently overgrown by N. rhodnii. No physiological effect on the bugs was noticed.

Brecher, G. & Wigglesworth, V. B. (1944). The transmission of Actinomyces rhodnii Erikson in Rhodnius prolixus Stal (Hemiptera) and its influence on the growth of the host. Parasitology, 35: 220-224.

Carcavallo, R. U. (1970). Epidemiologia de la trypanosomiasis americana y las possibilidades del control biologico. pp. 381-390. In K. S. Singh and B. K. Tandan (eds.), H. D. Srivastava commemorative volume. Indian Veterinary Research Institute, Izatnagar, India.

Protozoan, bacterial, and fungal parasites of triatomids are reviewed. Most are symbiotic. /From Rev. Appl. Entomol. B, 1974, 62(4): 698/

Chatton, E. (1911). Microsporidies considérées comme causes d'erreurs dans l'étude du cycle évolutif des trypanosomides chez les insectes. Bull. Soc. Pathol. Exot., 4: 662-664.

Cuba, C. A. C. (1972). Development of T. rangeli metacyclic forms from Triatomid bugs' salivary glands ingested by Rhodnius ecuadoriensis. Trans. R. Soc. Trop. Med. Hyg., 66: 944-945.

When metacyclic trypanosomes were ingested, they developed in the digestive tract and invaded the haemocoel. After multiplication, the parasites invaded the salivary glands and completed their development.

Cuba, C. et al. (1972). Hallazgo de Rhodnius ecuadoriensis Lent and León, 1958 infectado naturalmente por trypanosomas semejantes a Trypanosoma rangeli Tejera, 1920 en caseríos del Distrito de Cascas, Contumazá, Dpto. de Cajamarca, Perú. Rev. Inst. Med. Trop São Paulo, 14(3): 191-202. (S, e).

Seven per cent. of the R. ecuadoriensis individuals examined contained a flagellate similar to T. rangeli.

D'Alessandro, A. et al. (1971).   Distribution of triatomine-transmitted trypanosomiasis in Colombia and new records of the bugs and infections.   J. Med. Entomol., 8: 159-172.

The distribution and host range of Trypanosoma rangeli and T. cruzi are reviewed.

Dias, J. C. P. & Leão, A. E. A. (1967).   Parasitismo de fungos (Beauveria bassiana) sôbre Triatomíneos brasileiros criados en laboratório.   Atas. Soc. Biol. Rio de J., 11(3): 85-87.

Laboratory colonies of Triatoma infestans, T. vitticeps, and Panstrongylus megistus became infected with B. bassiana after the fungal spores were sprinkled on them. Although some nymph and adult mortality was observed, the colonies were not greatly affected.   The fungus was reisolated from body cavity, body surface, and excreta of infected bugs.

Duncan, J. T. (1926).   On a bactericidal principle present in the alimentary canal of insects and arachnids.   Parasitology, 18: 238-252.

In experimental tests, gut contents of Rhodnius prolixus were bactericidal for Bacillus mesentericus and B. subtilis and not for B. mycoides, Staphlococcus albus, and S. aureus. No pathological conditions were noted.

Gomez, I. (1967).   Nuevas observaciones acerca de la acción patógena del Trypanosoma rangeli Tejera, 1920 sôbre Rodnius prolixus Stal, 1859.   Rev. Inst. Med. Trop São Paulo, 9: 5-10. (Pt, e).

Six per cent. of bugs fed on blood infected with T. rangeli survived to adulthood, while 62% of the controls survived.   After taking an infected meal, there is an increase in the haemolymphatic cells.

Grewal, M. S. (1957).   Pathogenicity of Trypanosoma rangeli Tejera, 1920 on the invertebrate host.   Exp. Parasitol., 6: 123-130.

Infected Rhodnius prolixus were sluggish and unnaturally light in colour.   The parasite apparently interferes with the molting process.   Mortality was highest among nymphs fed on animals with heaviest infections.

Grewal, M. S. (1969).   Studies on Trypanosoma rangeli, a South American human trypanosome. Res. Bull. Panjab Univ. Sci., 20: 449-486.

Haemolymph infection of Rhodnius prolixus depends upon the age of the bugs and on the concentration of the trypanosomes in their blood meals.   Younger bugs develop haemolymph infection within 10 days of an infective meal.   Crithidial and trypanosome forms from the haemolymph invade the salivary glands and transform into metacyclic trypanosomes there.

Halff, L. A. (1956).   Untersuchungen über die Abhängigkeit der Entwicklung der Reduviide Triatoma infestans Klug von ihrem Darmsymbioten.   Acta Trop., 13: 225-253.   (G, e, f).

A symbiotic bacterium was observed in the midgut of T. infestans.   It appears to have a similar structure and function to Nocardia rhodnii in Rhodnius prolixus.

Hanson, W. L. et al. (1968).   Experimental infection of Triatoma infestans and Rhodnius prolixus with Trypanosomatidae of the genera Crithidia and Blastocrithidia.   J. Protozool., 15(2): 346-349.

The experimental infections demonstrated that the protozoa could survive and in most cases reproduce in the bugs for as long as 2 months after infection.   The trypanosomes have a loose host specificity.   No pathogenic conditions were noted.

Herrer, A. et al. (1972).   Presencia de Trypanosoma rangeli Tejera, 1920, en el Perú.
I. El insecto vector, Rhodnius ecuadoriensis Lent & León, 1958.   Rev. Biol. Trop., 20: 141-
149.   (S, e).

   Metacyclic forms of T. rangeli were seen in the salivary glands of R. ecuadoriensis.
This strain of T. rangeli was not infective for R. prolixus.

Jenkins, D. W. (1964).   Pathogens, parasites, and predators of medically important
arthropods.   Bull. WHO, 30(Suppl.), 150 pp.

   A review of the literature through 1963 is given.

Kramer, J. P. (1972).   Octoporea carloschagasi n. sp., a Microsporidian associate of
Trypanosoma cruzi in Panstrongylus megistus.   Z. Parasitenkd., 39: 221-224.

Lake, P. & Friend, W. G. (1968).   The use of artificial diets to determine some of the
effects of Nocardia rhodnii on the development of Rhodnius prolixus.   J. Insect. Physiol.,
14(4): 543-562.

   The bug cannot develop normally on rabbit blood in the absence of its normal symbiont,
N. rhodnii.   The symbiont apparently supplies its host with pantothenic acid, folic
acid, pyridoxine, and thiamine, and may also provide nicotinic acid and riboflavin.

Lucena, D. T. de & Vergetti, J. G. (1973).   Infecção natural de Panstrongylus megistus
(Burmeister, 1835) por Trypanosoma rangeli (Tejera, 1920), no interior do Estado de Alagoas.
Rev. Inst. Med. Trop. São Paulo, 15: 171-178.   (Pt, e).

   A male P. megistus was infected by T. rangeli in Alagoas State, Brazil.   Identification
was based on general morphology and measurements of the crithidial and metacyclic stages.

Marchette, N. J. & Hatié, C. (1965).   Microbial isolates from the digestive tract of
Triatoma protracta (Uhler) (Reduviidae).   J. Invertebr. Pathol., 7: 45-48.

   Nocardia (Mycobacterium) sp., N. rhodochrous, and Corynebacterium sp., and an unidentified yeast were isolated from the digestive tracts of fourth instar nymphs.   The
bacteria were most numerous in the foregut.   It is not known if these organisms are
true symbiotes or if they perform a vital physiological function.

Mayer, M. & Pifano, F. (1949).   Sôbre una schizogregrarine de Eutriatoma maculata de
Venezuela.   Arch. Venez. Patol. Trop. Parasitol. Med., 1(2): 335-340.

Nyirady, S. A. (1973).   The germfree culture of three species of Triatominae: Triatoma
protracta (Uhler), Triatoma rubida (Uhler) and Rhodnius prolixus Stal.   J. Med. Entomol.,
10(5): 417-448.

   Germfree reduviids that ingested Escherichia coli or Streptococcus faecalis died within
3 days of feeding.   Dissection of the dead bugs revealed that blood had escaped from
the stomach and filled the body cavity.   Bacillus thuringiensis var. thuringiensis was
fed to adults and fifth instar nymphs of R. prolixus and to adults and fifth instar
nymphs of T. protracta.   All died within 1 week of feeding.

Silva, E. O. R. & Amaral, A. D. F. (1971).   Sôbre o encontro de um parasita do gênero
Blastocrithidia en exemplares de T. infestans criados em colônias de laboratório.   Rev.
Saúde Publica, 5(2): 298.   (Pt)

   Trypanosomatids were found in 10% of the laboratory colony of T. infestans maintained
in São Paulo State, Brazil.

Steinhaus, E. A. (1946). Insect microbiology; an account of the microbes associated with insects and ticks, with special reference to the biologic relationships involved. Comstock Publ. Co., Ithaca, N.Y., 763 pp.

General reference.

Steinhaus, E. A. & Marsh, G. A. (1962). Report of diagnoses of diseased insects 1951-1961. Hilgardia, 33: 349-490.

Although no evidence of microbial death was found, abundant crystal inclusions were found in 3 adults and 8 nymphs of Rhodnius prolixus from a laboratory colony.

Tejera, E. (1920). Un nouveau flagellé de Rhodnius prolixus, Trypanosoma (ou Crithidia) rangeli n. sp. Bull. Soc. Pathol. Exot., 13: 527-530.

T. rangeli is described as a new species from R. prolixus collected in Venezuela.

Tobie, E. J. (1965). Biological factors influencing transmission of Trypanosoma rangeli by Rhodnius prolixus. J. Parasitol., 51: 837-841.

When the first instar nymphs of R. prolixus are exposed to a meal containing T. rangeli, survival during the nymphal stage decreases. Of those exposed, 34% had died by the end of the fifth instar, while 16% of the controls had died. The infection apparently interferes with the molting process. T. rangeli was more pathogenic to females than to males.

Tobie. E. J. (1968). Fate of some culture flagellates in the hemocoel of Rhodnius prolixus. J. Parasitol., 54: 1040-1046.

Eighteen species of flagellates were inoculated into R. prolixus adults. Trypanosoma rangeli multiplied and survived for an indefinite time. T. cruzi and other flagellates for which R. prolixus is not a natural host were killed by the haemocytes. Crithidia fasciculata multiplied and was pathogenic.

Tobie, E. J. (1970). Observations on the development of Trypanosoma rangeli in the hemocoel of Rhodnius prolixus. J. Invertebr. Pathol., 15: 118-125.

The salivary glands are not always invaded when T. rangeli is injected in the haemocoel of R. prolixus. The parasite can live and multiply in the haemocoel. Crithidial forms enter the plasmotocytes and multiply, eventually the cells rupture.

Watkins, R. (1971a). Histology of Rhodnius prolixus infected with Trypanosoma rangeli. J. Invertebr. Pathol., 17: 59-66.

T. rangeli infections appear to inhibit Nocardia rhodnii, a symbiote needed for normal development. This would explain failure to molt by some infected individuals. Also, infections result in nerve damage and hypertrophy of tracheal cells. The physiological effects of infection are summarized.

Watkins, R. (1971b). Trypanosoma rangeli: effect on excretion in Rhodnius prolixus. J. Invertebr. Pathol., 17: 67-71.

Excretion is reduced in infected bugs. Several reasons for this are presented, including: tissue damage, lack of diuretic hormone, presence of an inhibitor, and changes in osmotic pressure.

Zeledón, R. and de Monge, E. (1966).   Natural immunity of the bug <u>Triatoma infestans</u> to the protozoan <u>Trypanosoma rangeli</u>.   <u>J. Invertebr. Pathol</u>., <u>8</u>(3): 420-424.

<u>Rhodnius prolixus</u> from El Salvador and <u>T. infestans</u> from Chile were artificially infected with two strains of <u>T. rangeli</u>.   A high degree of parasitism was found in the body cavity of <u>R. prolixus</u> (up to 43×10^6 flagellates/ml), while <u>T. infestans</u> was only occasionally invaded.   <u>T. infestans</u> disposes of the infection mainly by phagocytosis.

XX. PATHOGENS OF ACARINA (MITES)

Elizabeth West Davidson

Division of Agriculture
Arizona State University
Tempe, AZ 85281, USA

PATHOGENS OF ACARINA (MITES)

Host	Host stage infected	Pathogen or parasite	% Incidence	Locality	Lab. or field study	Reference
Cepheus tegeocranus Oribatid mite	?	Actino-bacillus Ray-fungus	"Very scarce"	?	Field	Warren (1947)
Ceratoppia bipilis	?	Nosema helminthorum	3	-	Lab.	Dissanaike (1958)
Dermanyssus gallinae Fowl mite	Adult	"Streptotrichian" similar to actinomycosis	25, females 4, males	Kent England	Field	Warren (1940)
Eutrombicula alfreddugesi	All	Yeast-like spherical microorganisms 6-12 µm in diameter	100	USA	Field	Kroman et al. (1969)
Haemogamasus hirsutus	Female	Acarinocola hirsutus Nematode	1 specimen	Warwickshire, England	Field	Warren (1941)
Haemogamasus liponysoides	?	Eastern equine encephalitis virus	-	-	Lab.	Clark et al. (1966)
Haemogamasus reidi potential vector of plague, typhus, and tularemia	?	Hepatozoon griseisciure Haemogregarine of grey squirrel	-	-	Lab.	Redington & Jachowski (1971)
Haemogamasus reidi	All	Hepatozoon griseisciure	-	-	Lab., Field	Redington & Jachowski (1972)
Haemolaelaps centrocarpus	?	Hepatozoon balfouri	20	USA	Lab.	Furman (1966)
Haemolaelaps longipes	?	Hepatozoon balfouri	100	USA	Lab.	"

PATHOGENS OF ACARINA (MITES) (continued)

Host	Host stage infected	Pathogen or parasite	% Incidence	Locality	Lab. or field study	Reference
Hannemania sp.	All	Yeast-like spherical microorganisms 6-12 μm in diameter	100	USA	Field	Kroman et al. (1969)
Hermannia sp.	Both sexes	Acoccodium ventriculi	–	England	Field	Warren (1944)
Laelaps ecidninus	?	Hepatozoon perniciosum Haemogregarine				Miller (1908)
Leptotrombidium akamushi Chigger	Females and offspring	Rickettsia tsutsugamushi Scrub typhus	100 transm. to F_1	–	Lab.	Rapmund et al. (1969)
Leptotrombidium arenicola Chigger	Females and offspring	Rickettsia tsutsugamushi	F_1-transm. to 1/39 F_2-transm. to 3/49	–	Lab.	Rapmund et al. (1972)
Leptotrombidium fletcheri	All	"	Varied	–	Lab.	Roberts et al. (1975)
Leptotrombidium panamense	All	Yeast-like spherical micro-organisms 6-12 μm in diameter	100	USA	Field	Kroman et al. (1969)
Liponyssus bacoti	?	Litomosoides carinii	–	–	Lab.	Williams & Brown (1945)
Liponyssus bacoti	?	Litomosoides carinii	–	–	Lab.	Williams & Brown (1946)
Liponyssus bacoti	?	"	–	–	Lab.	Hughes (1950)
Liponyssus bacoti	13-day-old females	Litomosoides carinii	–	–	Lab.	Quraishi et al. (1966)

PATHOGENS OF ACARINA (MITES) (continued)

Host	Host stage infected	Pathogen or parasite	% Incidence	Locality	Lab. or field study	Reference
Liponyssus (Ornithonyssus) bacoti Tropical rat mite suspect vector of scrub and murine typhus	?	Litomosoides carinii Cotton rat worm	-	-	Lab.	Bertram et al. (1946)
Liponyssus bacoti	?	Litomosoides carinii	-	-	Lab.	Bertram (1947)
Liponyssus saurarum	Eggs, adults	Karyolysis lacertae Haemogregarine	-	-	Lab.	Reichenow (1921)
Neotrombicula harperi	All	Yeast-like spherical microorganisms 6-12 μm in diameter	100	USA	Field	Kroman et al. (1969)
Ornithonyssus bacoti	Both sexes	Litomosoides carinii	-	-	Lab.	Williams & Kershaw (1961)
Ornithonyssus bacoti	?	Eastern equine encephalitis virus	-	-	Lab.	Clark et al. (1966)
Urodinychus sp.	Both sexes	Paramoncystis simplex	-	England	Field	Warren (1944)
Urodinchus sp.	Both sexes	Acoccidium tubulorum	-	England	Field	"
Xenillus tegeocranus	?	Nosema helminthorum	3	-	Lab.	Dissanaike (1958)
Mite colonies	All	Fungi	-	-	Lab.	Audy & Lavoipierre (1964)

ABSTRACTS

Mary Ann Strand

Audy, J. R. & Lavoipierre, M. M. J. (1964).   The laboratory rearing of parasitic acarina.
Bull. World Health Organ., 31: 583-586.

 Fungi may cause mortality in mite colonies when humidity is kept above 75%.

Bertram, D. S. (1947).   The period required by Litomosoides carinii to reach the infective
stage in Liponyssus bacoti and the duration of the mite's infectivity.   Ann. Trop. Med.
Parasitol., 41: 253-261.   (RAEb 39: 22).

 L. carinii reached the infective stage 14 days after the host mites became infected and
 remained infective for 22 more days.   The infection was transmitted by the mites to rats
 on the 3rd through the 5th meal (15-25 days).

Bertram, D. S. et al. (1946).   The biology and maintenance of Liponyssus bacoti Hirst, 1913,
and an investigation into its role as a vector of Litomosoides carinii to cotton rats and white
rats, together with some observations on the infection in the white rats.   Ann. Trop. Med.
Parasitol., 40: 228-254.   (RAEb 37: 139)

 The microfilariae are taken up by nymph and adult mites, but no developmental forms were
 found in the nymphs.   Of 11 adult female mites exposed, late developmental forms were
 found in 5 after 20-33 days.   Filariae transversed the gut wall of the mite and are
 transmitted as the mite feeds.

Clark, G. M. et al. (1966).   Observations on the ability of Haemogamasus liponyssoides Ewing
and Ornithonyssus bacoti (Hirst) (Acarina, Gamasina) to retain eastern equine encephalitis
virus: Preliminary report.   Am. J. Trop. Med. Hyg., 15: 107-112.

 The mites became infected with EEE after feeding on blood of infected hosts.   The virus
 persisted in the H. liponyssoides for 24 hours at room temperature or longer at higher
 temperatures.   In O. bacoti, the virus persisted 7 days at room temperature.   No
 increase in virus was seen in the mites.

Dissanaike, A. S. (1958).   Experimental infection of tapeworms and oribatid mites with
Nosema helminthorum.   Exp. Parasitol., 7: 306-318.   (A 33: 27738).

 The experimental infection of 2 mites, Ceratoppia bipilis and Xenillus tegeocranus, with
 N. helminthorum is reported.   The microsporidian is a hyper-parasite of Moniezia of
 lambs.   Its spores are similar to, though smaller than, N. helminthorum from cestodes.

Furman, D. P. (1966).   Hepatozoon balfouri (Laveran, 1905): Sporogonic cycle, pathogenesis,
and transmission by mites to jerboa hosts.   J. Parasitol., 52(2): 373-382.

 H. balfouri was successfully transferred to 2 species of laboratory reared mites,
 Haemolaelaps longipes and Haemolaelaps centrocarpus.   Heavy infections were pathogenic
 and often fatal.   Nearly all of the H. longipes mites and about 20% of the
 H. centrocarpus became infected when fed infective blood.   No difference in the time
 required for parasite development was noted in the 2 species of mites.

Hughes, T. E. (1950). Some stages of Litomosoides carinii in Liponyssus bacoti. Ann. Trop. Med. Parasitol., 44: 285-290. (BA 28: 25690)

The microfilariae penetrate the gut wall within 24 hrs and undergo successive moults in the surrounding parenchyma. Those lingering in the gut lumen are ingested by phagocytes.

Kroman, R. A. et al. (1969). The presence of yeastlike cells of probable symbiotic nature in trombiculid mites. Proc. 2nd Int. Cong. Acarol., pp. 309-315.

Yeastlike cells were observed in the ventriculi of larval and adult chiggers. In some adults they were also distributed in the parenchyma-like tissue throughout the body. Their presence in egg and deutova indicates they are not food inclusions.

Lipa, J. J. (1971). Microbial control of mites and ticks. pp. 357-373. In Burges, H. D. & Hussey, N. W. (ed.). Microbial control of insects and mites. Academic Press, New York.

This article mainly deals with mites which are agricultural pests.

Miller, W. M. (1908). Hepatozoon perniciosum (n.g., n.sp.) a haemogregarine pathogenic in the intermediate host, a mite (Laelaps ecidninus). Bull. U.S. Pub. Health Mar. Hosp. Serv. Hyg. Lab., 46: 1-51.

Neither abstract nor reference seen.
(Reference from Clark et al., 1966)

Quraishi, M. A. H. et al. (1966). The effect of X-irradiation on the survival of Liponyssus bacoti and on its susceptibility to infection with Litomosoides carinii. Ann. Trop. Med. Parasitol., 60: 57-62.

When exposed to radiation prior to feeding on an infected host, the intensity of infection of mites surviving irradiation was greater than the unirradiated controls. However mortality of the irradiated mites was five times higher than the controls. When exposed to radiation after infection, the intensity of infection of the survivors was lower than unirradiated controls.

Rapmund, G. et al. (1972). Transovarial transmission of Rickettsia tsutsugamushi in Leptotrombidium (Leptotrombidium) arenicola Traub (Acarina: Trombiculidae). J. Med. Entomol., 9: 71-72.

Transovarial transmission was demonstrated through 2 generations of the mites. From one initially infected female, 1 of 63 offspring transmitted the rickettsia and 3 of 49 of her offspring also were infective.

Rapmund, G. et al. (1969). Transovarial development of scrub typhus rickettsiae in a colony of vector mites. Trans. R. Soc. Trop. Med. Hyg., 63: 251-258.

The transovarial infection rate over 3 generations of Leptotrombidium akamushi was 98%. Infected female mites produced only female offspring, however, the role of the infection in this occurrence is not known.

Redington, B. C. & Jachowski, L. A. Jr (1971). Syngamy and sporogony of Hepatozoon griseisciuri Clark, 1958 (Sporozoa: Haemogregarinidae), in its natural vector, Haemogamasus reidi Ewing, 1925 (Acari: Mesostigmata). J. Parasitol., 57: 953-960.

The sporogonic cycle in the mite is described. Microgametes are not formed prior to zygote formation as is the case with many other hemogregarines.

Redington, B. C. & Jachowski, L. A. Jr (1972). Role of Haemogamasus reidi (Acari: Mesostigmata) in the life cycle of the grey squirrel protozoan, Hepatozoon griseisciuri (Sporozoa: Haemogregarinidae). J. Parasitol., 58: 401-403.

All mites exposed to blood infected with the sporozoan, which lived longer than 5 days post exposure, were found to harbour oocysts of the parasite. Heavy infections appeared to have a deleterious effect on some mites. Death or delayed moulting was observed. Although all feeding stages of the mites are suitable hosts for this sporozoan under experimental conditions, there are factors involved in the natural transmission of the parasite which result in only the adult female acting as host.

Reichenow, E. (1921). Die Hämococcidien der Eidechsen. Arch. Protistenkd., 42: 179-292. (RAEb 10: 224), also published as: Los Hemococcidios de los Lacertidos. Trab. Mus. Nac. Ciencias Nat. (Madrid), ser. zool., 40, 153 pp. (RAEb 8: 208).

A study of the blood parasite Karyolysus lacertae from the lizard Lacerta muralis is reported. The parasite is transmitted by the mite Liponyssus saurarum. The life cycle of the mite and development of K. lacertae are described.

Roberts, L. W. et al. (1975). J. Med. Entomol., 12: 345-348. Distribution of Rickettsia tsutsugamushi in organs of Leptotrombidium (Leptotrombidium) fletcheri (Prostigmata: Trombiculidae).

Warren, E. (1940). The genital system of Dermanyssus gallinae. Ann. Natal Mus., 9: 409-459.

A "streptotrichian"[*] attacks the gut epithelium and ovaries of the fowl mite. The parasite causes degeneration of ova. The infected mites were collected near Kent, England.

Warren, E. (1941). On the occurrence of nematodes in the haemocoel of certain gamasid mites. Ann. Natal Mus., 10: 79-94. (BA 17: 17517).

Nematodes, which have been found in the haemocoel of 5 species of gamasid mites, are described and placed in a new genus Acarinocola. The nematodes apparently reproduce parthenogenically in the haemocoel of the mites, but a free-living sexual generation may occur.

Warren, E. (1944). Observations on the anatomy and histology of a myrmecophilous mite (Urodinychus sp.) and an account of certain of its sporozoon parasites. Ann. Natal Mus., 10(3): 359-406.

---

[*] Probably an actinomycete.

Paramonocystis simplex, described as a new genus and species of gregarine, attacks the giant gland-cells of the posterior caeca of the ventriculus of the mite. The parasite may in turn be attacked by actinomycetes. The Malpighian tubules of the mites are, also, attacked by a haplosporidian, Acoccidium tubulorum, a newly described species. An allied sporozoon, which attacks the epithelium of the ventriculus of Hermannia sp. is described as A. ventriculi, also a new species.

Warren, E. (1947). On the genital system and gut of the oribatid mite, Cepheus tegeocranus (Herm.), and the reaction of these organs to a ray-fungus parasite. Ann. Natal Mus., 11: 1-36. (BA 23: 19978)

A fungus attacks the epithelium of the gut-caeca. No definite spores or hyphae were found. The growth phase consisted of "zoogleae" and short radiating chained filaments. It was tentatively identified as Actinobacillus,[*] a streptotrichian. Infection in the genital system results in degeneration of the male organs.

Williams, R. W. & Brown, H. W. (1945). The development of Litomosoides carinii filariid parasite of the cotton rat in the tropical rat mite. Science, 102: 482-483.

All stages of development of the filariid were recovered from Liponyssus bacoti fed on infected rats.

Williams, R. W. & Brown, H. W. (1946). The transmission of Litomosoides carinii, filariid parasite of the cotton rat, by the tropical rat mite, Liponyssus bacoti. Science, 103: 224.

Laboratory evidence demonstrated that the mite is the vector of the filariid to cotton rats.

Williams, P. & Kershaw, W. E. (1961). Studies on the intake of microfilariae by their insect vectors, their survival, and their effect on the survival of their vectors. X. The survival of the tropical rat mite, the vector of filariasis in the cotton rat. Ann. Trop. Med. Parasitol., 55: 217-230.

The rate of mortality of mites infected with Litomosoides carinii is greater than that of an uninfected population. Developing filaria do not adversely affect infected female mites but may be lethal when they migrate from the mites. Infected male mites have a higher mortality rate than infected females.

---

[*] Probably an actinomycete.

XXI. PATHOGENS OF ACARINA (TICKS) [a]

H. Hoogstraal

NAMRU 3
FPO New York 09527, USA

---

[a] The literature for this table and bibliography covers the period 1961–1975.

PATHOGENS OF ACARINA (TICKS)

Host	Host stage infected	Pathogen	% Incidence	Locality	Lab. or field study	Reference
Argas persicus	Nymphs, adults	Viral etiology assumed	"Regularly observed"	USSR (lab. colony)	Lab.	Sidorov & Shcherbakov (1973)
Dermacentor andersoni	Nymphs, larvae, adults	Proteus mirabilis	100 mortality of naturally infected	USA (lab. colony)	Lab.	Brown et al. (1970)
"	Adults	Wolbachia-like symbiotes	Massive injection caused mortality	USA	Lab.	Burgdorfer et al. (1973)
"	-	Serratia marcescens	-	USA	Lab.	Steinhaus (1959)
Dermacentor marginatus	Nymphs, larvae, adults	Rickettsia prowazeki	-	USSR	-	Reháček (1965)
"	Nymphs, adults	Aspergillus parasiticus	49 adults	Czechoslovakia	Field	Samšiňaková et al. (1974)
"	Nymphs, adults	Beauveria bassiana				
"	Nymphs, adults	B. tenella				
"	Nymphs, adults	Cephalosporium coccorum				
"	Adults	Paecilomyces fumosoroseus				
"	Nymphs, adults	Facultative parasitic and saprophytic fungi				
Dermacentor pictus	Nymphs, larvae, adults	Rickettsia prowazeki	-	USSR	-	Reháček (1965)

PATHOGENS OF ACARINA (TICKS) (continued)

Host	Host stage infected	Pathogen	% Incidence	Locality	Lab. or field study	Reference
Dermacentor reticulatus	Nymphs, adults	Aspergillus parasiticus		Czechoslovakia	Field	Samšiňaková et al. (1974)
"	Nymphs, adults	Beauveria bassiana				
"	Nymphs, adults	B. tenella				
"	Nymphs, adults	Cephalosporium coccorum	49 adults			
"	Adults	Paecilomyces fumosoroseus				
"	Nymphs, adults	Facultative parasitic and saprophytic fungi				
Ixodes ricinus	Adults	Aspergillus parasiticus		Czechoslovakia	Field	Samšiňaková et al. (1974)
"	Adults	B. bassiana				
"	Adults	B. tenella				
"	Adults	C. coccorum	53			
"	Adults	P. fumosoroseus				
"	Adults	Facultative parasitic and saprophytic fungi				
Ornithodoros (A.) lahorensis	Nymphs, adults	Viral etiology assumed	High	USSR (lab. colony)	Lab.	Sidorov & Shcherbakov (1973)
Ornithodoros moubata	Adults	Wolbachia persica	-		Lab.	Weyer (1973)
"	Nymphs, adults	Viral etiology assumed	Rare	USSR (lab. colony)	Lab.	Sidorov & Shcherbakov (1973)

PATHOGENS OF ACARINA (TICKS) (continued)

Host	Host stage infected	Pathogen	% Incidence	Locality	Lab. or field study	Reference
Ornithodoros tartakovskyi	Nymphs, adults	Viral etiology assumed	"Regularly observed"	USSR (lab. colony)	Lab.	Sidorov & Shcherbakov (1973)
Ornithodoros tholozani	Nymphs, adults	Viral etiology assumed	"Regularly observed"	USSR (lab. colony)	Lab.	Sidorov & Shcherbakov (1973)
Ornithodoros verrucosus	Nymphs, adults	Viral etiology assumed	"Regularly observed"	USSR (lab. colony)	Lab.	Sidorov & Shcherbakov (1973)
Ixodidae (several species)	All	Aspergillus fumigatus	-	-	Field, lab.	Lipa (1971)
"	All	Beauveria bassiana	-	-	Field, lab.	
"	All	Penicillium insectivora	-	-	Field, lab.	
Unidentified tick	-	Torrubiella sp. (?)	-	Korea	Field	Steinhaus & Marsh (1962)

ABSTRACTS

Mary Ann Strand

Brown, R. S. et al. (1970).   An endemic disease among laboratory populations of <u>Dermacentor</u> <u>andersoni</u> (= <u>D. venustus</u>) (Acarina: Ixodidae).   <u>J. Invertebr. Pathol</u>., <u>16</u>(1): 142-143.

A bacterium, <u>Proteus mirabilis</u>, was isolated from dead engorged ticks.   Infected nymphs turned from grey to black and died within six days after repletion.   The disease mechanism is apparently triggered in otherwise healthy ticks by ingestion of blood. Transovarial transmission was demonstrated.

Burgdorfer, W. et al. (1973).   Isolation and characterization of symbiotes from the Rocky Mountain wood tick, <u>Dermacentor andersoni</u>.   <u>J. Invertebr. Pathol</u>., <u>22</u>(3): 424-434.

<u>Wolbachia</u>-like symbiotes in the Rocky Mountain wood tick were isolated by injection of ovarial tissues into five-day-old chick embryos.   Symbiotes thus cultivated were injected intracoelomically into adult ticks and produced infestations in hemocytes, hypodermal tissues, salivary glands, and in connective tissues surrounding midgut, Malpighian tubules, and ovary.   The massive invasion of tissues by infected symbiotes invariably caused death.

Černý, V. et al. (1974).   The entomogenous fungi associated with <u>Ixodes ricinus</u> (L.). (Abstr.) <u>Proc. 3rd Int. Cong. Parasitol. (Munich</u>), <u>2</u>: 952-953.

See Samšiňaková et al., 1974.

Lipa, J. J. (1971).   Microbial control of mites and ticks.   In: H. D. Burges and N. W. Hussey, ed.   <u>Microbial control of insects and mites</u>.   New York, Academic Press, pp. 357-373.

In this review article, ticks are treated briefly.   <u>Aspergillus fumigatus</u>, <u>Penicillium</u> <u>insectivorum</u>, and <u>Beauveria bassiana</u> have been reported as pathogens of several species of ticks by Russian authors.

Rehácek, J. (1965).   Development of animal viruses and rickettsiae in ticks and mites. <u>Ann. Rev. Entomol</u>., <u>10</u>: 1-24

<u>Rickettsia prowazeki</u> is reported to kill <u>Dermacentor marginatus</u> and <u>D. pictus</u>.   Infected females die prematurely and egg production is reduced.   Large numbers of rickettsia are found in all of their organs.

Samšiňaková, A. (1974).   Entomogenous fungi associated with the tick <u>Ixodes ricinus</u> (L.). <u>Folia Parasitol. (Praha</u>), <u>21</u>(1): 38-48.

The ticks <u>I. ricinus</u>, <u>Dermacentor marginatus</u>, and <u>D. reticulatus</u> were infected with 17 species of fungi in nature.   Five were identified as obligate parasites, five as facul- tative parasites, and seven as saprophytes.   The relationship between fungi and ticks was markedly influenced by the season.   Overwintering ticks rarely died of fungal infections, only 6.2% of unengorged and 4.8% of the engorged specimens.   During the summer, infected engorged females ranged from 45.1-53.7%.   The influence of biotype on the death rate before and after oviposition was not significant.

Sidorov, V. E. & Shcherbakov, S. V. (1973). /Mass epizootics among Alveonasus lahorensis Neumann ticks./ Med. Parazitol. Parazit. Bolezn., 42(1): 47-51.

Two-year investigation of mass epizootic in laboratory colony of Ornithodoros (A.) lahorensis (chief species colonized at Gamaleya Institute) but "regularly" observed also in laboratory colonies of Argas (Persicargas) persicus (Oken), Ornithodoros (Pavlovskyella) tholozani (Laboulbene and Negnin (= "O. papillipes"), O. (P.) tartakovskyi Olenev, O. (Alectorobius) verrucosus Olenev, Sassuchin, and Feniuk, and "very rarely" in O. (O.) moubata (Murray). Three types of ulcers described on different body areas, particularly deformities of mouthparts: (1) cuticular ulceration, (2) deformities and tumourlike outgrowths and papillomas, with or without ulceration, but with changes in shape, size, and number of mouthparts, and (3) "gangerene" of extremities. Infection source thought to be rabbits (hosts of ticks) from one breeding farm.

Steinhaus, E. A. (1959). Serratia marcescens Bizio as an insect pathogen. Hilgardia, 28: 351-380.

Death of ticks was occasionally observed in experiments where Serratia marcescens was used as a test organism.

Steinhaus, E. A. & Marsh, G. A. (1962). Report of diagnoses of diseased insects 1951-1961. Hilgardia, 33: 349-490.

An unidentified tick collected in Korea was diagnosed as having a mycosis probably caused by Torrubiella sp.

Weyer, F. (1973). Versuche zur Übertragung von Wolbachia persica auf Kleiderläuse. Z. Angew. Zool., 60: 77-93.

W. persica was successfully transferred into the gut of the tick Ornithodoros moubata. In all cases the organisms multiplied and had a damaging effect on the hosts which died in a few weeks. Multiplication of the rickettsia took place in the hemolymph of the ticks, and they were found in excreted fluids and in oviposited eggs, but not in nymphs or the F_1 generation.

PATHOGEN–HOST LIST [a]

Mary Ann Strand & Ping Chin

Boyce Thompson Institute
Yonkers, NY 10701, USA

[a] Full citations for " References " are given in the host–pathogen section listed under " Host group ".

XXII. VIRUSES

Pathogen	Host group	Host	References
Baculovirus from Aedes sollicitans	Culicidae	Aedes aegypti	Clark & Fukuda (1971)
Baculovirus from Aedes sollicitans	Culicidae	Aedes nigromaculis	Clark (1972)
Baculovirus	Culicidae	Aedes sollicitans	Clark et al. (1969)
Baculovirus	Culicidae		Clark & Fukuda (1971)
Baculovirus	Culicidae	Aedes taeniorhynchus	Chapman (1974)
Baculovirus from Aedes sollicitans	Culicidae	Aedes tormentor	Clark & Fukuda (1971)
Baculovirus from Aedes sollicitans	Culicidae	Aedes triseriatus	Federici & Anthony (1972)
Baculovirus from Aedes sollicitans	Culicidae	Aedes triseriatus	Clark & Fukuda (1971)
Baculovirus from Aedes sollicitans	Culicidae	Aedes triseriatus	Federici & Lowe (1972)
Baculovirus	Culicidae	Aedes triseriatus	Chapman (1974)
Baculovirus	Culicidae	Anopheles crucians	Chapman (1974)
Baculovirus (?) unconfirmed or possibly parvovirus	Culicidae	Anopheles subpictus	Das Gupta & Ray (1957)
Baculovirus	Culicidae	Culex pipiens quinquefasciatus	Chapman (1974)
Baculovirus	Culicidae	Culex salinarius	Clark & Fukuda (1971)
Baculovirus	Culicidae	Psorophora continnis	Clark & Fukuda (1971)
Baculovirus of Aedes sollicitans	Culicidae	Psorophora ferox	Clark & Fukuda (1971)
Baculovirus	Culicidae	Psorophora ferox	Chapman (1974)
Baculovirus from Aedes sollicitans	Culicidae	Psorophora varipes	Clark & Fukuda (1971)
Baculovirus	Culicidae	Uranotaenia sapphirina	Hazard (1972)
Baculovirus	Culicidae	Wyeomyia smithii	Hall & Fish (1974)
Cytoplasmic poly-hedrosis virus	Culicidae	Aedes sollicitans	Clark & Fukuda (1971)
Cytoplasmic poly-hedrosis virus and baculovirus	Culicidae	Aedes sollicitans	Clark & Fukuda (1971), Chapman (1974)
Cytoplasmic poly-hedrosis virus	Culicidae	Aedes sticticus	Chapman (1972)

XXII. VIRUSES (continued)

Pathogen	Host group	Host	References
Cytoplasmic poly-hedrosis virus of Aedes sollicitans	Culicidae	Aedes taeniorhynchus	Clark & Fukuda (1971)
Cytoplasmic poly-hedrosis virus	Culicidae	Aedes taeniorhynchus	Chapman (1972)
Cytoplasmic poly-hedrosis virus	Culicidae		Federici (1973)
Cytoplasmic poly-hedrosis virus	Culicidae	Aedes thibaulti	Chapman (1972)
Cytoplasmic poly-hedrosis virus	Culicidae	Aedes triseriatus	Clark et al. (1969)
Cytoplasmic poly-hedrosis virus	Culicidae	Anopheles bradleyi	Chapman (1972)
Cytoplasmic poly-hedrosis virus	Culicidae	Anopheles crucians	Chapman (1972)
Cytoplasmic poly-hedrosis virus	Culicidae	Anopheles quadrimaculatus	Anthony et al. (1973)
Cytoplasmic poly-hedrosis virus	Culicidae	Anopheles stephensi	Bird et al. (1972), Davies et al. (1971)
Cytoplasmic poly-hedrosis virus	Simuliidae	Cnephia mutata	Bailey et al. (1975)
Cytoplasmic poly-hedrosis virus	Culicidae	Culex peccator	Chapman (1972)
Cytoplasmic poly-hedrosis virus	Culicidae	Culex restuans	Clark et al. (1969)
CPV of Culiseta melanura Culex salinarius Culex territans	Culicidae	Culex salinarius	Clark & Fukuda (1971)
Cytoplasmic poly-hedrosis virus	Culicidae	Culex salinarius	Clark et al. (1969)
Cytoplasmic poly-hedrosis virus	Culicidae	Culex territans	Clark et al. (1969)
CPV of Culiseta melanura Culex salinarius Culex territans	Culicidae	Culex territans	Clark & Fukuda (1971)
Cytoplasmic poly-hedrosis virus	Culicidae	Culiseta inornata	Chapman (1972)
Cytoplasmic poly-hedrosis virus	Culicidae	Culiseta melanura	Clark & Fukuda (1971), Chapman (1972)

XXII.   VIRUSES (continued)

Pathogen	Host group	Host	References
Cytoplasmic poly-hedrosis virus	Culicidae	Orthopodomyia signifera	Clark et al. (1969)
Cytoplasmic poly-hedrosis virus	Simuliidae	Prosimulium mixtum	Bailey et al. (1975)
Cytoplasmic poly-hedrosis virus	Culicidae	Psorophora continnis	Clark et al. (1969)
Cytoplasmic poly-hedrosis virus	Culicidae	Psorophora ferox	Chapman (1972)
Cytoplasmic poly-hedrosis virus from Aedes sollicitans	Culicidae	Psorophora ferox	Clark & Fukuda (1971)
Cytoplasmic poly-hedrosis virus	Culicidae	Uranotaenia sapphirina	Clark et al. (1969)
Cytoplasmic poly-hedrosis virus	Culicidae		Hazard (1972)
Eastern equine encephalitis virus	Acarina mites	Haemogamasus liponysoides	Clark et al. (1966)
Eastern equine encephalitis virus	Acarina mites	Ornithonyssus bacoti	Clark et al. (1966)
Entomopoxvirus	Culicidae	Aedes aegypti	Lebedeva & Zelenko (1972)
Chilo iridescent virus (Type 6)	Culicidae	Aedes aegypti	Fukuda (1971)
Sericesthis iridescent virus (Type 2)	Culicidae	Aedes aegypti	Day & Mercer (1964)
Iridescent virus (Type 5)	Culicidae	Aedes annulipes	Weiser (1965)
Iridescent virus (Type 12)	Culicidae	Aedes cantans	Tinsley et al. (1971)
Iridescent virus (Type 4)	Culicidae	Aedes cantans	Weiser (1965)
Iridescent virus (Type 14)	Culicidae	Aedes detritus	Hasan et al. (1960, 1971)
Iridescent virus	Culicidae	Aedes detritus	Service (1968)
Iridescent virus (Type 15)	Culicidae	Aedes detritus	Vago et al. (1969)
Iridescent virus (unconfirmed)	Culicidae	Aedes dorsalis	Chapman et al. (1969)
Iridescent virus	Culicidae	Aedes fulvus pallens	Chapman et al. (1966)

XXII.   VIRUSES (continued)

Pathogen	Host group	Host	References
Chilo iridescent virus (Type 6)	Culicidae	Aedes sierrensis	Fukuda (1971)
Chilo iridescent virus (Type 6)	Culicidae	Aedes sollicitans	Fukuda (1971)
Iridescent viruses from Aedes taeniorhynchus	Culicidae	Aedes sollicitans	Woodard & Chapman (1968)
Iridescent virus	Culicidae	Aedes sticticus	Chapman et al. (1969)
Iridescent virus (Type 11)	Culicidae	Aedes stimulans	Anderson (1970)
Chilo iridescent virus (Type 6)	Culicidae	Aedes taeniorhynchus	Fukuda (1971)
Iridescent virus from Aedes vexans	Culicidae	Aedes taeniorhynchus	Woodard & Chapman (1968)
Iridescent virus from Psorophora ferox	Culicidae	Aedes taeniorhynchus	Woodard & Chapman (1968)
Regular mosquito iridescent virus (Type 3)	Culicidae	Aedes taeniorhynchus	Lowe et al. (1970)
Regular mosquito iridescent virus (Type 3)	Culicidae		Matta & Lowe (1970)
Regular mosquito iridescent virus (Type 3)	Culicidae		Stoltz (1971)
Regular mosquito iridescent virus (Type 3) and turquoise MIV	Culicidae		Stoltz (1973)
Regular mosquito iridescent virus (Type 3)	Culicidae		Stoltz & Summers (1974)
Regular mosquito iridescent virus (Type 3)	Culicidae	Aedes taeniorhynchus	Clark et al. (1965)
Regular mosquito iridescent virus (Type 3)	Culicidae		Hall & Anthony (1971)
Regular mosquito iridescent virus (Type 3)	Culicidae		Hall & Lowe (1971)

XXII.   VIRUSES (continued)

Pathogen	Host group	Host	References
Regular mosquito iridescent virus (Type 3) and turquoise mosquito iridescent virus	Culicidae		Hall & Lowe (1972)
Regular mosquito iridescent virus (Type 3)	Culicidae		Linley & Nielsen (1968a, b)
Regular mosquito iridescent virus (Type 3) and turquoise mosquito iridescent virus	Culicidae	Aedes taeniorhynchus	Anthony & Hall (1970)
Regular mosquito iridescent virus (Type 3) and turquoise MIV	Culicidae	Aedes taeniorhynchus	Wagner et al. (1973)
RMIV TMIV	Culicidae	Aedes taeniorhynchus	Woodard & Chapman (1968)
RMIV	Culicidae	Aedes taeniorhynchus	
Iridescent virus	Culicidae	Aedes vexans	Chapman et al. (1966)
Iridescent virus	Culicidae		Woodard & Chapman (1968)
Iridescent virus from Psorophora ferox	Culicidae	Aedes vexans	Woodard & Chapman (1968)
Chilo iridescent virus (Type 6)	Culicidae	Anopheles albimanus	Fukuda (1971)
Chilo iridescent virus (Type 3)	Culicidae	Anopheles quadrimaculatus	Fukuda (1971)
Chilo iridescent virus (Type 3)	Culicidae	Culex peccator	Fukuda (1971)
Chilo iridescent virus (Type 3)	Culicidae	Culex salinarius	Fukuda (1971)
Culicoides iridescent virus (CuIV)	Cerato- pogonidae	Culicoides sp. prob. prob. arboricola	Chapman et al. (1968, 1969), Chapman (1973)
Chilo iridescent virus (Type 6)	Culicidae	Culiseta inornata	Fukuda (1971)
Chilo iridescent virus (Type 6)	Culicidae	Culiseta melanura	Fukuda (1971)
Iridescent virus	Culicidae	Psorophora continnis	Chapman (1974)
Chilo iridescent virus (Type 6)	Culicidae	Psorophora ferox	Fukuda (1971)

XXII.   VIRUSES (continued)

Pathogen	Host group	Host	References
Iridescent virus	Culicidae	Psorophora ferox	Chapman et al. (1966), Federici (1970), Woodard & Chapman (1968)
Iridescent virus	Culicidae	Psorophora horrida	Chapman et al. (1969)
Chilo iridescent virus (Type 3)	Culicidae	Psorophora varipes	Fukuda (1971)
Iridescent virus	Culicidae	Psorophora varipes	Chapman et al. (1969)
Iridescent virus	Simuliidae	Simulium ornatum	Weiser (1968)
Non-occluded virus	Culicidae	Anopheles stephensi	Davies et al. (1971)
Non-occluded virus	Culicidae	Culex tarsalis	Richardson et al. (1974)
Non-occluded virus	Culicidae	Culiseta inornata	Federici (1974)
Parvovirus (densonucleosis)	Culicidae	Aedes aegypti	Lebedeva et al. (1973)
Tetragonal virus	Culicidae	Aedes sierrensis	Kellen et al. (1963)
Tetragonal virus	Culicidae	Anopheles crucians	Chapman et al. (1970)
Tetragonal virus	Culicidae	Anopheles freeborni	Kellen et al. (1963)
Tetragonal virus	Culicidae	Culex salinarius	Clark & Chapman (1969)
Tetragonal virus	Culicidae		Chapman et al. (1970)
Tetragonal virus	Culicidae	Culex salinarius	Stoltz et al. (1974)
Tetragonal virus of Culex salinarius	Culicidae	Culex tarsalis	Clark & Chapman (1969)
Tetragonal virus (referred to as a CPV)	Culicidae	Culex tarsalis	Kellen et al. (1963, 1966)
Viral etiology assumed	Acarina ticks	Argas persicus	Sidorov & Shcherbakov (1973)
Intranuclear virus-like particles	Glossina	Glossina f. fuscipes	Jenni & Steiger (1974)
Viruslike particles	Glossina	Glossina morsitans centralis	Jenni (1973)
Viral etiology assumed	Acarina ticks	Ornithodoros (A.) lahorensis	Sidorov & Shcherbakov (1973)
Viral etiology assumed	Acarina ticks	Ornithodoros moubata	Sidorov & Shcherbakov (1973)
Viral etiology assumed	Acarina ticks	Ornithodoros tartakovskyi	Sidorov & Shcherbakov (1973)
Viral etiology assumed	Acarina ticks	Ornithodoros tholozani	Sidorov & Shcherbakov (1973)
Viral etiology assumed	Acarina ticks	Ornithodoros verrucosus	Sidorov & Shcherbakov (1973)

XXIII.   RICKETTSIAE

Pathogen	Host group	Host	References
Rickettsia akari	Anoplura - Mallophaga	Pediculus humanus	Jenkins (1973)
Rickettsia conorii	Anoplura - Mallophaga	Pediculus humanus	Jenkins (1973)
Symbiotes (Rickettsia) lectularius	Cimicidae	Cimex lectularius	Arkwright et al. (1921), Louis et al. (1973) Chang & Musgrave (1973), Chang (1974)
Rickettsia prowazeki	Acarina ticks	Dermacentor marginatus	Reháček (1965)
Rickettsia prowazeki	Acarina ticks	Dermacentor pictus	Reháček (1965)
Rickettsia prowazeki	Anoplura - Mallophaga	Pediculus humanus	Jenkins (1973), Reháček (1965)
Rickettsia rickettsi	Anoplura - Mallophaga	Pediculus humanus	Jenkins (1973)
Rickettsia prowazeki	Anoplura - Mallophaga	Polyplax spinulosus	Mooser et al. (1931)
Rickettsia tsutsugamushi Scrub typhus	Acarina mites	Leptotrombidium akamushi Chigger	Rapmund et al. (1969)
Rickettsia tsutsugamushi	Acarina mites	Leptotrombidium arenicola Chigger	Rapmund et al. (1972)
Rickettsia tsutsugamushi	Anoplura - Mallophaga	Pediculus humanus	Jenkins (1973)
Rickettsia typhi	Anoplura - Mallophaga	Pediculus humanus	Jenkins (1973)
Rickettsiella blattae	Blattidae	Blatta orientalis	Huger (1964)
Wolbachia persica	Acarina ticks	Ornithodoros moubata	Weyer (1973)
Wolbachia persica	Anoplura - Mallophaga	Pediculus humanus	Jenkins (1973), Weyer (1973)
Wolbachia- like symbiotes	Acarina ticks	Dermacentor andersoni	Burgdorfer et al. (1973)

XXIII.   RICKETTSIAE (continued)

Pathogen	Host group	Host	References
Rickettsia	Phlebotominae	Lutzomyia vexator vexator (Coq.)	Hertig (1936)
Filamentous rickettsiae	Siphonaptera	Ctenocephalides canis	Jenkins (1964e)
Filamentous rickettsiae	Siphonaptera	Nosopsyllus fasciatus	Jenkins (1964u)
Filamentous rickettsiae	Siphonaptera	Pulex irritans	Jenkins (1964e)
Rickettsia-like organism	Glossina	Glossina brevipalpis	Pinnock & Hess (1974)
Rickettsia-like organism	Glossina	Glossina fuscipes	Pinnock & Hess (1974)
Rickettsia-like organism	Glossina	Glossina morsitans	Pinnock & Hess (1974)
Rickettsia-like organism	Glossina	Glossina pallidipes	Pinnock & Hess (1974)
Cocci - (?) Rickettsiae	Ceratopogonidae	Culicoides sanguisuga	Hertig & Wolbach (1924)

XXIV. BACTERIA

Pathogen	Host group	Host	References
Aerobacter aerogenes	Blattidae	Blaberus craniifer	Briscoe et al. (1961, 1963)
Aerobacter aerogenes	Blattidae	Blattella germanica	Henry (1965)
Aerobacter cloacae	Blattidae	Blaberus craniifer	Briscoe et al. (1961, 1963)
Aeromonas punctata	Culicidae	Anopheles annulipes	Kalucy & Daniel (1972)
Aeromonas sp.	Glossina	Glossina morsitans	Bauer (1974)
Bacillus alvei	Blattidae	Blattella germanica	Henry (1965)
Bacillus alvei-circulans-brevis	Culicidae	Aedes aegypti	Singer (1974)
Bacillus alvei-circulans-brevis	Culicidae	Culex pipiens fatigans	Singer (1973, 1974, 1975)
Bacillus anthracis	Cimicidae	Cimex lectularius	Rosenholz & Owsjannikowa (1921)
Bacillus cereus	Blattidae	Blattella germanica	Henry (1965)
Bacillus cereus	Culicidae	Culex pipiens	Zharov (1969)
Bacillus cereus var. mycoides	Culicidae	Culex pipiens fatigans	Thomas & Poinar (1973)
Bacillus cereus	Musca	Musca domestica	Briggs (1960), Thomas & Poinar (1973)
Bacillus lentimoribus	Blattidae	Blattella germanica	Henry (1965)
Bacillus pediculi	Anoplura Mallophaga	Pediculus humanus	Arkwright & Bacot (1921)
Bacillus sphaericus	Culicidae	Aedes aegypti	Singer (1974, 1975), Kellen et al. (1965), Kellen & Myers (1964)
Bacillus sphaericus	Culicidae	Aedes atropalpus	Kellen et al. (1965)
Bacillus sphaericus	Culicidae	Aedes sierrensis	Kellen et al. (1965) Kellen & Myers (1964)
Bacillus sphaericus	Culicidae	Aedes squaniger	Kellen et al. (1965)
Bacillus sphaericus	Culicidae	Aedes triseriatus	Kellen et al. (1965)
Bacillus sphaericus (SSII-1)	Culicidae	Anopheles spp.	Singer (1975)
Bacillus sphaericus	Culicidae	Culex erythrothorax	Kellen et al. (1965), Kellen & Myers (1964)
Bacillus sphaericus	Culicidae	Culex partecips	Kellen & Myers (1964)
Bacillus sphaericus	Culicidae	Culex peus	Kellen & Myers (1964)
Bacillus sphaericus	Culicidae	Culex pipiens	Kellen et al. (1965)
Bacillus sphaericus var. fusiformis	Culicidae	Culex pipiens	Goldberg et al. (1974)

XXIV.   BACTERIA (continued)

Pathogen	Host group	Host	References
Bacillus sphaericus var. fusiformis	Culicidae	Culex pipiens fatigans	Rogoff et al. (1969)
Bacillus sphaericus (SSII-1)	Culicidae	Culex pipiens fatigans	Davidson & Singer (1973), Singer (1973, 1974a)
Bacillus sphaericus (SS-1)	Culicidae	Culex pipiens fatigans	Singer (1975)
Bacillus sphaericus	Culicidae	Culex tarsalis	Goldberg et al. (1974), Kellen & Myers (1964)
Bacillus sphaericus (SSII-1)	Culicidae	Culex sp.	Singer (1975)
Bacillus sphaericus	Culicidae	Culiseta incidens	Reeves (1970), Kellen & Myers (1964)
Bacillus thuringiensis	Culicidae	Aedes aegypti	Singer (1974)
Bacillus thuringiensis (Toxin)	Culicidae	Aedes aegypti	Davidson & Singer (1973)
Bacillus thuringiensis (Bakthane-L-69)	Culicidae	Aedes aegypti	Shaikh & Morrison (1966)
Bacillus thuringiensis (BA 068)	Culicidae	Aedes aegypti	Reeves (1970)
Bacillus thuringiensis	Culicidae	Aedes dorsalis	Singer (1975), Briggs (1973b)
Bacillus thuringiensis (BA 068)	Culicidae	Aedes nigromaculus	Reeves (1970)
Bacillus thuringiensis (BA 068)	Culicidae	Aedes sierrensis	Reeves (1970)
Bacillus thuringiensis (Bakthane L-69)	Culicidae	Aedes stimulans	Shaikh & Morrison (1966)
Bacillus thuringiensis (BA 068)	Culicidae	Aedes triseriatus	Reeves (1970)
Bacillus thuringiensis (BA 068)	Culicidae	Aedes spp.	Reeves & Garcia (1971a, b)
Bacillus thuringiensis	Culicidae	Anopheles spp.	Lavrent'yev et al. (1965)
Bacillus thuringiensis	Blattidae	Blattella germanica	Henry (1965)
Bacillus thuringiensis	Blattidae	Blatta orientalis	Yarnykh & Tonkozhenko (1969)
Bacillus thuringiensis (various preparations)	Anoplura Mallophaga	Bovicola bovis	Gingrich et al. (1974)
Bacillus thuringiensis	Anoplura Mallophaga	Bovicola crassipes	Gingrich et al. (1974)
Bacillus thuringiensis	Anoplura Mallophaga	Bovicola limbatus	Gingrich et al. (1974)

XXIV.   BACTERIA (continued)

Pathogen	Host group	Host	References
Bacillus thuringiensis	Anoplura Mallophaga	Bovicola ovis	Gingrich et al. (1974)
Bacillus thuringiensis var. thuringiensis	Culicidae	Culex pipiens fatigans	Rogoff et al. (1969)
Bacillus thuringiensis	Culicidae	Culex pipiens fatigans	Singer (1974)
Bacillus thuringiensis (entobacterin, exotoxin)	Culicidae	Culex modestus	Saubenova et al. (1973)
Bacillus thuringiensis (exotoxin)	Culicidae	Culex pipiens molestus	Gurgenidze (1970), Yarnykh & Tonhozhenko (1969)
Bacillus thuringiensis dendrolimus	Culicidae	Culex pipiens	Zharov (1969)
Bacillus thuringiensis	Culicidae	Culex pipiens	Zharov (1969)
Bacillus thuringiensis (toxin)	Culicidae	Culex tarsalis	Ahmad et al. (1971), Goldberg et al. (1974)
Bacillus thuringiensis (BA 068)	Culicidae	Culex spp.	Reeves & Garcia (1971a, b)
Bacillus thuringiensis	Culicidae	Culex spp.	Lavent'yev et al. (1965)
Bacillus thuringiensis	Anoplura Mallophaga	Lipeorus caponis	Hoffman & Gingrich (1968)
Bacillus thuringiensis	Anoplura Mallophaga	Menacanthus stramineus	Hoffman & Gingrich (1968)
Bacillus thuringiensis	Anoplura Mallophaga	Menopon gallinae	Hoffman & Gingrich (1968)
Bacillus thuringiensis	Musca	Musca domestica	Briggs (1960), Burgerjon & Galichet (1965), Cantwell et al. (1964), Connor & Hansen (1967), Feigin (1963), Galichet (1966), Gingrich (1965), Greenwood 1964, Harvey & Howell (1965), Millar (1965)
Bacillus thuringiensis	Blattidae	Periplaneta americana	Yarnykh & Tonkozhenko (1969)
Bacillus thuringiensis	Reduviidae	Rhodnius prolixus	Nyirady (1973)
Bacillus thuringiensis	Blattidae	Shelfordella tartara	Ryan & Nicholas (1972), Yarnykh & Tonkozhenko (1969)
Bacillus thuringiensis var. thuringiensis Heimpel & Angus	Stomoxys	Stomoxys calcitrans	Gingrich (1965)
Bacillus thuringiensis	Reduviidae	Triatoma protracta	Nyirady (1973)

XXIV.   BACTERIA (continued)

Pathogen	Host group	Host	References
Bacillus thuringiensis	Reduviidae	Triatoma rubida	Nyirady (1973)
Bacillus sp.	Culicidae	Culex annuterostria	Singer (1975)
Bacillus sp.	Culicidae	Culex pipiens fatigans	Singer (1975)
Bacillus spp.	Culicidae	Culex pipiens fatigans	Fulton et al. (1974), Singer (1975)
Bacillus sp.	Culicidae	Culex pipiens	Briggs (1964)
Bacillus sp.	Culicidae	Culex restuans	Fulton et al. (1974)
Bacillus sp.	Culicidae	Culex salinarius	Fulton et al. (1974)
Bacillus	Glossina	Glossina fuscipes	Carpenter (1912), Carpenter (1913)
Bacillus spp.	Musca	Musca domestica	Rogoff et al. (1969)
Bacterium delendae-muscae	Musca	Musca domestica	Roubaud & Descazeaux (1923)
Bacterium mathisi (unrecognized taxon)	Glossina	Glossina morsitans	Roubaud & Treillard (1935)
Bacterium mathisi	Glossina	Glossina palpalis	Roubaud & Treillard (1936)
Bacterium mathisi (not a recognized taxon)	Musca	Musca domestica	Roubaud & Treillard (1935)
Borrelia glossinae	Glossina	Glossina palpalis	Novy & Knapp (1906)
Brevibacterium (Bacterium) tegumenticola (classification doubtful)	Cimicidae	Cimex lectularius	Steinhaus (1941)
Brucella (=Micrococcus) melitensis (Hughes) Meyer & Shaw	Stomoxys	Stomoxys calcitrans	Kennedy (1906)
Cicadomyces sp.	Glossina	Glossina morsitans	Nogge (1974)
Cicadomyces sp.	Glossina	Glossina palpalis	Roubaud (1919)
Cicadomyces sp.	Glossina	Glossina palpalis	Wallace (1931)
Cicadomyces sp.	Glossina	Glossina tachinoides	Roubaud (1919)
Cillopasteurella delendae-muscae (Roubaud & Descazeaux) Prévot (=Bacterium delende-muscae)	Stomoxys	Stomoxys calcitrans	Roubaud & Descazeaux (1923)
Citrobacter sp.	Blattidae	Leudophaea maderae	Thomas & Poinar (1973)
Clostridium sp.	Culicidae	Mosquito (?)	Thomas & Poinar (1973)
Couccus (?)	Cimicidae	Cimex lectularius	Steinhaus (1941)

XXIV.   BACTERIA (continued)

Pathogen	Host group	Host	References
Corynebacterium paurometabolum	Cimicidae	Cimex lectularius	Steinhaus (1941)
Corynebacterium sp.	Culicidae	Aedes aegypti	Micks & Ferguson (1973)
Corynebacterium sp.	Blattidae	Periplaneta americana	Ryan & Nicholas (1972)
Corynebacterium sp.	Reduviidae	Triatoma protracta	Marchette & Hatie (1965)
Diplococcus pneumaniae	Blattidae	Blaberus craniifer	Anderson et al. (1973)
Diplococcus (possible symbiote)	Reduviidae	Rhodnius prolixus	Bewig & Schwartz (1954)
Diplococcus (possible symbiote)	Reduviidae	Triatoma infestans	Bewig & Schwartz (1954)
Eberthella typhosa (not a recognized taxon)	Anoplura Mallophaga	Pediculus humanus	Jenkins (1973)
Escherichia coli	Blattidae	Blaberus craniifer	Anderson et al. (1973)
Escherichia coli	Blattidae	Blattella germanica	Ulewicz & Zawistowski (1973)
Escherichia coli	Anoplura Mallophaga	Pediculus humanus	Jenkins (1973)
Escherichia coli	Blattidae	Periplaneta americana	Ryan & Nicholas (1972)
Escherichia coli	Reduviidae	Rhodnius prolixus	Nyirady (1973)
Escherichia coli	Reduviidae	Triatoma protracta	Nyirady (1973)
Escherichia coli	Reduviidae	Triatoma rubida	Nyirady (1973)
Escherichia freundii	Blattidae	Blaberus craniifer	Briscoe et al. (1961, 1963)
Flavobacterium near lutescens	Culicidae	Aedes freeborni	Steinhaus & Marsh (1962)
Francisella (Bacterium) tularensis	Cimicidae	Cimex lectularius	Francis (1927)
Haffnia (Erwinia?)	Blattidae	Leucophaea maderae	Thomas & Poinar (1973)
Hymenolepis diminuta	Siphonaphera	Nosopsyllus fasciatus	Jenkins (1964h)
Hymenolepis murina	Siphonaphera	Nosopsyllus fasciatus	Jenkins (1964h)
Micrococcus luteus (=Sarcina flava)	Cimicidae	Cimex lectularius	Steinhaus (1941)
Mycobacterium sp.	Culicidae	Aedes aegypti	Mikhnovka et al. (1972)
Mycobacterium sp.	Culicidae	Culex pipiens molestus	Mikhnovka et al. (1972)
Nocardia rhodni	Reduviidae	Rhodnius prolixus	Bewig & Schwartz (1954), Baines (1956), Lake & Friend (1968), Auden (1974). For more references, see Jenkins (1964)
Nocardia rhodnii (symbiote)	Reduviidae	Triatoma infestans	Bewig & Schwartz (1954)
Nocardia (Mycobacterium) sp.	Reduviidae	Triatoma protracta	Marchette & Hatie (1965) (gut flora)

XXIV.   BACTERIA (continued)

Pathogen	Host group	Host	References
Nocardia (M.) rhodochrous	Reduviidae	Triatoma protracta	Marchette & Hatie (1965)
Paracolobactrum aerugenoides	Blattidae	Blaberus craniifer	Briscoe et al. (1961)
Paracolobactrum intermedium	Blattidae	Blaberus craniifer	Briscoe et al. (1961, 1963)
Pasturella tularensis	Anoplura Mallophaga	Pediculus humanus	Price (1957)
Proteus mirabilis	Acarina ticks	Dermacentor andersoni	Brown et al. (1970)
Proteus morganii	Blattidae	Blaberus craniifer	Bricoe et al. (1961)
Proteus vulgaris	Anoplura Mallophaga	Pediculus humanus	Jenkins (1973)
Pseudomonas aeruginosa	Culicidae	Aedes vexans	Povazhna (1974)
Pseudomonas aeruginosa	Blattidae	Blaberus craniifer	Anderson et al. (1973)
Pseudomonas aeruginosa	Blattidae	Blaptica dubia	Sauerlander & Ehrhardt (1961), Sauerlander & Kohler (1961)
Pseudomonas aeruginosa	Blattidae	Blattella germanica	Henry (1965)
Pseudomonas aeruginosa	Culicidae	Culex pipiens fatigans	Thomas & Poinar (1973)
Pseudomonas aeruginosa	Culicidae	Culex pipiens	Povazhna (1974)
Pseudomonas aeruginosa	Blattidae	Periplaneta americana	Sauerlander & Ehrhardt (1961), Sauerlander & Kohler (1961), Zuberi et al. (1969)
Pseudomonas aeruginosa	Culicidae	Mosquito (?)	Thomas & Poinar (1973)
Pseudomonas sp.	Culicidae	Aedes aegypti	Mikhnovka et al. (1972)
Pseudomonas sp.	Culicidae	Culex pipiens molestus	Mikhnovka et al. (1972)
Pseudomonas sp.	Cerato- pogonidae	Culicoides salinarius	Becker (1958)
Pseudomonas sp.	Glossina	Glossina morsitans	Bauer (1974)
Pseudomonas sp.	Phlebotominae	Lutzomyia vexator occidentis	Chaniotis & Anderson (1968)
Pseudomonas (near septica)	Musca	Musca domestica	Amonkar et al. (1967)
Salmonella avium (not a recognized taxon)	Cimicidae	Cimex lectularius	Jenkins (1964)
Salmonella choleraesuis	Blattidae	Blaberus craniifer	Briscoe et al. (1961, 1963)
Salmonella enteritidis	Anoplura Mallophaga	Pediculus humanus	Jenkins (1973)

XXIV.   BACTERIA (continued)

Pathogen	Host group	Host	References
Salmonella enteritidis	Siphonaphera	Xenopsylla cheopis	Vaschenok et al. (1971)
Salmonella typhosa	Blattidae	Blaberus craniifer	Anderson et al. (1973)
Serratia near marcescens	Culicidae	Anopheles punctipennis	Briggs (1974)
Serratia marcescens	Blattidae	Blaptica dubia	Sauerlander & Ehrhardt (1961)
Serratia marcescens	Blattidae	Blattella germanica	Henry (1965)
Serratia marcescens	Acarina ticks	Dermacentor andersoni	Steinhaus (1959)
Serratia marcescens	Musca	Musca domestica	Steinhaus (1951)
Serratia marcescens	Blattidae	Periplaneta americana	Sauerlander & Ehrhardt (1961), Steinhaus & Marsh (1962)
Shigella dysenteriae	Anoplura Mallophaga	Pediculus humanus	Jenkins (1973)
Spirochaeta ctenocephali	Siphonaphera	Ctenopephalides felis	Jenkins (1964q)
Spirochaeta phlebotomi (=Treponema phlebotomi)	Phlebotominae	Phlebotomus perniciosus	Pringault (1921)
Staphylococcus aureus	Anoplura Mallophaga	Pediculus humanus	Jenkins (1973)
Staphylococcus aureus	Blattidae	Supella suppellectilium	Lord et al. (1964)
Staphylococcus muscae	Musca	Musca domestica	Glaser (1924)
Streptococcus faecalis	Blattidae	Leucophaea maderae	Thomas & Poinar (1973)
Streptococcus faecalis	Reduviidae	Rhodnius prolixus	Nyirady (1973), Brecher & Wigglesworth (1944)
Streptococcus faecalis	Reduviidae	Triatoma protracta	Nyirady (1973)
Streptococcus faecalis	Reduviidae	Triatoma rubida	Nyirady (1973)
Strickeria jurgensi	Anoplura Mallophaga	Pediculus humanus	Stempell (1916)
Treponema macfiei	Glossina	Glossina palpalis	Macfie (1915)
Treponema macfiei	Glossina	Glossina tachinoides	Macfie (1914b)
Vibrio sp.	Culicidae	Aedes aegypti	Chapman et al. (1967)
Vibrio sp.	Culicidae	Aedes fulvus pallens	Chapman et al. (1967)
Vibrio sp.	Culicidae	Aedes sollicitans	Chapman et al. (1969)
Vibrio sp.	Culicidae	Aedes taeniorhynchus	Chapman et al. (1967)
Vibrio sp.	Culicidae	Aedes triseriatus	Chapman et al. (1967)

XXIV.   BACTERIA (continued)

Pathogen	Host group	Host	References
Vibrio sp.	Culicidae	Anopheles barberi	Chapman et al. (1969)
Vibrio sp.	Culicidae	Anopheles bradleyi	Chapman et al. (1969)
Vibrio sp.	Culicidae	Culex pipiens fatigans	Fulton et al. (1974), Chapman et al. (1967, 1969)
Vibrio sp.	Culicidae	Culex restuans	Chapman et al. (1967)
Vibrio sp.	Culicidae	Culex salinarius	Chapman et al. (1967, 1969)
Vibrio sp.	Culicidae	Culex territans	Chapman et al. (1967)
Vibrio sp.	Culicidae	Culiseta inornata	Chapman et al. (1967)
Yersinia enterocolitica	Anoplura Mallophaga	Pediculus humanus	Jenkins (1973)
Yersinia (Pasteurella) pestis	Siphonaphera	Ceratophyllus gallinae	Jenkins (1964g, v)
Yersinia (Pasteurella) pestis	Siphonaphera	Ceratophyllus laevicepts	Kondrashkina et al. (1968)
Yersinia (Pasteurella) pestis	Siphonaphera	Ceratophyllus tesquorum	Kondrashkina et al. (1968)
Yersinia (Pasteurella) pestis	Siphonaphera	Ctenocephalides canis	Jenkins (1964g, v)
Yersinia (Pasteurella) pestis	Siphonaphera	Ctenocephalides felis	Jenkins (1964g, v)
Yersinia (Pasteurella) pestis	Siphonaphera	Diamanus montanus	Jenkins (1964g, v)
Yersinia (Pasteurella) pestis	Siphonaphera	Leptopsylla segnis	Jenkins (1964g, v)
Yersinia (Pasteurella) pestis	Siphonaphera	Neopsylla setosa	Konkrashkina et al. (1968)
Yersinia (Pasteurella) pestis	Siphonaphera	Nosopsyllus fasciatus	Jenkins (1964g, v)
Yersinia (Pasteurella) pestis	Siphonaphera	Pulex irritans	Jenkins (1964g, v)
Yersinia (Pasteurella) pestis	Siphonaphera	Xenopsylla cheopis	Kartman & Quan (1964), Jenkins (1964g, v), Kondrashkina et al. (1968)
Yersinia pestis	Anoplura Mallophaga	Linognathus sp.	Jettman (1925)
Yersinia pseudotuberculosis	Anoplura Mallophaga	Pediculus humanus	Jenkins (1973)
Azotobacteraceae	Simuliidae	Simulium damnosum	Burton et al. (1973)
Bacillaceae	Simuliidae	Simulium damnosum	Burton et al. (1973)

XXIV.   BACTERIA (continued)

Pathogen	Host group	Host	References
Long, rod shaped bacterium	Culicidae	Aedes aegypti	Briggs (1973b)
Bacteria (?)	Culicidae	Anopheles gambiae	Briggs (1967)
Non-spore forming short-rod(s) bacteria	Culicidae	Anopheles gambiae	Briggs (1973b)
Identification uncertain (bacteria)	Cerato-pogonidae	Culicoides nubeculosus	Steinhaus (1946)
A non-spore forming bacterium described in error as Bacillus lutzae	Musca	Musca domestica	Brown (1927), Brown & Heffron (1929)
Bacteroids	Glossina	Glossina austeni	Hueber & Davey (1974)
Bacteroids	Glossina	Glossina morsitans	Ma & Denlinger (1974)
Coccoid bacterium	Glossina	Glossina sp.	Vey (1974a)
Enterobacteriaceae	Culicidae	Culex pipiens	Muspratt (1964)
Micrococcaceae	Simuliidae	Simulium damnosum	Burton et al. (1973)
Pseudomondaceae	Simuliidae	Simulium damnosum	Burton et al. (1973)
Spore forming bacillus	Culicidae	Aedes aegypti	Briggs (1972)
Spore-former	Culicidae	Aedes aegypti	Mikhnovka et al. (1972)
Aerobic spore-former	Culicidae	Aedes triseriatus	Briggs (1966)
Spore-former	Culicidae	Anopheles gambiae	Briggs (1972)
Spore-former	Culicidae	Culex annuterostria	Briggs (1974)
Spore-former	Culicidae	Culex pipiens fatigans	Briggs (1973b)
Spore-former	Culicidae	Culex pipiens molestus	Mikhnovka et al. (1972)
Spirochaete	Glossina	Glossina tachinoides	Macfie (1914a)
Bacterial symbiont	Cerato-pogonidae	Culicoides sp.	Mayer (1934)
Hereditary bacteria symbiont	Cerato-pogonidae	Dasyhelea obscura	Keilin (1921a, 1927)
Intracellular symbiont	Cerato-pogonidae	Dasyhelea sp.	Stammer, in Buchner (1930)
Intracellular symbiont	Cerato-pogonidae	Dasyhelea versicolor	Stammer, in Buchner (1930)
Symbionts	Cerato-pogonidae	Culicoides nubeculosus	Lawson (1951)
Bacterium (symbiote)	Reduviidae	Triatoma infestans	Halff (1956)
Bacteria-like organism	Glossina	Glossina pallidipes	Rogers (1973)

XXV.   PROTOZOA OTHER THAN MICROSPORIDA

Pathogen	Host group	Host	References
Acanthamoeba hyalina	Blattidae	Blaberus craniifer	Briscoe (1971)
Acoccidium tubulorum	Acarina mites	Urodinychus sp.	Warren (1944)
Acoccidium ventriculi Haplosporidian protos	Acarina mites	Hermannia sp.	Warren (1944)
Actinocephalus parvus	Siphonaphora	Ceratophyllus gallinae	Jenkins (1964p, y)
Adelina sp.	Phlebotominae	Sergentomyia minuta minuta (Rond.)	Rioux et al. (1972)
Agrippina bona	Siphonaphora	Nosopsyllus fasciatus	Jenkins (1964w)
Akiba (Leucocytozoon) caulleryi	Ceratopogonidae	Culicoides arakawai	Akiba (1960), Bennett et al. (1965)
Allantoccystis dasyhelei	Ceratopogonidae	Dasyhelea obscura	Keilin (1920b, 1927)
Balantidium blattarum	Blattidae	Blaberus craniifer	Briscoe (1971)
Blastocrithidia spp.	Reduviidae	Rhodnius prolixus	Hanson et al. (1968)
Blastocrithidia spp.	Reduviidae	Triatoma infestans	Rocha de Silva & Amaral (1971), Hanson et al. (1968)
Carchesium	Culicidae	Culex pipiens	Larson (1967)
Caudospora alaskansis	Simuliidae	Prosimulium alpestre	Jamnback (1970)
Caudospora brevicauda	Simuliidae	Cnephia mutata	Ezenwa (1974c), Frost & Nolan (1972), Jamnback (1970)
Caudospora nasiae	Simuliidae	Simulium adersi	Jamnback (1970)
Caudospora pennsylvanica	Simuliidae	Prosimulium magnum	Beaudoin & Wills (1965)
Caudospora simulii	Simuliidae	Prosimulium fuscum	Frost & Nolan (1970)
Caudospora simulii	Simuliidae	Prosimulium mixtum	Frost & Nolan (1970)
Caudospora simulii	Simuliidae	Prosimulium mixtum/fuscum	Ezenwa (1974d), Frost (1970)
Caudospora simulii	Simuliidae	Simuliidae	Briggs (1966)
Caudospora simulii	Simuliidae	S. (Eusimulium) latipes	Vavra (1968), Briggs (1972)
Caulleryella	Culicidae	Aedes taeniorhynchus	Chapman (1974)
Caulleryella	Culicidae	Culiseta melanura	Chapman (1974)
Caulleryella	Culicidae	Orthyopodomyia signifera	Chapman (1974)

XXV.   PROTOZOA OTHER THAN MICROSPORIDA (continued)

Pathogen	Host group	Host	References
Cocconema pediculusvestimenti	Anoplura-Mallophaga	Pediculus humanus	Popov & Manuilova (1926)
Crithidia fasciculata	Culicidae	Aedes aegypti	Kramer (1964), Sinha (1972)
Crithidia fasciculata	Culicidae	Aedes sollicitans	Chapman et al. (1967)
Crithidia fasciculata	Culicidae	Anopheles freeborni	Clark et al. (1964)
Crithidia fasciculata	Culicidae	Anopheles gambiae	Brooker (1971A, 1972)
Crithidia fasciculata	Culicidae	Anopheles hackeri	Sandosham et al. (1962)
Crithidia fasciculata	Culicidae	Anopheles quadrimaculatus	Brooker (1970, 1971b)
Crithidia fasciculata	Culicidae	Culex apicalis	Clark et al. (1964)
Crithidia fasciculata	Culicidae	Culex boharti	Clark et al. (1964)
Crithidia fasciculata	Culicidae	Culex peus	Clark et al. (1964)
Crithidia fasciculata	Culicidae	Culex pipiens	Clark et al. (1964) Kramer (1964)
Crithidia fasciculata	Culicidae	Culex pipiens fatigans	Sinha (1972)
Crithidia fasciculata	Culicidae	Culex tarsalis	Clark et al. (1964)
Crithidia fasciculata	Culicidae	Culiseta incidans	Clark et al. (1964)
Crithidia fasciculata	Culicidae	Culiseta inornata	Clark et al. (1964)
Crithidia ctenocephali	Siphonaphora	Ctenocephalides canis	Jenkins (1964r)
Crithidia pulicis	Siphonaphora	Pulex irritans	Jenkins (1964s, t)
Crithidia sp.	Siphonaphora	Ctenocephalides felis	Jenkins (1964s, t)
Crithidia sp.	Culicidae	Mansonia titillans (Wik.)	Page (1972)
Crithidia spp.	Reduviidae	Rhodnius prolixus	Hanson et al. (1968)
Crithidia spp.	Reduviidae	Triatoma infestans	Hanson et al. (1968)
Gregarine, Diplocystis sp.	Phlebotominae	Phlebotomus caucasicus Marzinowsky	Lisova (1962)
Gregarine, Diplocystis sp.	Phlebotominae	Phlebotomus chinensis Newst.	Lisova (1962)
Gregarine, Diplocystis sp.	Phlebotominae	Phlebotomus sergenti	Lisova (1962)
Gregarine, Diplocystis sp.	Phlebotominae	Phlebotomus papatasii	Lisova (1962)
Endamoeba blattae	Blattidae	Blaberus giganteus	Hoyte (1961a)
Endamoeba blattae	Blattidae	Blatta orientalis	Hoyte (1961a)

XXV.   PROTOZOA OTHER THAN MICROSPORIDA (continued)

Pathogen	Host group	Host	References
Endamoeba blattae	Blattidae	Blattella germanica	Hoyte (1961a)
Endamoeba blattae	Blattidae	Periplaneta americana	Hoyte (1961a)
Endamoeba blattae	Blattidae	Periplaneta australasiae	Hoyte (1961a)
Endamoeba thomsoni	Blattidae	Blaberus craniifer	Briscoe (1971)
Endolimax blattae	Blattidae	Blaberus craniifer	Briscoe (1971)
Endolimax blattae	Blattidae	Blatta orientalis	Warhurst (1963, 1967)
Endolimax blattae	Blattidae	Periplaneta americana	Warhurst (1963, 1967)
Epistylis sp.	Culicidae	Aedes trivittatus	Larson (1967)
Epistylis sp.	Culicidae	Aedes vexans	Larson (1967)
Epistylis sp.	Culicidae	Anopheles punctipennis	Larson (1967)
Epistylis sp.	Culicidae	Culex duttoni	World Health Organization (1968)
Epistylis sp.	Culicidae	Culex pipiens	Larson (1967)
Epistylis sp.	Culicidae	Culex restuans	Larson (1967)
Possible Eugregarinid	Culicidae	Aedes aegypti	Briggs (1972)
Fusiona geusi	Blattidae	Pycnoscelus surinamensis	Stejskal (1965)
Glugea sp.	Ceratopogonidae	Dasyhelea sp.	Keilin (1927)
Gregarina blaberae	Blattidae	Blabera fusca	Tuzet et al. (1965)
Gregarina blattarum	Blattidae	Blaberus craniifer	Briscoe (1971)
Gregarina blattarum	Blattidae	Periplaneta americana	Desportes (1966)
Haemogregarina triatomae	Reduviidae	Eutriatoma (Triatoma) rubrovaria	Steinhaus (1946)
Haplosporidium simulii	Simuliidae	Simulium venustum	Beaudoin & Wills (1968)
Helicosporida	Culicidae	Culex territans	Chapman (1974)
Helicosporidium parasiticum	Culicidae	Culex fatigans	Kellen & Lindegren (1973)
Helicosporidium parasiticum	Ceratopogonidae	Dasyhelea obscura	Keilin (1921a, b, 1927)
Hepatozoon balfouri	Acarina mites	Haemolaelaps centrocarpus	Furman (1966)
Hepatozoon balfouri	Acarina mites	Haemolaelaps longipes	Furman (1966)

XXV.   PROTOZOA OTHER THAN MICROSPORIDA (continued)

Pathogen	Host group	Host	References
Hepatozoon griseisciure	Acarina mites	Haemogamasus reidi	Redington & Jachowski (1972)
Hepatozoon griseisciure Haemogregarine of grey squirrel	Acarina mites	Haemogamasus reidi potential vector of plague, typhus, and tularaemia	Redington & Jachowski (1971)
Hepatozoon perniciosum Haemogregarine	Acarina mites	Laelaps ecidninus	Miller (1908)
Hepatozoon pettiti	Glossina	Glossina palpalis	Buxton (1955)
Haemogregarine (Hepatozoon sp.)	Glossina	Glossina palpalis	Chatton & Roubaud (1913)
Herpetomonas ctenocephali	Siphonaphora	Ctenocephalides canis	Jenkins (1964i, l)
Herpetomonas ctenocephali	Siphonaphora	Ctenocephalides felis	Jenkins (1964i)
Herpetomonas ctenophali	Siphonaphora	Pulex irritans	Jenkins (1964i)
Herpetomonas ctenopsyllae	Siphonaphora	Leptopsylla segnis	Jenkins (1964m)
Herpetomonas muscae-domesticae	Musca	Musca domestica	Flu (1911), Ross & Hussain (1924)
Herpetomonas muscarum	Musca	Musca domestica	Kramer (1961), Laird (1959)
Herpetomonas pattoni	Siphonaphora	Nosopsyllus fasciatus	Jenkins (1964m)
Karyolysis lacertae Haemogregarine	Acarina mites	Liponyssus saurarum	Reichenow (1921)
Lankesteria barretti n.sp.	Culicidae	Aedes triseriatus	Vavra (1969)
Lankesteria clarki sp.n.	Culicidae	Aedes sierrensis	Sanders et al. (1973)
Lankesteria culicis	Culicidae	Aedes aegypti	Walsh et al. (1969), Barrett (1968), Barrett et al. (1971), Briggs (1973), Gentile et al. (1971), Hati & Ghosh (1963), Hayes & Haverfield (1971), Kramer (1964), McCray et al. (1970), Sheffield et al. (1971), Sinha (1972), Stapp & Casten (1971)

XXV.   PROTOZOA OTHER THAN MICROSPORIDA (continued)

Pathogen	Host group	Host	References
Lankesteria culicis	Culicidae	Aedes sierrensis	Weiser (1968)
Lankesteria sp.	Phlebotominae	Lutzomyia vexator occidentis (Fchld. & Hertig)	Ayala (1971, 1973)
Gregarine, Lankesteria sp.	Phlebotominae	Phlebotomus longipes Parrot & Martin	Ashford (1974)
Legerella fasciatus	Siphonaphora	Nosopsyllus fasciatus	Jenkins (1964p)
Legerella parva	Siphonaphora	Ceratophyllus gallinae	Jenkins (1964p)
Leishmania (Herpetomonas) donovani	Cimicidae	Cimex hemipterus rotundatus	Patton (1913)
Leishmania (Herpetomonas) donovani	Cimicidae	Cimex lectularius	Patton (1913)
Leishmania (Herpetomonas) tropica	Cimicidae	Cimex hemipterus rotundatus	Patton et al. (1921)
Leptomonas ctenocephali	Siphonaphora	Ctenocephalides canis	Jenkins (1964b, x)
Leptomonas ctenocephali	Siphonaphora	Xenopsylla cheopis	Amonkar & Mushi (1965)
Leptomonas pulicis	Siphonaphora	Pulex irritans	Jenkins (1964)
Leptomonas muscae-domesticae	Musca	Musca domestica	Flu (1911)
Leptomonas pediculi	Anoplura-Mallophaga	Pediculus humanus	Fantham (1912)
Leptomonas sp.	Glossina	Glossina morsitans	Buxton (1955)
Leptomonas sp.	Siphonaphora	Spilopsyllus cuniculi	Rothschild (1969)
Lophomonas blattarum	Blattidae	Blaberus craniifer	Briscoe (1971)
Lophomonas blattarum	Blattidae	Blaberus giganteus	Hoyte (1961a)
Lophomonas blattarum	Blattidae	Blatta orientalis	Hoyte (1961a)
Lophomonas blattarum	Blattidae	Blattella germanica	Hoyte (1961a)
Lophomonas blattarum	Blattidae	Periplaneta americana	Hoyte (1961a)
Lophomonas blattarum	Blattidae	Periplaneta australasiae	Hoyte (1961a)
Lophomonas striata	Blattidae	Blaberus giganteus	Hoyte (1961a)
Lophomonas striata	Blattidae	Blatta orientalis	Hoyte (1961a)
Lophomonas striata	Blattidae	Blattella germanica	Hoyte (1961a)
Lophomonas striata	Blattidae	Periplaneta americana	Hoyte (1961a)
Lophomonas striata	Blattidae	Periplaneta australasiae	Hoyte (1961a)

Pathogen	Host group	Host	References
Machadoella triatomae	Reduviidae	Eutriatoma maculata	Mayer & Pifano (1949)
Machadoella triatomae	Reduviidae	Triatoma dimidiata	Steinhaus (1946)
Malpighiella refringens	Siphonaphora	Nosopsyllus fasciatus	Jenkins (1964n)
Monocercomonoides opthopterorum	Blattidae	Blaberus giganteus	Hoyte (1961a)
Monocercomonoides opthopterorum	Blattidae	Blatta orientalis	Hoyte (1961a)
Monocercomonoides opthopterorum	Blattidae	Blattella germanica	Hoyte (1961a)
Monocercomonoides opthopterorum	Blattidae	Periplaneta americana	Hoyte (1961a)
Monocercomonoides opthopterorum	Blattidae	Periplaneta australasiae	Hoyte (1961a)
Monocercomonas sp.	Blattidae	Blaberus sp.	Perez-Reyes (1966)
Monocystis chagasi Adler & Mayrink	Phlebotominae	Lutzomyia flaviscutellata (Mang.)	Lewis, Lainson & Shaw (1970)
Monocystis chagasi	Phlebotominae	Lutzomyia longipalpis	Adler & Mayrink (1961)
Monocystis chagasi	Phlebotominae	Lutzomyia longipalpis	Coelho & Falcão (1964)
Monocystis chagasi	Phlebotominae	Lutzomyia sallesi (Galvão & Coutinho)	Coelho & Falcão (1964)
Monocystis mackiei	Phlebotominae	Phlebotomus argentipes Ann. & Brun.	Shortt & Swaminath (1927)
Monocystis mackiei	Phlebotominae	Phlebotomus papatasii	Missiroli (1929, 1932)
Gregarine, Monocystis (?)	Phlebotominae	Phlebotomus rodhaini Parrot	Barnley (1968)
Myxosporidium heibergi	Glossina	Glossina palpalis	Dutton et al. (1907)
Nephridiophaga blattellae	Blattidae	Blattella germanica	Woolever (1966)
Nyctotherus ovalis	Blattidae	Blaberus craniifer	Briscoe (1971)
Nyctotherus ovalis	Blattidae	Blaberus giganteus	Hoyte (1961a, b, c)
Nyctotherus ovalis	Blattidae	Blatta orientalis	Hoyte (1961a, b, c)
Nyctotherus ovalis	Blattidae	Blattella germanica	Hoyte (1961a, b, c)
Nyctotherus ovalis	Blattidae	Periplaneta americana	Hoyte (1961a, b, c)
Nyctotherus ovalis	Blattidae	Periplaneta australasiae	Hoyte (1961a)
Octosporea carloschagasi	Reduviidae	Panstrogylus megistus	Kramer (1972)

XXV.   PROTOZOA OTHER THAN MICROSPORIDA (continued)

Pathogen	Host group	Host	References
Octosporea muscae-domesticae	Reduviidae	Panstrogylus megistus	Chatton (1911)
Octosporea muscae-domestica	Musca	Musca autumnalis	Kramer (1973)
Octosporea muscae-domestica	Musca	Musca domestica	Flu (1911), Kramer (1964b), Kramer (1965), Kramer (1966), Laird (1959)
Parahaemoproteus canachites	Ceratopogonidae	Culicoides sphagnumensis	Fallis & Bennett (1961), Bennett et al. (1965)
Parahaemoproteus nettionis	Ceratopogonidae	Culicoides near piliferus	Fallis & Bennett (1961), Bennett et al. (1965)
Parahaemoproteus sp.	Ceratopogonidae	Culicoides crepuscularis	Fallis & Bennett (1961), Bennett et al. (1965)
Parahaemoproteus sp.	Ceratopogonidae	Culicoides stilobezzioides	Fallis & Bennett (1961), Bennett et al. (1965)
Paramonocystis simplex	Acarina mites	Urodinychus sp.	Warren (1944)
"Ectoparasitic cysts" c.f. Perezella	Ceratopogonidae	Culicoides cubitalis	Kettle & Lawson (1952)
Perezella sp.	Ceratopogonidae	Culicoides odibilis	Becker (1958)
Perezella sp.	Ceratopogonidae	Culicoides pulicaris	Becker (1958)
Perezella sp.	Ceratopogonidae	Culicoides riethi	Becker (1958)
Perezella sp.	Ceratopogonidae	Culicoides salinarius	Becker (1958)
Polymastix periplanetae	Blattidae	Periplaneta americana	Qadri & Rao (1963)
Polymastix sp.	Blattidae	Periplaneta americana	Warhurst (1966)
Retortomonas blattae	Blattidae	Blaberus craniifer	Briscoe (1971)
Rhizomastix periplanetae	Blattidae	Periplaneta americana	Rao (1963)
Schizocystis gregarinoides	Ceratopogonidae	Ceratopogon sp.	Léger (1900, 1906) Weiser (1963a, b)
Spiroglugea octospora	Ceratopogonidae	Ceratopogon sp.	Léger & Hesse (1922), Kudo (1924), Jírovec (1936), Weiser (1961)
Steinina rotundata	Siphonaphora	Ceratophyllus gallinae	Jenkins (1964a)
Stylocystis riouxi	Ceratopogonidae	Dasyhelea lithotelmatica	Tuzet & Ormières (1964)

XXV.   PROTOZOA OTHER THAN MICROSPORIDA (continued)

Pathogen	Host group	Host	References
Taeniocystis mira	Ceratopogonidae	Ceratopogon solstitialis	Léger (1906)
Taeniocystis parva	Ceratopogonidae	Forcypomyia sp.	Foerster (1938)
Tetrahymena gelei	Culicidae	Aedes sinensis	Weiser (1968)
Tetrahymena pyriformis	Culicidae	Aedes aegypti	Grassmick & Rowley (1973)
Tetrahymena pyriformis	Culicidae	Culex pipiens fatigans	Hati & Ghosh (1963)
Tetrahymena pyriformis	Culicidae	Culex tarsalis	Grassmick & Rowley (1973)
Tetrahymena pyriformis (= Probalantidium knowlesi, Balantidium knowlesi Leptoglena knowlesi)	Ceratopogonidae	Culicoides peregrinus	Ghosh (1925), Abe (1927), Grasse & Boissezon (1929), Jenkins (1964)
Tetrahymena pyriformis	Blattidae	Periplaneta americana	Seaman & Clement (1970), Seaman & Robert (1968), Seaman & Tosney (1967), Seaman et al. (1972)
Trichomonas vaginalis	Blattidae	Blabera fusca	Gorbert et al. (1974)
Tetrahymena	Culicidae	Aedes atlanticus	Chapman (1974)
Tetrahymena sp.	Culicidae	Aedes sierrensis	Sanders (1972)
Tetrahymena	Culicidae	Aedes sticticus	Chapman (1974)
Tetrahymena	Culicidae	Culex territans	Chapman (1974)
Toxoplasma gondii	Anoplura-Mallophaga	Pediculus humanus	Dutkiewicz (1966)
Toxoglugea vibrio	Ceratopogonidae	Ceratopogon sp.	Léger & Hesse (1922), Kudo (1924), Jírovec (1936), Weiser (1961)
Trypanosoma lewisi	Siphonaphora	Nosopsyllus fasciatus	Jenkins (1964f, o)
Trypanosoma rangeli	Cimicidae	Cimex lectularius	Grewal (1957, 1959)
Trypanosoma rangeli	Reduviidae	Panstrogylus megistus	Lucena & de Vergetti (1973)
Trypanosoma rangeli	Reduviidae	Rhodnius brethesi	D'Alessandro et al. (1971)
Trypanosoma rangeli	Reduviidae	Rhodnius ecuadoriensis	Cuba et al. (1972), Cuba (1972), Herrer et al. (1972)

XXV.   PROTOZOA OTHER THAN MICROSPORIDA (continued)

Pathogen	Host group	Host	References
Trypanosoma rangeli	Reduviidae	Rhodnius pallescens	D'Alessandro et al. (1971)
Trypanosoma rangeli	Reduviidae	Rhodnius prolixus	Tejera (1920), D'Alessandro et al. (1971), Grewal (1957, 1969), Zeledon & de Monge (1966), Gomez (1967), Tobie (1965, 1968, 1970), Watkins (1971a, b)
Trypanosoma rangeli	Reduviidae	Rhodnius robustus	D'Alessandro et al. (1971)
Trypanosoma rangeli	Reduviidae	Triatoma dimidiata	D'Alessandro et al. (1971)
Trypanosoma rangeli	Reduviidae	Triatoma infestans	Zeldon & de Monge (1966)
Vorticella convallaria	Culicidae	Aedes trivittatus	Larson (1967)
Vorticella convallaria	Culicidae	Aedes vexans	Larson (1967)
Vorticella convallaria	Culicidae	Culex pipiens	Larson (1967)
Vorticella convallaria	Culicidae	Culex restuans	Larson (1967)
Vorticella convallaria	Culicidae	Culex salinarius	Larson (1967)
Vorticella convallaria	Culicidae	Culex tarsalis	Larson (1967)
Vorticella convallaria	Culicidae	Culiseta inornata	Larson (1967)
Vorticella striata	Culicidae	Aedes trivittatus	Larson (1967)
Vorticella striata	Culicidae	Aedes vexans	Larson (1967)
Vorticella striata	Culicidae	Culex pipiens	Larson (1967)
Vorticella striata	Culicidae	Culex restuans	Larson (1967)
Vorticella striata	Culicidae	Culex salinarius	Larson (1967)
Vorticella striata	Culicidae	Culex territans	Larson (1967)
Vorticella sp.	Culicidae	Aedes nigromaculis	Larson (1967)
Vorticella sp.	Culicidae	Aedes trivittatus	Larson (1967)
Vorticella sp.	Culicidae	Aedes vexans	Larson (1967)
Vorticella sp.	Culicidae	Anopheles squamosus	Briggs (1968)

XXV. PROTOZOA OTHER THAN MICROSPORIDA (continued)

Pathogen	Host group	Host	References
Vorticella sp.	Culicidae	Culex pipiens	Schober (1967)
Vorticellidae	Culicidae	Anopheles gambiae	Briggs (1970)
Acephaline Gregarines	Phlebotominae	Lutzomyia camposi (Rodriquez)	McConnell & Correa (1964)
Acephaline Gregarines	Phlebotominae	Lutzomyia cruciata (Coq.)	McConnell & Correa (1964)
Acephaline Gregarines	Phlebotominae	Lutzomyia hartmanni (Fchld. & Hertig)	McConnell & Correa (1964)
Acephaline Gregarines	Phlebotominae	Lutzomyia panamensis (Shannon)	McConnell & Correa (1964)
Acephaline Gregarines	Phlebotominae	Lutzomyia sanguinaria (Fchld. & Hertig)	McConnell & Correa (1964)
Acephaline Gregarines	Phlebotominae	Lutzomyia trapidoi (Fchld. & Hertig)	McConnell & Correa (1964)
Acephaline Gregarines	Phlebotominae	Lutzomyia trinidadensis (Newst.)	McConnell & Correa (1964)
Acephaline Gregarines	Phlebotominae	Lutzomyia ylephiletor (Fchld. & Hertig)	McConnell & Correa (1964)
Acephaline Gregarine	Phlebotominae	Sergentomyia garnhami (Heisch, Guggisberg & Teesdale)	Lewis & Minter (1960)
Internal Ciliates	Culicidae	Aedes dupreei	Chapman et al. (1969)
Internal Ciliates	Culicidae	Aedes vexans	Chapman et al. (1967, 1969)
Internal Ciliates	Culicidae	Culex restuans	Chapman et al. (1969)
Ciliates	Ceratopogonidae	Culicoides austeni	Sharp (1928)
Internal Ciliates	Ceratopogonidae	Culicoides sp. prob. nanus	Chapman et al. (1969)
Ciliates	Phlebotominae	Lutzomyia shannoni	Lewis (1965a)
Internal Ciliates	Culicidae	Orthyopodmyia signifera	Chapman et al. (1969)
Ciliate	Simuliidae	Simulium damnosum	Lewis (1965), Marr & Lewis (1964)
Epibionts	Culicidae	Anopheles gambiae	Briggs (1973)
Flagellate	Culicidae	Anopheles funestus	Hazard (1972)
Flagellates (Leptomonas type)	Ceratopogonidae	Culicoides austeni	Sharp (1928)

XXV. PROTOZOA OTHER THAN MICROSPORIDA (continued)

Pathogen	Host group	Host	References
Flagellates (Strigomonas?)	Ceratopogonidae	Culicoides salinarius	Kremer et al.
Flagellate	Glossina	Glossina pallicera	Foster (1964)
Flagellate	Glossina	Glossina palpalis	Foster (1964)
Gregarine	Blattidae	Gromphadorhina portentosa	Bhatnagar & Edwards (1970)
Gregarine	Phlebotominae	Lutzomyia shannoni	McConnell & Correa (1964)
Gregarine	Phlebotominae	Lutzomyia shannoni (Dyar)	Garnham & Lewis (1959)
Haemogregarine	Glossina	Glossina palpalis	Macfie (1915)
Reptilian haemogregarine	Glossina	Glossina tachinoides	Lloyd et al. (1924)
Haplosporidia	Simuliidae	Simulium damnosum	Lewis (1965)
Hepatozoon	Phlebotominae	Phlebotomus papatasii	Adler & Theodor (1929)
Herpetomonad	Glossina	Glossina morsitans	Lloyd (1924)
Protozoan	Simuliidae	Prosimulium sp.	Briggs (1969)
Protozoan	Simuliidae	Simuliidae	Maitland & Penney (1967)
Protozoa	Simuliidae	Simulium taylori	Briggs (1968)

XXVI. MICROSPORIDA

Pathogen	Host group	Host	References
Amblyospora benigna	Culicidae	Culex apicalis	Kellen & Wills (1962a)
Amblyospora bolinasae	Culicidae	Aedes squamiger	Kellen & Wills (1962a)
Amblyospora californica	Culicidae	Culex tarsalis	Kellen & Lipa (1960), Chapman et al. (1967), Chapman (1966), Initial report, Tsai et al. (1969)
Amblyospora campbelli	Culicidae	Culiseta incidens	Kellen & Wills (1962a), Chapman (1966)
Amblyospora canadensis	Culicidae	Aedes canadensis	Kellen & Wills (1962a), Anderson (1968), Chapman et al. (1966), Hazard & Oldacre (1975), Bailey et al. (1967a), Wills & Beaudoin (1965)
Amblyospora gigantea	Culicidae	Culex erythrothorax	Kellen & Wills (1962a)
Amblyospora inimica	Culicidae	Culiseta inornata	Kellen & Wills (1962a), Hazard & Oldacre (1975), Chapman et al. (1966), Chapman (1966), Tsai et al. (1969)
Amblyospora keenani	Culicidae	Aedomyia squamipennis	Hazard & Oldacre (1975)
Amblyospora khaliulini	Culicidae	Aedes communis	Welch (1960), Weiser (1946), Noller (1920), Chapman et al. (1973), Hazard & Oldacre (1975), Khaliulin & Ivanov (1971)
Amblyospora minuta	Culicidae	Culex erraticus	Kudo (1924c), Chapman et al. (1967)
Amblyospora mojingensis	Culicidae	Anopheles eiseni	Hazard & Oldacre (1975)
Amblyospora noxia	Culicidae	Culex thriambus	Kellen & Wills (1962a)
Amblyospora opacita (doubtful parasite of this host)	Culicidae	Culex pipiens pipiens	Simmers (1974b)
Amblyospora opacita	Culicidae	Culex territans	Kudo (1922), Chapman et al. (1969), Anderson (1968), Hazard & Oldacre (1975)
Amblyospora opacita (doubtful parasite of this host)	Culicidae	Culiseta inornata	Simmers (1974b)

XXVI.   MICROSPORIDA (continued)

Pathogen	Host group	Host	References
<u>Amblyospora unica</u>	Culicidae	<u>Aedes melanimon</u>	Kellen & Wills (1962a), Chapman (1966)
<u>Amblyospora</u> sp.	Culicidae	<u>Aedes abserratus</u>	Anderson (1968), Hazard & Oldacre (1975)
<u>Amblyospora</u> sp.	Culicidae	<u>Aedes cantator</u>	Chapman et al. (1973)
<u>Amblyospora</u> sp.	Culicidae	<u>Aedes cantans</u>	Hazard & Oldacre (1975), Noller (1920), Weiser (1971)
<u>Amblyospora</u> sp.	Culicidae	<u>Aedes caspius</u>	Tour et al. (1971)
<u>Amblyospora</u> sp.	Culicidae	<u>Aedes cataphylla</u>	Kellen et al. (1965), Chapman et al. (1973)
<u>Amblyospora</u> sp.	Culicidae	<u>Aedes cinereus</u>	Chapman (1966), Anderson (1968), Hazard & Oldacre (1975)
<u>Amblyospora</u> sp.	Culicidae	<u>Aedes detritus</u>	Tour et al. (1971)
<u>Amblyospora</u> sp.	Culicidae	<u>Aedes dorsalis</u>	Kellen et al. (1965), Chapman (1966), Tsai et al. (1969)
<u>Amblyospora</u> sp.	Culicidae	<u>Aedes excrucians</u>	Chapman et al. (1973), Anderson (1968), Hazard & Oldacre (1975)
<u>Amblyospora</u> sp.	Culicidae	<u>Aedes fitchii</u>	Chapman et al. (1973)
<u>Amblyospora</u> sp.	Culicidae	<u>Aedes grossbecki</u>	Chapman et al. (1966)
<u>Amblyospora</u> sp.	Culicidae	<u>Aedes hexodontus</u>	Chapman et al. (1973), Kellen et al. (1965)
<u>Amblyospora</u> sp.	Culicidae	<u>Aedes increpitus</u>	Kellen et al. (1965)
<u>Amblyospora</u> sp.	Culicidae	<u>Aedes pullatus</u>	Chapman et al. (1973)
<u>Amblyospora</u> sp.	Culicidae	<u>Aedes punctor</u>	Chapman et al. (1973), Hazard & Oldacre (1975)
<u>Amblyospora</u> sp.	Culicidae	<u>Aedes riparius</u>	Chapman et al. (1973)
<u>Amblyospora</u> sp.	Culicidae	<u>Aedes sollicitans</u>	Kellen et al. (1966a), Hazard & Oldacre (1975)
<u>Amblyospora</u> sp.	Culicidae	<u>Aedes sticticus</u>	Chapman et al. (1966)
<u>Amblyospora</u> sp.	Culicidae	<u>Aedes stimulans</u>	Franz & Hagmann (1962), Anderson (1968)
<u>Amblyospora</u> sp.	Culicidae	<u>Aedes taeniorhynchus</u>	Hazard & Oldacre (1975), Kellen et al. (1966a)
<u>Amblyospora</u> sp.	Culicidae	<u>Aedes thibaulti</u>	Initial report
<u>Amblyospora</u> sp.	Culicidae	<u>Aedes ventrovittis</u>	Kellen et al. (1965)

XXVI.   MICROSPORIDA (continued)

Pathogen	Host group	Host	References
Amblyospora sp.	Culicidae	Coquillettidia perturbans	Chapman et al. (1967) Hazard & Oldacre (1975)
Amblyospora sp.	Culicidae	Culex annulirostris	Laird (1956)
Amblyospora sp.	Culicidae	Culex peccator	Chapman et al. (1969)
Amblyospora sp.	Culicidae	Culex peus	Kellen et al. (1965)
Amblyospora sp.	Culicidae	Culex pipiens pipiens	Initial report
Amblyospora sp.	Culicidae	Culex salinarius	Kellen et al. (1966a), Hazard & Oldacre (1975)
Amblyospora sp.	Culicidae	Culiseta annulata	Hazard & Oldacre (1975)
Amblyospora sp.	Culicidae	Culiseta impatiens	Tsai et al. (1969)
Amblyospora sp.	Culicidae	Culiseta particeps	Kellen et al. (1965)
Amblyospora sp.	Culicidae	Mansonia leberi	Hazard & Oldacre (1975)
Amblyospora sp.	Culicidae	Mansonia dyari	Hazard & Oldacre (1975)
Amblyospora sp.	Culicidae	Psorophora columbiae	Chapman et al. (1966), Hazard & Oldacre (1975)
Hyalinocysta chapmani	Culicidae	Culiseta melanura	Hazard & Oldacre (1975)
Nosema adie	Cimicidae	Cimex hemipterus rotundatus	Jenkins (1964)
Nosema aedis	Culicidae	Aedes aegypti	Kudo (1930)
Nosema algerae	Culicidae	Aedes aegypti	Canning & Hulls (1970), Hazard & Lofgren (1971)
Nosema algerae	Culicidae	Anopheles albimanus	Hazard (1970), Initial report
Nosema algerae	Culicidae	Anopheles balabacensis	Hazard (1970)
Nosema algerae	Culicidae	Anopheles crucians	Chapman et al. (1970a)
Nosema algerae	Culicidae	Anopheles gambiae	Fox & Weiser (1959), Hazard (1971), Canning & Hulls (1970)
Nosema algerae	Culicidae	Anopheles melas	Hazard (1970), Hazard & Lofgren (1971)
Nosema algerae	Culicidae	Anopheles quadrimaculatus	Fox & Weiser (1959), Canning & Hulls (1970)
Nosema algerae	Culicidae	Anopheles stephensi	Vavra & Undeen (1970), Canning & Hulls (1970), Hazard (1970)
Nosema algerae	Culicidae	Culex fatigans (pipiens quinque-fasciatus)	Hazard & Lofgren (1971)
Nosema algerae	Culicidae	Culex pipiens pipiens	

XXVI.   MICROSPORIDA (continued)

Pathogen	Host group	Host	References
Nosema algerae	Culicidae	Culex pipiens pipiens	Canning & Hulls (1970)
Nosema algerae	Culicidae	Culex salinarius	Hazard & Lofgren (1971)
Nosema ctenocephali/pulicis	Siphonaphora	Ctenocephalides canis	Jenkins (1964j)
Nosema ctenocephali	Siphonaphora	Ctenocephalides felis	Jenkins (1964j, k)
Nosema culicis (doubtful generic placement)	Culicidae	Culex pipiens pipiens	Bresslau & Bushkielm (1919)
Nosema helminthorum	Acarina mites	Ceratoppea bipilis	Dissanaike (1958)
Nosema helminthorum	Acarina mites	Xenillus tegeocranus	Dissanaike (1958)
Nosema kingi[*]	Musca	Musca domestica	Kramer (1964a)
Nosema lutzi (may be an invalid sp.)	Culicidae	Aedes aegypti	Lutz & Splendore (1908)
Nosema periplanetae	Blattidae	Blattella germanica	Selmair (1962)
Nosema sphaeromiadis	Ceratopogonidae	Sphaeromias sp.	Weiser (1957, 1961, 1963)
Nosema stricklandi	Simuliidae	Simulium sp.	Briggs (1972)
Nosema sp.	Culicidae	Anopheles albimanus	Initial report
Nosema sp. (doubtful generic placement)	Culicidae	Culex salinarius	Chapman et al. (1967)
Nosema n. sp.	Ceratopogonidae	Culicoides nanus	Chapman (1973)
Nosema sp.	Blattidae	Periplaneta americana	Fisher & Sanborn (1964)
Parathelohania africana	Culicidae	Anopheles gambiae	Hazard & Anthony (1974)
Parathelohania anomala	Culicidae	Anopheles ramsayi	Sen (1941)
Parathelohania anophelis	Culicidae	Anopheles quadrimaculatus	Kudo (1924c), Chapman (1966), Hazard & Anthony (1974)

XXVI. MICROSPORIDA (continued)

Pathogen	Host group	Host	References
Parathelohania barra	Culicidae	Aedes australis	Pillai (1968)
Parathelohania chagrasensis	Culicidae	Aedomyia squamipennis	Hazard & Oldacre (1975)
Parathelohania illinoisensis	Culicidae	Anopheles punctipennis	Kudo (1921), Hazard & Anthony (1974), Anderson (1968), Fantham et al. (1941)
Parathelohania indica	Culicidae	Anopheles hyrcanus	Kudo (1929)
Parathelohania legeri	Culicidae	Anopheles maculipennis	Hesse (1904a), Missiroli (1929), Weiser (1961)
Parathelohania legeri (doubtful parasite of this host)	Culicidae	Anopheles punctipennis	Simmers (1974a)
Parathelohania legeri (doubtful parasite of this host)	Culicidae	Psorophora ciliata	Simmers (1974a)
Parathelohania obesa	Culicidae	Anopheles crucians	Chapman et al. (1966a) Hazard & Weiser (1968)
Parathelohania obesa	Culicidae	Anopheles quadrimaculata	Kudo (1924c), Wills & Beaudoin (1965), Chapman et al. (1966), Hazard & Weiser (1968)
Parathelohania obscura	Culicidae	Anopheles varuna	Kudo (1929)
Parathelohania octolagenella	Culicidae	Anopheles pretoriensis	Hazard & Anthony (1974)
Parathelohania periculosa	Culicidae	Anopheles franciscanus	Kellen & Wills (1962a)
Parathelohania sp.	Culicidae	Anopheles albimanus	Hazard & Oldacre (1975)
Parathelohania sp.	Culicidae	Anopheles annularis	Kudo (1929)
Parathelohania sp.	Culicidae	Anopheles barbirostris	Kudo (1929)
Parathelohania sp.	Culicidae	Anopheles bradleyi	Chapman et al. (1966)
Parathelohania sp.	Culicidae	Anopheles earlei	Initial report
Parathelohania sp.	Culicidae	Anopheles funestus	Hazard & Oldacre (1975)
Parathelohania sp.	Culicidae	Anopheles labranchiae atroparvus	Tour et al. (1971)
Parathelohania sp.	Culicidae	Anopheles nili	Hazard & Oldacre (1975)
Parathelohania sp.	Culicidae	Anopheles pharoensis	Hazard & Oldacre (1975)

XXVI.   MICROSPORIDA (continued)

Pathogen	Host group	Host	References
Parathelohania sp.	Culicidae	Anopheles pseudopunctipennis pseudopunctipennis	Camey-Pacheco (1965) Initial report
Parathelohania sp.	Culicidae	Anopheles sinensis	Initial report
Parathelohania sp.	Culicidae	Anopheles subpictus	Kudo (1929)
Parathelohania sp.	Culicidae	Anopheles triannulatus	Hazard & Oldacre (1975)
Parathelohania sp.	Culicidae	Anopheles vagus	Sen (1941)
Parathelohania sp.	Culicidae	Anopheles walkeri	Laird (1961)
Pilosporella chapmani	Culicidae	Aedes triseriatus	Hazard & Oldacre (1975)
Pilosporella fishi	Culicidae	Wyeomyia vanduzeei	Hazard & Oldacre (1975)
Pleistophora caecorum	Culicidae	Culiseta inornata	Chapman & Kellen (1967)
Pleistophora chapmani	Culicidae	Culex territans	Clark & Fukuda (1971)
Pleistophora chapmani	Culicidae	Culiseta inornata	Initial report
Pleistophora collessi	Culicidae	Culex gelidus	Laird (1959a)
Pleistophora collessi	Culicidae	Culex tritaeniorhynchus summorosus	Laird (1959a)
Pleistophora culicis	Culicidae	Aedes aegypti	Garnham (1959)
Pleistophora culicis	Culicidae	Aedes triseriatus	Chapman (1974a)
Pleistophora culicis	Culicidae	Anopheles albimanus	Chapman et al. (1970a), Weiser & Coluzzi (1972)
Pleistophora culicis	Culicidae	Anopheles dureni	Reynolds (1966)
Pleistophora culicis	Culicidae	Anopheles franciscanus	Chapman et al. (1970a)
Pleistophora culicis	Culicidae	Anopheles gambiae	Garnham (1951), Canning (1957b), Weiser & Coluzzi (1972)
Pleistophora culicis	Culicidae	Anopheles stephensi	Garnham (1956), Reynolds (1966), Weiser & Coluzzi (1972)
Pleistophora culicis	Culicidae	Culex fatigans	Reynolds (1966)
Pleistophora culicis	Culicidae	Culex pipiens pipiens	Weiser (1947)
Pleistophora culicis	Culicidae	Culex salinarius	Chapman (1974)
Pleistophora culicis	Culicidae	Culex territans	Chapman et al. (1969)
Pleistophora culicis	Culicidae	Culiseta corigrarcolata	Briggs (1972)
Pleistophora culicis	Culicidae	Culiseta longiareolata	Weiser & Coluzzi (1964)
Pleistophora debaiseuxi	Simuliidae	Eusimulium latipes	Rubtsov (1966b)

XXVI.   MICROSPORIDA (continued)

Pathogen	Host group	Host	References
Pleistophora debaiseuxi	Simuliidae	Simulium sp.	Maurand & Manier (1968)
Pleistophora leasi	Simuliidae	Simulium ornatum	Gassouma (1972)
Pleistophora milesi	Culicidae	Aedes notoscriptus	Pillai (1974)
Pleistophora milesi	Culicidae	Maorigoeldia arggropus	Pillai (1974)
Pleistophora multispora	Simuliidae	Simulium corbis	Ezenwa (1974d)
Pleistophora multispora	Simuliidae	Simulium decorum	Ezenwa (1973)
Pleistophora multispora	Simuliidae	Simulium morsitans	Briggs (1972)
Pleistophora multispora	Simuliidae	Simulium venustum	Ezenwa (1973, 1974d)
Pleistophora simulii	Simuliidae	Boophthora erythrocephala	Rubtsov (1966b)
Pleistophora simulii	Simuliidae	Eusimulium latipes	Ezenwa (1974c, d)
Pleistophora simulii	Simuliidae	Simulium bezzii	Maurand (1967)
Pleistophora simulii	Simuliidae	Simulium corbis	Ezenwa (1974d)
Pleistophora simulii	Simuliidae	Simulium equinum	Maurand & Manier (1967)
Pleistophora simulii	Simuliidae	Simulium monticola	Maurand (1967)
Pleistophora simulii	Simuliidae	Simulium morsitans	Maurand (1967), Rubtsov (1966b)
Pleistophora simulii	Simuliidae	Simulium ornatum	Maurand (1967)
Pleistophora simulii	Simuliidae	Simulium tuberosum	Ezenwa (1974d)
Pleistophora simulii	Simuliidae	Simulium venustum	Frost (1970), Ezenwa (1973), Maurand (1967)
Pleistophora simulii	Simuliidae	Simulium sp.	Briggs (1972)
Pleistophora simulii	Simuliidae	Simulium sp.	Manier & Maurand (1966), Maurand (1966), Maurand & Manier (1968)
Pleistophora stegomyiae	Culicidae	Aedes aegypti	Marchoux et al. (1903)
Pleistophora tillingbournei	Simuliidae	Simulium ornatum	Gassouma (1972), Gassouma & Ellis (1973)
Pleistophora sp.	Culicidae	Aedes canadensis	Chapman et al. (1967)
Pleistophora sp.	Culicidae	Aedes vexans	Chapman et al. (1967)
Pleistophora sp.	Culicidae	Anopheles funestus	Hazard (1972)
Pleistophora sp.	Culicidae	Anopheles gambiae	Hazard (1972)

XXVI.   MICROSPORIDA (continued)

Pathogen	Host group	Host	References
Pleistophora sp.	Culicidae	*Culex fatigans*	Initial report
Pleistophora sp.	Culicidae	*Culex nigripalpus*	Initial report
Pleistophora sp.	Culicidae	*Culex restuans*	Initial report
Pleistophora sp.	Culicidae	*Culex salinarius*	Chapman et al. (1969)
Pleistophora spp.	Ceratopogonidae	*Culicoides* sp. prob. *nanus*	Chapman et al. (1967, 1968, 1969), Chapman (1973)
Pleistophora sp.	Culicidae	*Orthopodomyia signifera*	Chapman et al. (1967)
Pleistophora sp.	Simuliidae	*Psorophora ciliata*	Initial report
Pleistophora sp.	Culicidae	*Toxorhynchites rutulus septentrionalis*	Chapman et al. (1967)
Pleistophora sp.	Culicidae	*Uranotaenia sapphirina*	Initial report
Stempellia lunata	Culicidae	*Culex pilosus*	Hazard & Savage (1970)
Stempellia magna (doubtful parasite of this host)	Culicidae	*Aedes detritus*	Tour et al. (1971)
Stempellia magna (doubtful parasite of this host)	Culicidae	*Aedes sierrensis*	Clark & Fukuda (1967)
Stempellia magna (doubtful parasite of this host)	Culicidae	*Anopheles punctipennis*	Simmers (1974c)
Stempellia magna (doubtful parasite of this host)	Culicidae	*Culex pipiens pipiens*	Kudo (1921)
Stempellia magna	Culicidae	*Culex restuans*	Kudo (1922), Wills & Beaudoin (1965), Chapman et al. (1967), Anderson (1968), Bailey et al. (1967b)
Stempellia magna (doubtful parasite of this host)	Culicidae	*Culiseta inornata*	Simmers (1974c)
Stempellia milleri	Culicidae	*Culex fatigans*	Hazard & Fukuda (1974)
Stempellia milleri	Culicidae	*Culex pipiens pipiens*	Hazard & Fukuda (1974)
Stempellia milleri	Culicidae	*Culex restuans*	Initial report
Stempellia milleri	Culicidae	*Culex salinarius*	Hazard & Fukuda (1974)
Stempellia milleri	Culicidae	*Culex tarsalis*	Hazard & Fukuda (1974)
Stempellia milleri	Culicidae	*Culex territans*	Hazard & Fukuda (1974)
Stempellia rubtosovi	Simuliidae	*Odagmia caucasica*	Issi (1966)
Stempellia simulii	Simuliidae	*Simulium bezzii*	Maurand (1967)

XXVI.   MICROSPORIDA (continued)

Pathogen	Host group	Host	References
Stempellia simulii	Simuliidae	Simulium sp.	Maurand & Manier (1968)
Stempellia tuzetae	Culicidae	Aedes detritus	Tour et al. (1971)
Stempellia sp.	Culicidae	Mansonia dyari	Initial report
Stempellia sp.	Culicidae	Psorophora ferox	Chapman et al. (1967)
Stempellia sp.	Culicidae	Psorophora horrida	Chapman et al. (1969)
Stempellia sp.	Culicidae	Uranotaenia sapphirina	Initial report, Chapman et al. (1969)
Thelohania avacuolata	Simuliidae	Simulium ornatum	Gassouma (1972)
Thelohania barbata (nomen nudum)	Culicidae	Aedes cantans	Anderson (1968)
Thelohania barbata (nomen nudum)	Culicidae	Aedes vexans	Weiser (1961)
Thelohania bertrami	Simuliidae	Simulium ornatum	Gassouma (1972)
Thelohania bracteata	Simuliidae	Eusimulium latipes	Ezenwa (1974c, d)
Thelohania bracteata	Simuliidae	Simulium decorum	Ezenwa (1973)
Thelohania bracteata	Simuliidae	Simulium equinum	Maurand & Manier (1967)
Thelohania bracteata	Simuliidae	Simulium monticola	Maurand (1967)
Thelohania bracteata	Simuliidae	Simulium ornatum	Maurand (1967)
Thelohania bracteata	Simuliidae	Simulium variegatum	Maurand (1967)
Thelohania bracteata	Simuliidae	Simulium venustum	Ezenwa (1974d), Liu et al. (1971), Liu & Davies (1972a, b, c, d)
Thelohania bracteata	Simuliidae	Eusimulium sp.	Briggs (1972)
Thelohania bracteata	Simuliidae	Simulium sp.	Manier & Maurand (1966), Maurand (1966), Maurand & Manier (1968)
Thelohania canningi	Simuliidae	Simulium ornatum	Gassouma (1972), Gassouma & Ellis (1973)
Thelohania fibrata	Simuliidae	Eusimulium latipes	Rubtsov (1966b)
Thelohania fibrata	Simuliidae	Simulium venustum	Steinhaus & Marsh (1962)
Thelohania fibrata	Simuliidae	Eusimulium sp.	Steinhaus & Marsh (1962)
Thelohania fibrata	Simuliidae	Simulium sp.	Briggs (1972), Maurand & Boriux (1969)
Thelohania grassi (doubtful generic placement)	Culicidae	Anopheles maculipennis	Missiroli (1929)
Thelohania minuta	Simuliidae	Simulium ornatum	Gassouma (1972), Gassouma & Ellis (1973)
Thelohania pyriformis (doubtful generic placement)	Culicidae	Anopheles sp.	Kudo (1924c)

XXVI. MICROSPORIDA (continued)

Pathogen	Host group	Host	References
Thelohania simulii	Simuliidae	Simulium ornatum	Gassouma (1972)
Thelohania tabani	Tabanidae	Tabanus atratus	Gingrich (1965)
Thelohania varians	Simuliidae	Eusimulium latipes	Rubtsov (1966b)
Thelohania varians	Simuliidae	Odagmia ornata	Rubtsov (1966b)
Thelohania sp.	Culicidae	Culex apicalis	Kellen et al. (1965)
Thelohania sp. (doubtful generic placement)	Culicidae	Culex pipiens pipiens	Iturbe & Gonzales (1921)
Thelohania sp. (doubtful generic placement)	Culicidae	Culex territans	Kellen et al. (1966a), Chapman et al. (1967)
Thelohania sp.	Simuliidae	Prosimulium flaveantennus	Jamnback (1970)
Thelohania sp.	Simuliidae	Simulium adersi	Briggs (1969)
Thelohania sp.	Simuliidae	Simulium arcticum	Steinhaus & Marsh (1962)
Thelohania sp.	Simuliidae	Simulium canadensis	Steinhaus & Marsh (1962)
Thelohania sp.	Simuliidae	Simulium ornatum	Briggs (1968)
Thelohania sp.	Simuliidae	Simulium piperi	Steinhaus & Marsh (1962)
Thelohania sp.	Simuliidae	Simuliidae	Maitland & Penney (1967), Thomas & Poinar (1973)
Weiseria laurenti	Simuliidae	Prosimulium inflatum	Doby & Saquez (1964)
Weiseria sommeranae	Simuliidae	Gymnopais sp.	Jamnback (1970)
Weiseria spinosa	Culicidae	Culex pipiens pipiens	Gol'berg (1971)
Undetermined microsporidium	Culicidae	Culex secutor	Briggs (1972)
Microsporida	Phlebotominae	Lutzomyia lainsoni	Ward & Killick-Kendrick (1974)
Microsporidia	Simuliidae	Odagmia ornata	Shipitsina (1963)
Microsporida, Adelina sp.	Phlebotominae	Phlebotomus perniciosus Newst.	Rioux et al. (1972)
Microsporidia	Simuliidae	Prosimulium hirtipes	Anderson & Dicke (1960)
Microsporidia	Simuliidae	Simulium arakawae	Steinhaus & Marsh (1962)
Microsporidia	Simuliidae	Simuliidae	Anderson & Dicke (1960), Bobrova (1971)
Microsporidia	Simuliidae	Simulium damnosum	Briggs (1968)
Microsporidia	Simuliidae	Simulium reptans	Shipitsina (1963)
Microsporida (to be described)	Tabanidae	Tabanus lineola	Hazard & Knell (1973)
Microsporida (to be described)	Tabanidae	Tabanus subsimilis	Harlan (1973)
Microsporidia	Simuliidae	Tetanopteryx maculata	Shipitsina (1963)

XXVII.　FUNGI OTHER THAN <u>COELOMOMYCES</u>

Pathogen	Host group	Host	References
<u>Absidia repens</u>	Glossina	<u>Glossina fusca congolensis</u>	Vey (1971)
<u>Actinobacillus</u> Ray-fungus	Acarina mites	<u>Cerheus tegeocranus</u> Oribatid mite	Warren (1947)
<u>Amoebidium parasiticum</u>	Culicidae	<u>Aedes aegypti</u>	Kuno (1973)
<u>Amoebidium parasiticum</u>	Culicidae	<u>Aedes detritus</u>	Manier et al. (1964)
<u>Amoebidium parasiticum</u>	Culicidae	<u>Culex theileri</u>	Manier et al. (1964)
<u>Amoebidium</u> sp.	Simuliidae	<u>Simulium argyreatum</u>	Manier (1964)
<u>Amoebidium</u> sp.	Simuliidae	<u>Simulium fasciatum</u>	Manier (1971)
<u>Aspergillus flavipes</u>	Glossina	<u>Glossina fusca congolensi</u>	Vago & Meynadier (1973)
<u>Aspergillus flavus</u>	Blattidae	<u>Blatella germanica</u>	Henry (1965)
<u>Aspergillus flavus</u>	Tabanidae	<u>Chrysops relictus</u>	Koval & Andreeva (1971), Andreeva (1972)
<u>Aspergillus flavus</u>	Cimicidae	<u>Cimex lectularius</u>	Cockbain & Hastie (1961)
<u>Aspergillus flavus</u>	Musca	<u>Musca domestica</u>	Amonkar & Nair (1965), Beard & Walton (1965), Thomas & Poinar (1973)
<u>Aspergillus flavus</u>	Cimicidae	<u>Paracimex</u> sp.	Usinger (1966)
<u>Aspergillus flavus</u>	Tabanidae	<u>Tabanus autumnalis</u>	Andreeva (1972)
<u>Aspergillus fumigatus</u>	Tabanidae	<u>Chrysops relictus</u>	Koval & Andreeva (1971), Andreeva (1972)
<u>Aspergillus fumigatus</u>	Acarina ticks	Ixodidae	Lipa (1971)
<u>Aspergillus fumigatus</u>	Tabanidae	<u>Tabanus autumnalis</u>	Andreeva (1972)
<u>Aspergillus glaucus</u>	Culicidae	<u>Anopheles</u> sp.	Speer (1927), Galli-Valerio & Rochaz de Jongh (1905)
<u>Aspergillus glaucus</u>	Culicidae	<u>Culex</u> sp.	Galli-Valerio & Rochaz de Jongh (1905)
<u>Aspergillus niger</u>	Culicidae	<u>Anopheles</u> sp.	Galli-Valerio & Rochaz de Jongh (1905), Speer (1927)
<u>Aspergillus niger</u>	Tabanidae	<u>Chrysops relictus</u>	Koval & Andreeva (1971), Andreeva (1972)
<u>Aspergillus niger</u>	Culicidae	<u>Culex</u> sp.	Galli-Valerio & Rochaz de Jongh (1905), Speer (1927)
<u>Aspergillus niger</u>	Tabanidae	<u>Tabanus autumnalis</u>	Andreeva (1972)
<u>Aspergillus ochraceus</u>	Glossina	<u>Glossina</u> sp.	Vey (1974b)
<u>Aspergillus parasiticus</u>	Culicidae	<u>Anopheles subpictus</u>	Hati & Ghosh (1965)

Pathogen	Host group	Host	References
<u>Aspergillus parasiticus</u>	Culicidae	<u>Culex gelidus</u>	Hati & Ghosh (1965)
<u>Aspergillus parasiticus</u>	Culicidae	<u>Culex pipiens fatigans</u>	Hati & Ghosh (1965)
<u>Aspergillus parasiticus</u>	Acarina ticks	<u>Dermacentor marginatus</u>	Samšiňaková et al. (1974)
<u>Aspergillus parasiticus</u>	Acarina ticks	<u>Dermacentor reticulatus</u>	Samšiňaková et al. (1974)
<u>Aspergillus parasiticus</u>	Acarina ticks	<u>Ixodes ricinus</u>	Samšiňaková et al. (1974)
<u>Aspergillus parasiticus</u>	Musca	<u>Musca domestica</u>	Marchionatto (1945)
<u>Aspergillus</u> sp.	Culicidae	<u>Anopheles</u> sp.	Christophers (1952)
<u>Aspergillus</u> sp.	Culicidae	<u>Culex tritaeniorhynchus summorosus</u>	Laird (1959)
<u>Aspergillus</u> sp.	Culicidae	<u>Culex</u> sp.	Speer (1927)
Fungus, <u>Aspergillus</u> sp.	Phlebotominae	<u>Lutzomyia</u> spp.	Hertig & Johnson (1961), Hertig (1964)
<u>Aspergillus</u> sp.	Simuliidae	<u>Simulium damnosum</u>	Lewis (1965)
<u>Beauveria bassiana</u>	Culicidae	<u>Aedes aegypti</u>	Clark et al. (1967, 1968)
<u>Beauveria bassiana</u>	Culicidae	<u>Aedes nigromaculis</u>	Clark et al. (1967, 1968)
<u>Beauveria bassiana</u>	Culicidae	<u>Aedes sierrensis</u>	Clark et al. (1967, 1968)
<u>Beauveria bassiana</u>	Culicidae	<u>Anopheles albimanus</u>	Clark et al. (1967, 1968)
<u>Beauveria bassiana</u>	Culicidae	<u>Anopheles maculipennis</u>	Roubaud & Toumanoff (1930)
<u>Beauveria bassiana</u>	Culicidae	<u>Anopheles quadrimaculatus</u>	Charles (1939)
<u>Beauveria bassiana</u>	Culicidae	<u>Culex apicalis</u>	Dyl'ko (1971)
<u>Beauveria bassiana</u>	Culicidae	<u>Culex exilis</u>	Dyl'ko (1971)
<u>Beauveria bassiana</u> (possibly <u>Entomophthora culicis</u>)	Culicidae	<u>Culex pipiens</u>	Dyé (1905)
<u>Beauveria bassiana</u>	Culicidae	<u>Culex pipiens</u>	Roubaud & Toumanoff (1930), Clark et al. (1967, 1968)
<u>Beauveria bassiana</u>	Culicidae	<u>Culex tarsalis</u>	Clark et al. (1967, 1968)
<u>Beauveria bassiana</u>	Acarina ticks	<u>Dermacentor marginatus</u>	Samšiňaková et al. (1974)
<u>Beauveria bassiana</u>	Acarina ticks	<u>Dermacentor reticulatus</u>	Samšiňaková et al. (1974)

XXVII.   FUNGI OTHER THAN <u>COELOMOMYCES</u> (continued)

Pathogen	Host group	Host	References
<u>Beauveria bassiana</u>	Acarina ticks	<u>Ixodes ricinus</u>	Samšiňaková et al. (1974)
<u>Beauveria bassiana</u>	Acarina ticks	Ixodidae	Lipa (1971)
<u>Beauveria bassiana</u> (=<u>B. cinerea</u> from <u>theobaldiae</u>)	Culicidae	Mosquitos	Morquer (1933)
<u>Beauveria bassiana</u>	Musca	<u>Musca domestica</u>	Dresner (1950)
<u>Beauveria bassiana</u>	Reduviidae	<u>Panstrogylus megistus</u>	Dias & Leão (1967)
<u>Beauveria bassiana</u>	Reduviidae	<u>Triatoma infestans</u>	Dias & Leão (1967)
<u>Beauveria bassiana</u>	Reduviidae	<u>Triatoma vitticeps</u>	Dias & Leão (1967)
<u>Beauveria densa</u>	Tabanidae	<u>Chrysops relictus</u>	Andreeva (1972), Koval & Andreeva (1971)
<u>Beauveria densa</u>	Tabanidae	<u>Tabanus autumnalis</u>	Andreeva (1972)
<u>Beauveria tenella</u>	Culicidae	<u>Aedes aegypti</u>	Pinnock et al. (1973), Sanders (1972)
<u>Beauveria tenella</u>	Culicidae	<u>Aedes dorsalis</u>	Pinnock et al. (1973), Sanders (1972)
<u>Beauveria tenella</u>	Culicidae	<u>Aedes hexodontus</u>	Pinnock et al. (1973)
<u>Beauveria tenella</u>	Culicidae	<u>Aedes sierrensis</u>	Pinnock et al. (1973), Sanders (1972)
<u>Beauveria tenella</u>	Culicidae	<u>Culex pipiens</u>	Pinnock et al. (1973), Sanders (1972)
<u>Beauveria tenella</u>	Culicidae	<u>Culex tarsalis</u>	Pinnock et al. (1973)
<u>Beauveria tenella</u>	Culicidae	<u>Culeseta incidens</u>	Pinnock et al. (1973)
<u>Beauveria tenella</u>	Culicidae	<u>Culiseta</u> sp.	Sanders (1972)
<u>Beauveria tenella</u>	Acarina ticks	<u>Dermacentor marginatus</u>	Samšiňaková et al. (1974)
<u>Beauveria tenella</u>	Acarina ticks	<u>Dermacentor reticulatus</u>	Samšiňaková et al. (1974)
<u>Beauveria tenella</u>	Acarina ticks	<u>Ixodes ricinus</u>	Samšiňaková et al. (1974)
<u>Caphalosporium</u>, possibly <u>C. coccorum</u>	Culicidae	<u>Culex pipiens</u>	Service (1969)
<u>Carouxella scalaris</u>	Ceratopogonidae	<u>Dasyhelea lithotelmatica</u>	Manier et al. (1961)
<u>Cephalosporium coccorum</u>	Acarina ticks	<u>Dermacentor marginatus</u>	Samšiňaková et al. (1974)
<u>Cephalosporium coccorum</u>	Acarina ticks	<u>Dermacentor reticulatus</u>	Samšiňaková et al. (1974)
<u>Cephalosporium coccorum</u>	Acarina ticks	<u>Ixodes ricinus</u>	Samšiňaková et al. (1974)
<u>Cladosporium</u> ?	Culicidae	<u>Culex pipiens fatigans</u>	Laird (1959)

XXVII.   FUNGI OTHER THAN <u>COELOMOMYCES</u> (continued)

Pathogen	Host group	Host	References
<u>Coelomomyces citerrii</u> Leão Pedroso	Phlebotominae	<u>Lutzomyia</u> sp. or spp.	Leão & Pedroso (1965)
<u>Coelomomyces milkoi</u>	Tabanidae	<u>Chrysops relictus</u>	Andreeva (1972)
<u>Coelomomyces milkoi</u>	Tabanidae	<u>Tabanus autumnalis</u>	Andreeva (1972)
<u>Coelomomyces</u> sp.	Simuliidae	<u>Simulium ornatum nitidifrons</u>	Briggs (1967)
<u>Coelomycidium simulii</u>	Simuliidae	<u>Simulium venustum</u>	Ezenwa (1974d)
<u>Coelomycidium simulii</u>	Simuliidae	<u>Simulium vittatum</u>	Ezenwa (1974d)
<u>Coelomycidium simulii</u>	Simuliidae	<u>Simulium</u> sp.	Maurand & Manier (1968), Loubes & Manier (1970)
<u>Coelomycidium simulii</u>	Simuliidae	Simuliidae	Rubtsov (1969)
<u>Coelomycidium</u> sp.	Culicidae	<u>Culex modestus</u>	Shcherban' & Gol'berg (1971)
<u>Coelomycidium</u> sp.	Culicidae	<u>Culex pipiens pipiens</u>	Shcherban & Gol'berg (1971)
<u>Conidiobolus coronatus</u> (=<u>Entomophthora coronata</u>)	Culicidae	<u>Aedes taeniorhynchus</u>	Lowe & Kennel (1972)
<u>Conidiobolus coronatus</u> (=<u>Entomophthora coronata</u>)	Culicidae	<u>Culex pipiens fatigans</u> (=<u>C. pipiens quinquefasciatus</u>)	Lowe et al. (1968)
<u>Conidiobolus coronatus</u> (=<u>Entomophthora coronata</u>)	Culicidae	<u>Culex pipiens fatigans</u> (=<u>C. pipiens quinquefasciatus</u>)	Lowe & Kennel (1972)
<u>Culicinomyces clavosporus</u>	Culicidae	<u>Aedes atropalpus epactius</u>	Couch et al. (1974)
<u>Culicinomyces clavosporus</u>	Culicidae	<u>Anopheles punctipennis</u>	Couch et al. (1974)
<u>Culicinomyces clavosporus</u>	Culicidae	<u>Anopheles quadrimaculatus</u>	Couch et al. (1974)
<u>Culicinomyces clavosporus</u>	Culicidae	<u>Anopheles stephensi</u>	Couch et al. (1974)
<u>Culicinomyces clavosporus</u>	Culicidae	<u>Culex erraticus</u>	Couch et al. (1974)
<u>Culicinomyces clavosporus</u>	Culicidae	<u>Culex pipiens fatigans</u> (=<u>C. pipiens quinquefasciatus</u>)	Couch et al. (1974)
<u>Culicinomyces clavosporus</u>	Culicidae	<u>Culex restuans</u>	Couch et al. (1974)
<u>Culicinomyces clavosporus</u>	Culicidae	<u>Culex territans</u>	Couch et al. (1974)
<u>Culicinomyces clavosporus</u>	Culicidae	<u>Culiseta melanura</u>	Couch et al. (1974)
<u>Culicinomyces clavosporus</u>	Culicidae	<u>Psorophora confinnis</u>	Couch et al. (1974)
<u>Culicinomyces clavosporus</u>	Culicidae	<u>Uranotaenia sapphirina</u>	Couch et al. (1974)
<u>Culicinomyces</u> sp.	Culicidae	<u>Aedes australis</u>	Sweeney (1952a)
<u>Culicinomyces</u> sp.	Culicidae	<u>Aedes rupestris</u>	Sweeney & Panter (1974)
<u>Culiconomyces</u> sp.	Culicidae	<u>Aedes</u> sp.	Sweeney et al. (1973)

XXVII.   FUNGI OTHER THAN <u>COELOMOMYCES</u> (continued)

Pathogen	Host group	Host	References
<u>Culicinomyces</u> sp.	Culicidae	<u>Anopeheles amictus hilli</u>	Sweeney (1975a), Sweeney et al. (1973)
<u>Culicinomyces</u> sp.	Culicidae	<u>Anopheles annulipes</u>	Sweeney (1975a)
<u>Culicinomyces</u> sp.	Culicidae	<u>Culex pipiens fatigans</u>	Sweeney (1975a, b)
<u>Culicinomyces</u> sp.	Culicidae	<u>Culex</u> sp.	Sweeney et al. (1973)
<u>Grubyella ochoterenai</u>	Cerato-pogonidae	<u>Culicoides phlebotomus</u>	Ciferri (1929)
<u>Entomophthora aquatica</u>	Culicidae	<u>Aedes canadensis</u>	Anderson & Ringo (1969)
<u>Entomophthora aquatica</u>	Culicidae	<u>Culiseta morsitans</u>	Anderson & Ringo (1969)
<u>Entomophthora conglomerata</u> (=<u>Empusa conglomerata</u>, <u>Empusa thaxteri</u>)	Culicidae	<u>Aedes communis</u> (=<u>Culex nemorosus</u>)	Lakon (1919), Sorokin (1877)
<u>Entomophthora conglomerata</u> (=<u>Empusa conglomerata</u>, <u>Empusa thaxteri</u>)	Culicidae	<u>Culex pipiens</u>	Brumpt (1941), Lakon (1919)
<u>Entomophthora conglomerata</u>	Culicidae	<u>Culex pipiens</u>	Il'chenko (1968)
<u>Entomophthora conglomerata</u>	Culicidae	<u>Culex pipiens</u>	Kupriyanova (1966a, b)
<u>Entomophthora conglomerata</u>	Culicidae	<u>Culex pipiens</u>	Gol'berg (1969)
<u>Entomophthora conglomerata</u> (=<u>Empusa conglomerata</u>, <u>Empusa thaxteri</u>)	Culicidae	<u>Culex</u> sp.	Thaxter (1888)
<u>Entomophthora culicis</u>	Culicidae	<u>Aedes detrius</u>	Marshall (1938)
<u>Entomophthora culicis</u>	Culicidae	<u>Anopheles hispaniola</u>	López-Neyra & Guardiola Mira (1938)
<u>Entomophthora culicis</u>	Culicidae	<u>Anopheles maculipennis</u>	López-Neyra & Guardiola Mira (1938)
<u>Entomophthora culicis</u>	Culicidae	<u>Culex pipiens</u>	Braun (1855), López-Neyra & Guardiola Mira (1938), Marshall (1938), Speer (1927)
<u>Entomophthora culicis</u>	Culicidae	<u>Culex pipiens</u>	Gol'berg (1973)
<u>Entomophthora culicis</u>	Culicidae	<u>Culex</u> sp.	Nowakowski (1883), Thaxter (1888)
<u>Entomophthora culicis</u>	Culicidae	<u>Culex</u> sp.	Christophers (1952)
<u>Entomophthora</u> (<u>Entomophaga</u>) <u>destruens</u>	Culicidae	<u>Culex pipiens</u>	Novák (1965, 1967, 1971), Service (1969), Weiser & Batko (1966), Weiser & Novák (1964)
<u>Entomophthora destruens</u>	Culicidae	<u>Culex pipiens</u>	Gol'berg (1973)
<u>Entomophthora fresenius</u>	Stomoxys	<u>Stomoxys calcitrans</u>	Roubaud (1911)
<u>Entomophthora gracilis</u>	Culicidae	Mosquitos	Lakon (1919), Picard (1914), Thaxter (1888)

XXVII.   FUNGI OTHER THAN <u>COELOMOMYCES</u> (continued)

Pathogen	Host group	Host	References
<u>Entomophthora henrici</u>	Culicidae	<u>Culex pipiens</u>	Brumpt (1941), Molliard (1918)
<u>Entomophthora kansana</u>	Musca	<u>Musca domestica</u>	Hutchison (1962)
<u>Entomophthora muscae</u>	Musca	<u>Musca domestica</u>	Güssow (1913), Hesse (1913), Steinhaus (1951), Steinhaus & Marsh (1962), Thomas & Poinar (1973), Yeager (1939)
<u>Entomophthora muscae</u>	Glossina	<u>Glossina palpalis</u>	Vanderyst (1923)
<u>Entomophthora muscae</u>	Glossina	<u>Glossina</u> sp.	Roubaud (1911)
<u>Entomophthora</u> (=<u>Empusa</u>) <u>muscae</u> Cohn	Stomoxys	<u>Stomoxys calcitrans</u>	Surcouf (1923)
<u>Entomophthora ovispora</u>	Cerato-pogonidae	Ceratopogonidae	Gol'berg (1969)
<u>Entomophthora papatasii</u> (=<u>Empusa papatasii</u>)	Phlebotominae	<u>Phlebotomus papatasii</u>	Marett (1915), Larrousse (1921)
<u>Entomophthora papilliata</u> (=<u>Lamia apiculata</u>)	Culicidae	Mosquitos	Lakon (1919), Marshall (1952)
<u>Entomophthora rhizospora</u>	Culicidae	Mosquitos	Brumpt (1941), Howard et al. (1912)
<u>Entomophthora schroeteri</u> (=<u>E. rimosa</u>)	Culicidae	Mosquitos	Brumpt (1941), Schröter (1886)
<u>Entomophthora sphaerosperma</u> (=<u>Tarichium sphaerospermum</u>, <u>Empusa radicans</u>, <u>Entomophthora radicans</u>, <u>Entomophthora phytonomi</u>)	Culicidae	Mosquitos	Brumpt (1941), Fresenius (1856), Howard et al. (1912), Thaxter (1888)
<u>Entomophthora variabilis</u>	Culicidae	Mosquitos	Lakon (1919), Thaxter (1888)
<u>Entomophthora</u> sp.	Culicidae	<u>Culex pipiens</u>	Gol'berg (1969, 1970a, b, 1973)
<u>Entomophthora</u> spp.	Culicidae	<u>Culex pipiens</u>	Teernstra-Eeken & Engel (1967)
<u>Entomophthora</u> ?	Culicidae	<u>Culex pipiens</u>	Oda & Kuhlon (1973)
<u>Entomophthora</u> sp.	Phlebotominae	<u>Phlebotomus ariasi</u>	Rioux & Golvan (1964), Rioux et al. (1966)
<u>Entomophthora</u> sp.	Simulidae	<u>Simulium damnosum</u>	Lewis (1965)
<u>Fusarium avenaceum</u>	Tabanidae	<u>Chrysops relictus</u>	Andreeva (1972), Koval & Andreeva (1971)
<u>Fusarium avenaceum</u>	Tabanidae	<u>Tabanus autumnalis</u>	Andreeva (1972)
<u>Fusarium oxysporum</u>	Culicidae	<u>Aedes detrius</u>	Hasan & Vago (1972)
<u>Fusarium oxysporum</u>	Culicidae	<u>Culex pipiens pipiens</u>	Hasan & Vago (1972)

XXVII. FUNGI OTHER THAN <u>COELOMOMYCES</u> (continued)

Pathogen	Host group	Host	References
<u>Fusarium semitectum</u> var. majus	Glossina	<u>Glossina</u> sp.	Doidge (1950)
<u>Fusarium solani agrillacea</u>	Tabanidae	<u>Chrysops relictus</u>	Andreeva (1972), Koval & Andreeva (1971)
<u>Fusarium solani agrillacea</u>	Tabanidae	<u>Tabanus autumnalis</u>	Andreeva (1972)
<u>Fusarium</u> sp. ?	Culicidae	<u>Aedes aegypti</u>	Macfie (1917)
<u>Harpella melusinae</u>	Simuliidae	<u>Austrosimulium longicorne</u>	Crosby (1974)
<u>Harpella melusinae</u>	Simuliidae	<u>Austrosimulium multicorne</u>	Crosby (1974)
<u>Harpella melusinae</u>	Simuliidae	<u>Austrosimulium stewartense</u>	Crosby (1974)
<u>Harpella melusinae</u>	Simuliidae	<u>Austrosimulium tillvardianum</u>	Crosby (1974)
<u>Harpella melusinae</u>	Simuliidae	<u>Austrosimulium ungulatum</u>	Crosby (1974)
<u>Harpella melusinae</u>	Simuliidae	<u>Austrosimulium vexans</u>	Crosby (1974)
<u>Harpella melusinae</u>	Simuliidae	<u>Cnephia mutata</u>	Frost & Manier (1971)
<u>Harpella melusinae</u>	Simuliidae	<u>Prosimulium</u> sp.	Frost & Manier (1971)
<u>Harpella melusinae</u>	Simuliidae	<u>Simulium aureum</u>	Manier (1964)
<u>Harpella melusinae</u>	Simuliidae	<u>Simulium bezzii</u>	Manier (1969)
<u>Harpella melusinae</u>	Simuliidae	<u>Simulium equinum</u>	Moss (1970), Manier (1969)
<u>Harpella melusinae</u>	Simuliidae	<u>Simulium euryadminiculum</u>	Frost & Manier (1971)
<u>Harpella melusinae</u>	Simuliidae	<u>Simulium fasciatum</u>	Manier (1971)
<u>Harpella melusinae</u>	Simuliidae	<u>Simulium ornatum</u>	Manier (1969)
<u>Harpella melusinae</u>	Simuliidae	<u>Simulium variegatum</u>	Manier (1969)
<u>Harpella melusinae</u>	Simuliidae	<u>Simulium vittatum</u>	Frost & Manier (1971)
<u>Herpomyces stylopygae</u>	Blattidae	<u>Blatta orientalis</u>	Tavares (1966)
<u>Lagenidium giganteum</u> (=<u>L. culicidum</u>)	Culicidae	<u>Aedes aegypti</u>	Couch & Romney (1973), McCray et al. (1973a), Umphlett (1973)
<u>Lagenidium giganteum</u>	Culicidae	<u>Aedes atropalpus epactius</u>	Couch & Romney (1973)
<u>Lagenidium giganteum</u> (=<u>L. culicidum</u>)	Culicidae	<u>Aedes mediovittatus</u>	McCray et al. (1973a), Umphlett (1973)
<u>Lagenidium giganteum</u>	Culicidae	<u>Aedes nigromaculis</u>	McCray et al. (1973b)
<u>Lagenidium giganteum</u>	Culicidae	<u>Aedes polynesiensis</u>	Couch & Romney (1973)
<u>Lagenidium giganteum</u> (=<u>L. culicidum</u>)	Culicidae	<u>Aedes sollicitans</u>	McCray et al. (1973a), Umphlett (1973)

XXVII.   FUNGI OTHER THAN <u>COELOMOMYCES</u> (continued)

Pathogen	Host group	Host	References
<u>Lagenidium giganteum</u> (=<u>L. culicidum</u>)	Culicidae	<u>Aedes taeniorhynchus</u>	McCray et al. (1973a), Umphlett (1975)
<u>Lagenidium giganteum</u> (=<u>L. culicidum</u>)	Culicidae	<u>Aedes triseriatus</u>	McCray et al. (1973a), Umphlett (1973)
<u>Lagenidium giganteum</u>	Culicidae	<u>Aedes triseriatus</u>	Couch & Romney (1973)
<u>Lagenidium giganteum</u>	Culicidae	<u>Anopheles punctipennis</u>	Couch & Romney (1973)
<u>Lagenidium giganteum</u>	Culicidae	<u>Anopheles quadrimaculatus</u>	Couch & Romney (1973)
<u>Lagenidium giganteum</u>	Culicidae	<u>Anopheles stephensi</u>	Couch & Romney (1973)
<u>Lagenidium giganteum</u>	Culicidae	<u>Anopheles</u> sp.	Umphlett & Huang (1972), Umphlett (1973), Umphlett & McCray (1974)
<u>Lagenidium giganteum</u>	Culicidae	<u>Culiseta incidens</u>	Couch & Romney (1973)
<u>Lagenidium giganteum</u> (=<u>L. culicidum</u>)	Culicidae	<u>Culex nigripalpus</u>	McCray et al. (1973a), Umphlett (1973)
<u>Lagenidium giganteum</u>	Culicidae	<u>Culex pipiens fatigans</u> (=<u>C. pipiens quinquefasciatus</u>)	Couch & Romney (1973)
<u>Lagenidium giganteum</u> (=<u>L. culicidum</u>)	Culicidae	<u>Culex pipiens fatigans</u> (=<u>C. pipiens quinquefasciatus</u>)	McCray et al. (1973a), Umphlett (1973)
<u>Lagenidium giganteum</u>	Culicidae	<u>Culex restuans</u>	Umphlett & Huang (1972) Umphlett (1973), Couch & Romney (1973)
<u>Lagenidium giganteum</u>	Culicidae	<u>Culex tarsalis</u>	McCray et al. (1973b)
<u>Lagenidium giganteum</u> (=<u>L. culicidum</u>)	Culicidae	<u>Culex tarsalis</u>	Couch & Romney (1973), McCray et al. (1973a), Umphlett (1973)
<u>Lagenidium giganteum</u>	Culicidae	<u>Culex</u> sp.	Couch (1935, 1960)
<u>Lagenidium giganteum</u>	Culicidae	<u>Psorophora</u> sp.	Umphlett & Hugan (1972) Umphlett (1973)
<u>Lagenidium</u> sp.	Culicidae	<u>Armigeres dentatus</u>	Mattingly (1972a, b)
<u>Macrosporium</u> sp.	Musca	<u>Musca domestica</u>	Damodar et al. (1964)
<u>Metarrhizium anisopliae</u>	Culicidae	<u>Aedes aegypti</u>	Roberts (1970, 1974), Roberts & Mouchet (unpublished)
<u>Metarrhizium anisopliae</u>	Culicidae	<u>Aedes atropalpus</u>	Roberts (1970, 1974)
<u>Metarrhizium anisopliae</u>	Culicidae	<u>Aedes polynesiensis</u>	Mouchet (unpublished)
<u>Metarrhizium anisopliae</u>	Culicidae	<u>Aedes sollicitans</u>	Roberts (1970, 1974)
<u>Metarrhizium anisopliae</u>	Culicidae	<u>Aedes taeniorhynchus</u>	Roberts (1970, 1974)
<u>Metarrhizium anisopliae</u>	Culicidae	<u>Anopheles albimanus</u>	Roberts (1970, 1974)
<u>Metarrhizium anisopliae</u>	Culicidae	<u>Anopheles funestus</u>	Roberts (unpublished)

XXVII.   FUNGI OTHER THAN <u>COELOMOMYCES</u> (continued)

Pathogen	Host group	Host	References
<u>Metarrhizium anisopliae</u>	Culicidae	<u>Anopheles gambiae</u>	Roberts (unpublished)
<u>Metarrhizium anisopliae</u>	Culicidae	<u>Anopheles quadrimaculatus</u>	Roberts (1970, 1974)
<u>Metarrhizium anisopliae</u>	Culicidae	<u>Anopheles rufipes</u>	Roberts (unpublished)
<u>Metarrhizium anisopliae</u>	Culicidae	<u>Anopheles stephensi</u>	Roberts (1967, 1970, 1974)
<u>Metarrhizium anisopliae</u>	Tabanidae	<u>Chrysops relictus</u>	Andreeva (1972)
<u>Metarrhizium anisopliae</u>	Culicidae	<u>Culex pipiens fatigans</u>	Roberts & Mouchet (unpublished)
<u>Metarrhizium anisopliae</u>	Culicidae	<u>Culex pipiens pipiens</u>	Roberts (1967, 1970, 1974)
<u>Metarrhizium anisopliae</u>	Culicidae	<u>Culex restuans</u>	Roberts (1970, 1974)
<u>Metarrhizium anisopliae</u>	Culicidae	<u>Culiseta inornata</u>	Roberts (1970, 1974)
<u>Metarrhizium anisopliae</u>	Tabanidae	<u>Tabanus autumnalis</u>	Andreeva (1972)
<u>Monosporella unicuspidata</u>	Cerato-pogonidae	<u>Dasyhelea obscura</u>	Keilin (1920a, 1921a, 1927)
<u>Mucor hiemalis</u>	Blattidae	<u>Blabera fusca</u>	Heitor (1962)
<u>Mucor stolonifera</u>	Culicidae	<u>Anopheles</u> sp.	Bačinskij (1926)
<u>Mucor</u> sp.	Tabanidae	<u>Chrysops italicus</u>	Koval & Andreeva (1971)
<u>Oidium lactis</u>	Culicidae	<u>Anopheles</u> sp.	Bačinskij (1926)
<u>Oidium lactis</u>	Culicidae	<u>Culex</u> sp.	Bačinskij (1926)
<u>Paecilomyces fumosoroseus</u>	Acarina ticks	<u>Dermacentor marginatus</u>	Samšiňaková et al. (1974)
<u>Paecilomyces fumosoroseus</u>	Acarina ticks	<u>Dermacentor reticulatus</u>	Samšiňaková et al. (1974)
<u>Paecilomyces fumosoroseus</u>	Acarina ticks	<u>Ixodes ricinus</u>	Samšiňaková et al. (1974)
<u>Paramoebidium chattoni</u>	Simuliidae	<u>Simulium equinum</u>	Moss (1970), Manier (1969)
<u>Penella hovassi</u>	Simuliidae	<u>Prosimulium</u> sp.	Frost & Manier (1971)
<u>Panella hovassi</u>	Simuliidae	<u>Simulium monticola</u>	Frost & Manier (1971), Manier (1969)
<u>Panella chattoni</u>	Simuliidae	<u>Simulium ornatum</u>	Manier (1969)
<u>Panella hovassi</u>	Simuliidae	<u>Simulium vittatum</u>	Frost & Manier (1971)
<u>Penicillium glaucum</u>	Culicidae	<u>Anopheles</u> sp.	Bačinskij (1926)
<u>Penicillium glaucum</u>	Culicidae	<u>Culex</u> sp.	Bačinskij (1926)
Fungus, <u>Penicillium glaucum</u>	Phlebotominae	<u>Phlebotomus papatasii</u>	Zotov (1930)
<u>Penicillium insectivora</u>	Acarina ticks	Ixodidae	Lipa (1971)

XXVII.   FUNGI OTHER THAN <u>COELOMOMYCES</u> (continued)

Pathogen	Host group	Host	References
<u>Penicillium jant hinellum</u>	Glossina	<u>Glossina fusca congolensis</u>	Vago & Meynadier (1973)
<u>Penicillium lilacinum</u>	Glossina	<u>Glossina fusca congolensis</u>	Vey (1971)
<u>Polyscytalum</u> sp.	Culicidae	<u>Psorophora howardii</u>	Martini (1920), Speer (1927)
<u>Polyscytalum</u> sp.	Culicidae	<u>Psorophora lutzii</u>	Martini (1920)
<u>Polyscytalum</u>	Culicidae	<u>Psorophora</u> sp.	Howard et al. (1912)
<u>Pythium</u> sp. (near <u>P. adhaerens</u>)	Culicidae	<u>Aedes sierrensis</u>	Clark et al. (1966)
<u>Pythium</u> sp. (near <u>P. adhaerens</u>)	Culicidae	<u>Aedes triseriatus</u>	Clark et al. (1966)
<u>Pythium</u> sp. (near <u>P. adhaerens</u>)	Culicidae	<u>Anopheles freeborni</u>	Clark et al. (1966)
<u>Pythium</u> sp. (near <u>P. adhaerens</u>)	Culicidae	<u>Culex tarsalis</u>	Clark et al. (1966)
<u>Pythium</u> sp. (near <u>P. adhaerens</u>)	Culicidae	<u>Culiseta incidens</u>	Clark et al. (1966)
<u>Pythium</u> sp. (near <u>P. adhaerens</u>)	Culicidae	<u>Culiseta inornata</u>	Clark et al. (1966)
<u>Pythium</u> sp. (near <u>P. adhaerens</u>)	Culicidae	<u>Orthopodnyia californica</u>	Clark et al. (1966)
<u>Pythium</u> sp. (near <u>P. adhaerens</u>)	Culicidae	<u>Uranotaenia anhydor</u>	Clark et al. (1966)
<u>Rubetella inopinata</u>	Cerato-pogonidae	<u>Dasyhelea lithotelmatica</u>	Manier et al. (1961), Coluzzi (1966)
<u>Saprolegnia declina</u>	Culicidae	<u>Aedes berlandi</u>	Rioux & Achard (1956)
<u>Saprolegnia monica</u>	Culicidae	<u>Culex pipiens fatigans</u>	Hamlyn-Harris (1932)
<u>Saprolegnia</u> sp.	Culicidae	<u>Aedes geniculatus</u>	Marshall (1938)
<u>Saprolegnia</u> sp.	Culicidae	<u>Aedes rusticus</u>	Marshall (1938)
<u>Saprolegnia</u> sp.	Culicidae	<u>Aedes</u> sp.	Jettmar (1957)
<u>Saprolegnia</u> sp.	Culicidae	<u>Aedes</u> (<u>Stegomyia</u>) sp.	Martini (1920)
<u>Saprolegnia</u> sp.	Culicidae	<u>Anopheles</u> sp.	Jettmar (1947)
<u>Saprolegnia</u> sp.	Culicidae	<u>Culiseta annulata</u>	Marshall (1938)
<u>Saprolegnia</u> sp.	Culicidae	<u>Culiseta morsitans</u>	Marshall (1938)
Saprolegniaceae ?	Culicidae	<u>Anopheles maculipennis</u>	Chorine & Baranoff (1929)
<u>Smittium culicis</u>	Culicidae	<u>Aedes aegypti</u>	Williams & Lichtwardt (1972)
<u>Smittium culisetae</u>	Culicidae	<u>Aedes albopictus</u>	Williams & Lichtwardt (1972)

XXVII.   FUNGI OTHER THAN <u>COELOMOMYCES</u> (continued)

Pathogen	Host group	Host	References
<u>Smittium culicis</u>	Culicidae	<u>Aedes berlandi</u>	Manier (1969)
<u>Smittium culicis</u>	Culicidae	<u>Aedes caspius</u>	Manier (1969)
<u>Smittium culicis</u>	Culicidae	<u>Aedes detrius</u>	Manier (1969)
<u>Smittium culicis</u>	Culicidae	<u>Aedes geniculatus</u>	Manier (1969)
<u>Smittium</u> (=<u>Rubetella</u>) <u>culicis</u>	Culicidae	<u>Aedes melanimon</u>	Clark et al. (1963)
<u>Smittium culicis</u>	Culicidae	<u>Aedes stricticus</u>	Williams & Lichtwardt (1972)
<u>Smittium culicis</u>	Culicidae	<u>Anopheles atroparvus atroparvus</u>	Manier (1969)
<u>Smittium culicis</u>	Culicidae	<u>Anopheles claviger</u>	Manier (1969)
<u>Smittium culicis</u>	Culicidae	<u>Anopheles plumbeus</u>	Manier (1969)
<u>Smittium culicis</u>	Culicidae	<u>Culex hortensis</u>	Manier (1969)
<u>Smittium</u> (=<u>Orphella</u>) <u>culicis</u>	Culicidae	<u>Culex hortensis</u>	Tuzet & Manier (1947)
<u>Smittium culicis</u>	Culicidae	<u>Culex modestus</u>	Manier (1969)
<u>Smittium culicis</u>	Culicidae	<u>Culex pipiens pipiens</u>	Manier (1969)
<u>Smittium</u> (=<u>Rubetella</u>) <u>culicis</u>	Culicidae	<u>Culex theileri</u>	Manier et al. (1964)
<u>Smittium culicis</u>	Culicidae	<u>Culiseta</u> (=<u>Theobaldia</u>) <u>annulata</u>	Manier (1969)
<u>Smittium culicis</u>	Culicidae	<u>Uranotaenia unguiculata</u>	Manier (1969)
<u>Smittium culisetae</u>	Culicidae	<u>Aedes aegypti</u>	Williams & Lichtwardt (1972)
<u>Smittium culisetae</u>	Culicidae	<u>Aedes triseriatus</u>	M. E. Chapman (cited in Williams & Lichtwardt, 1972)
<u>Smittium culisetae</u>	Culicidae	<u>Aedes vexans</u>	Williams & Lichtwardt (1972)
<u>Smittium culisetae</u>	Culicidae	<u>Aedes</u> sp.	Williams & Lichtwardt (1972)
<u>Smittium culisetae</u>	Culicidae	<u>Culiseta impatiens</u>	Lichtwardt (1964), Williams & Lichtwardt (1972)
<u>Smittium culisetae</u>	Culicidae	<u>Culiseta incidens</u>	Lichtwardt (1964)
<u>Smittium culisetae</u>	Culicidae	<u>Culiseta inornata</u>	Lichtwardt (1964)
<u>Smittium</u> (=<u>Rubetella</u>) <u>inopinata</u>	Culicidae	<u>Aedes aegypti</u>	Coluzzi (1964)
<u>Smittium</u> (=<u>Rubetella</u>) <u>inopinata</u>	Culicidae	<u>Aedes vittatus</u>	Coluzzi (1964)
<u>Smittium</u> (=<u>Rubetella</u>) <u>inopinata</u>	Culicidae	<u>Anopheles gambiae</u>	Coluzzi (1964)

XXVII.   FUNGI OTHER THAN <u>COELOMYCES</u> (continued)

Pathogen	Host group	Host	References
<u>Smittium</u> (=<u>Rubetella</u>) <u>inopinata</u>	Culicidae	<u>Culex hortensis</u>	Coluzzi (1964)
<u>Smittium</u> (=<u>Rubetella</u>) <u>inopinata</u>	Culicidae	<u>Culex pipiens</u>	Coluzzi (1964)
<u>Smittium</u> (=<u>Rubetella</u>) <u>inopinata</u>	Culicidae	<u>Culiseta longiareolata</u>	Coluzzi (1964)
<u>Smittium simulii</u>	Culicidae	<u>Aedes aegypti</u>	Williams & Lichtwardt (1972)
<u>Smittium simulii</u>	Culicidae	<u>Aedes triseriatus</u>	M. E. Chapman (cited in Williams & Lichtwardt, 1972)
<u>Smittium simulii</u>	Simuliidae	<u>Simulium argus</u>	Lichtwardt (1964)
<u>Smittium simulii</u>	Simuliidae	<u>Simulium aureum</u>	Manier (1964)
<u>Smittium simulii</u>	Simuliidae	<u>Simulium bezzii</u>	Manier (1969)
<u>Smittium simulii</u>	Simuliidae	<u>Simulium ornatum</u>	Manier (1969)
<u>Smittium simulii</u>	Simuliidae	<u>Stipella vigilans</u>	Manier (1969)
<u>Smittium</u> (=<u>Rubetella</u>) sp.	Culicidae	<u>Culiseta incidens</u>	Clark et al. (1963)
<u>Smittium</u> sp.	Simuliidae	<u>Simulium equinum</u>	Moss (1970)
<u>Spicaria</u> sp.	Culicidae	<u>Anopheles quadrimaculatus</u>	Brown (1949)
<u>Stipella vigilans</u>	Simuliidae	<u>Simulium bezzii</u>	Manier (1969)
<u>Stipella vigilans</u>	Simuliidae	<u>Simulium ornatum</u>	Manier (1969)
<u>Stipella vigilans</u>	Simuliidae	<u>Stipella vigilans</u>	Manier (1969)
<u>Stipella</u> sp.	Simuliidae	<u>Simulium equinum</u>	Moss (1970)
<u>Torrubiella</u> sp. (?)	Acarina ticks	Unidentified tick	Steinhaus & Marsh (1962)
<u>Trenomyces helveticus</u>	Anoplura Mallophaga	<u>Bruelia cyclothorax</u>	Lunkashu (1970)
<u>Trenomyces helveticus</u>	Anoplura Mallophaga	<u>Bruelia gracilis</u>	Lunkashu (1970)
<u>Trenomyces helveticus</u>	Anoplura Mallophaga	<u>Bruelia nebulosa</u>	Lunkashu (1970)
<u>Trenomyces helveticus</u>	Anoplura Mallophaga	<u>Bruelia uniinosa</u>	Lunkashu (1970)
<u>Trenomyces helveticus</u>	Anoplura Mallophaga	<u>Picicola coutiguus</u>	Lunkashu (1970)
<u>Trenomyces helveticus</u>	Anoplura Mallophaga	<u>Sturnidoecus ruficeps</u>	Lunkashu (1970)
<u>Trenomyces histophtorus</u>	Anoplura Mallophaga	<u>Gonides gigas</u>	Chatton & Picard (1909)

XXVII.   FUNGI OTHER THAN <u>COELOMOMYCES</u> (continued)

Pathogen	Host group	Host	References
<u>Trenomyces histophtorus</u>	Anoplura Mallophaga	<u>Menopon gallinae</u>	Trinchieri (1910), Chatton & Picard (1909), Spegazzini (1917)
<u>Trenomyces platensis</u>	Anoplura Mallophaga	<u>Saemundssonia</u> sp.	Spegazzini (1917)
<u>Trenomyces</u> (5 species)	Anoplura Mallophaga	Many species of Mallophaga	Thaxter (1912)
<u>Trichoderma viride</u>	Culicidae	Mosquitos	Steinhaus (1949)
<u>Trichophyton</u> sp. ?	Culicidae	<u>Anopheles coustani</u>	Dyé (1905)
<u>Trichophyton</u> sp. ?	Culicidae	Mosquitos	Christophers (1952), Liston (1901)
Ascomycete	Glossina	<u>Glossina fuscipes</u>	Carpenter (1912)
Basidiomycete	Culicidae	<u>Aedes sollicitans</u>	Chapman et al. (1967, 1969)
Fungus	Culicidae	<u>Anopheles maculipennis</u>	Léon (1924)
"Fungal hyphae"	Cerato-pogonidae	<u>Culicoides furens</u>	Lewis (1958)
"Fungal hyphae" (lumen of oesophageal diverticulum; ? fungus ingested with raisin juice)	Cerato-pogonidae	<u>Culicoides nubeculosus</u>	Megahed (1956)
Facultative parasitic and saprophytic fungi	Acarina ticks	<u>Dermacentor reticulatus</u>	Samšiňaková et al. (1974)
Facultative parasitic and saprophytic fungi	Acarina ticks	<u>Dermacentor marginatus</u>	Samšiňaková et al. (1974)
Fungus	Glossina	<u>Glossina morsitans</u>	Nash (1933), Lester (1934)
Fungus	Glossina	<u>Glossina tachinoides</u>	Lester (1934)
Fungi Imperfecti	Glossina	<u>Glossina fuscipes</u>	Carpenter (1912)
Fungal mycelium	Glossina	<u>Glossina palpalis</u>	Van Hoof & Henrard (1937)
Fungus with septate hyphae	Glossina	<u>Glossina palpalis</u>	Macfie (1915)
Facultative parasitic and saprophytic fungi	Acarina ticks	<u>Ixodes ricinus</u>	Samšiňaková et al. (1974)
Fungus	Tabanidae	<u>Lutzomyia gomezi</u> (Nitz.)	McConnell & Correa (1964)
Fungus	Phlebotominae	<u>Lutzomyia lainsoni</u>	Ward & Killick-Kendrick (1974)
Fungus	Phlebotominae	<u>Lutzomyia lichyi</u> (Floch and Abonnenc), as <u>P. vexillarius</u> Fchld. & Hertig	McConnell & Correa (1964)

XXVII.   FUNGI OTHER THAN <u>COELOMOMYCES</u> (continued)

Pathogen	Host group	Host	References
Fungus	Phlebotominae	<u>Lutzomyia longipalpis</u> (Lutz & Neiva)	Sherlock & Sherlock (1959)
Fungus	Phlebotominae	<u>Lutzomyia longipalpis</u>	Adler & Mayrink (1961)
Fungus	Phlebotominae	<u>Lutzomyia ovallesi</u>	McConnell & Correa (1964)
Fungus	Phlebotominae	<u>Lutzomyia panamensis</u> (Shannon)	McConnell & Correa (1964)
Fungus	Phlebotominae	<u>Lutzomyia sanguinaria</u>	McConnell & Correa (1964)
Fungus	Phlebotominae	<u>Lutzomyia shannoni</u>	McConnell & Correa (1964), Lewis (1965a)
Fungus	Phlebotominae	<u>Lutzomyia trinidadensis</u>	McConnell & Correa (1964)
Fungus	Phlebotominae	<u>Lutzomyia vespertilionis</u> (Fchld. & Hertig)	McConnell & Correa (1964)
Fungus	Phlebotominae	<u>Lutzomyia vexator occidentis</u>	Chaniotis & Anderson (1968)
Fungus	Phlebotominae	<u>Lutzomyia ylephiletor</u>	McConnell & Correa (1964)
Fungus	Acarina mites	Mite colonies	Audy & Lavoipierre (1964)
Fungus	Phlebotominae	<u>Phlebotomus papatasii</u>	Adler & Theodor (1929)
Fungus	Phlebotominae	<u>Phlebotomus papatasii</u> (Scopoli)	Adler & Theodor (1927)
Fungus	Tabanidae	<u>Phlebotomus sergenti</u> Parrot	Adler & Theodor (1929)
Fungus	Tabanidae	<u>Phlebotomus</u> ? ?	?
Fungus	Tabanidae	<u>Sergentomyia africana</u> (Newst.)	Kirk & Lewis (1940)
Fungus	Tabanidae	<u>Sergentomyia ingrami</u> (Newst.)	Kirk & Lewis (1947)
Fungus ?	Simuliidae	<u>Simulium damnosum</u>	Lewis (1965), Briggs (1970, 1973)
Fungus ?	Simuliidae	<u>Simulium rostratum</u>	Bobrova (1971)
Fungus ?	Simuliidae	<u>Simulium vittatum</u>	Frost & Manier (1971)
Genistellaceae	Simuliidae	<u>Austrosimulium</u> spp.	Crosby (1974)
Laboulbeniales	Cerato-pogonidae	<u>Forcipomyia</u> sp.	Mayer (1934)
Phycomycete	Glossina	<u>Glossina brevipalpis</u>	Moggridge (1936)
Phycomycete	Glossina	<u>Glossina palpalis</u>	Swynnerton (1936)

XXVII. FUNGI OTHER THAN <u>COELOMOMYCES</u> (continued)

Pathogen	Host group	Host	References
Phycomycete	Glossina	<u>Glossina</u> sp.	Nash (1970)
Phycomycete	Simuliidae	<u>Simulium damnosum</u>	Briggs (1970, 1973)
"Streptotrichian" similar to actinomycosis	Acarina mites	<u>Dermanyssus gallinae</u> Fowl mite	Warren (1940)
Yeast	Culicidae	<u>Aedes aegypti</u>	Christophers (1952), Marchaux et al. (1903)
Yeast	Culicidae	<u>Anopheles maculipennis</u>	Laveran (1902)
Yeast-like spherical micro-organisms 6-12 μ in diameter	Acarina mites	<u>Eutrombicula alfreddugesi</u>	Kroman et al. (1969)
Yeast-like spherical micro-organisms 6-12 μ in diameter	Acarina mites	<u>Hannemania</u> sp.	Kroman et al. (1969)
Yeast-like spherical micro-organisms 6-12 μ in diameter	A .rina mites	<u>Leptotrombidium panamense</u>	Kroman et al. (1969)
Yeast-like spherical micro-organisms 6-12 μ in diameter	Acarina mites	<u>Neotrombicula harperi</u>	Kroman et al. (1969)
Yeast	Reduviidae	<u>Triatoma protracta</u>	Marchette & Hatie (1965)

XXVIII.   COELOMOMYCES

Pathogen	Host group	Host	References
Coelomomyces africanus	Culicidae	Anopheles funestus	Haddow (1942)
Coelomomyces africanus	Culicidae	Anopheles gambiae	Haddow (1942)
Coelomomyces africanus	Culicidae	Anopheles rufipes	Rodhain & Gayral (1971)
Coelomomyces anophelesicus	Culicidae	Anopheles subpictus	Iyengar (1935)
Coelomomyces anophelesicus	Culicidae	Anopheles vagus	Iyengar (1935)
Coelomomyces anophelesicus	Culicidae	Anopheles varuna	Iyengar (1935)
Coelomomyces ascariformis	Culicidae	Anopheles gambiae	Walker (1938) /validated by van Thiel (1962)/, Rodhain & Gayral (1971)
Coelomomyces ascariformis	Culicidae	Anopheles minimus	Manalang (1930)
Coelomomyces bisymmetricus	Culicidae	Anopheles crucians	Couch & Dodge (1947)
Coelomomyces cairnsensis	Culicidae	Anopheles farauti	Laird (1956a)
Coelomomyces ciferrii	Phlebotinae	Lutzomyia sp. or spp.	Leào & Pedroso (1965)
Coelomomyces cribrosus	Culicidae	Anopheles crucians	Couch & Dodge (1947)
Coelomomyces cribrosus	Culicidae	Anopheles punctipennis	Couch & Dodge (1947)
Coelomomyces cribrosus	Culicidae	Culex fraudatrix	Laird (1959b)
Coelomomyces cribrosus	Culicidae	Culex tritaeniorhynchus siamensis	Laird (1959a)
Coelomomyces cribrosus	Culicidae	Culex tritaeniorhynchus summorosus	Laird (1959b)
Coelomomyces dodgei	Culicidae	Anopheles crucians	Umphlett (1960) (unpubl.), Chapman & Glenn (1972)
Coelomomyces finlayae	Culicidae	Aedes notoscriptus	Laird (1959b)
Coelomomyces grassei	Culicidae	Anopheles gambiae	Rioux & Pech (1960, 1962)
Coelomomyces iliensis	Culicidae	Culex modestus	Dubitskii et al. (1970, 1973)
Coelomomyces indicus	Culicidae	Aedomyia catasticta	Laird (1956a)
Coelomomyces indicus	Culicidae	Anopheles aconitus	Iyengar (1935)
Coelomomyces indicus	Culicidae	Anopheles annularis	Iyengar (1935)
Coelomomyces indicus	Culicidae	Anopheles barbirostris	Iyengar (1935)
Coelomomyces indicus	Culicidae	Anopheles funestus	Muspratt (1946a)
Coelomomyces indicus	Culicidae	Anopheles gambiae	Muspratt (1946a, b, 1962), Madelin (1968)
Coelomomyces indicus	Culicidae	Anopheles hyrcanus var. nigerrimus	Iyengar (1935)

XXVIII. COELOMOMYCES (continued)

Pathogen	Host group	Host	References
Coelomomyces indicus	Culicidae	Anopheles jamesi	Iyengar (1935)
Coelomomyces indicus	Culicidae	Anopheles pharoensis	Gad & Sadek (1968)
Coelomomyces indicus	Culicidae	Anopheles pretoriensis	Muspratt (1946, 1962)
Coelomomyces indicus	Culicidae	Anopheles ramsayi	Iyengar (1935)
Coelomomyces indicus	Culicidae	Anopheles rivulorum	Muspratt (1946, 1962)
Coelomomyces indicus	Culicidae	Anopheles rufipes	Muspratt (1946, 1962)
Coelomomyces indicus	Culicidae	Anopheles squamosus	Muspratt (1946a)
Coelomomyces indicus (?)	Culicidae	Anopheles squamosus	Muspratt (1962)
Coelomomyces indicus	Culicidae	Anopheles subpictus	Laird (1959b), Iyengar (1961) (unpubl.), Gugnani et al. (1965)
Coelomomyces indicus	Culicidae	Anopheles varuna	Iyengar (1935)
Coelomomyces indicus	Culicidae	Culex simpsoni	Muspratt (1946a)
Coelomomyces keilini	Culicidae	Anopheles crucians	Couch & Dodge (1947)
Coelomomyces lativittatus	Culicidae	Anopheles crucians	Couch & Dodge (1947)
Coelomomyces lativittatus (var.?)	Culicidae	Anopheles earlei	Laird (1961)
Coelomomyces macleayae	Culicidae	Aedes polynesiensis	Pillai & Rakai (1970)
Coelomomyces macleayae	Culicidae	Aedes triseriatus	Couch (unpubl.)
Coelomomyces macleayae	Culicidae	Aedes (macleaya) sp.	Laird (1959b)
Coelomomyces macleayae	Culicidae	Toxorynchites rutilus septentrionalis	Nolan et al. (1973)
Coelomomyces milkoi[*]	Tabanidae	Chrysops relictus	Koval & Andryeyeva (1971)
Coelomomyces milkoi[*]	Tabanidae	Tabanus autumnalis	Koval & Andryeyeva (1971)
Coelomomyces pentangulatus	Culicidae	Culex erraticus	Couch (1945)
Coelomomyces psorophorae (?)	Culicidae	Aedes cinereus	Eckstein (1922)
Coelomomyces psorophorae var.	Culicidae	Aedes scatophagoides	Muspratt (1946)
Coelomomyces psorophorae var.	Culicidae	Aedes taeniorhynchus	Lum (1963)
Coelomomyces psorophorae var.	Culicidae	Aedes vexans	Laird (1961)

[*] Probably a member of the Entomophthorales, not Coelomomyces (M. Laird, personal communication).

XXVIII.   COELOMOMYCES (continued)

Pathogen	Host group	Host	References
Coelomomyces psorophorae	Culicidae	Aedes vexans	Zharov (1973)
Coelomomyces psorophorae (?)	Culicidae	Aedes vexans	Eckstein (1922)
Coelomomyces psorophorae var.	Culicidae	Culiseta inornata	Shemanchuk (1959)
Coelomomyces psorophorae	Culicidae	Psorophora ciliata	Laird (1961)
Coelomomyces psorophorae var.	Culicidae	Psorophora howardii	Couch & Dodge (1947)
Coelomomyces punctatus	Culicidae	Anopheles crucians	Chapman & Glenn (1972)
Coelomomyces punctatus	Culicidae	Anopheles quadrimaculatus	Couch & Dodge (1947)
Coelomomyces quadrangulatus var. lamborni	Culicidae	Aedes albopictus	Couch & Dodge (1947)
Coelomomyces quadrangulatus	Culicidae	Anopheles crucians	Couch & Dodge (1947)
Coelomomyces quadrangulatus var.	Culicidae	Anopheles georgianus	Couch & Dodge (1947)
Coelomomyces omorii	Culicidae	Culex tritaeniorhynchus summorosus	Laird et al. (1975)
Coelomomyces opifexi	Culicidae	Aedes australis	Pillai & Smith (1968), Pillai (1969), Pillai & Woo (1973)
Coelomomyces opifexi	Culicidae	Opifex fuscus	Pillai & Smith (1968)
Coelomomyces quadrangulatus	Culicidae	Anopheles punctipennis	Couch & Dodge (1947)
Coelomomyces quadrangulatus var. irregularis	Culicidae	Anopheles punctipennis	Couch & Dodge (1947)
Coelomomyces quadrangulatus var.	Culicidae	Anopheles quadrimaculatus	Couch & Dodge (1947)
Coelomomyces quadrangulatus var. parvus	Culicidae	Culex tritaeniorhynchus siamensis	Laird (1959a)
Coelomomyces quadrangulatus (var.?)	Culicidae	Anopheles walkeri	Laird (1961)
Coelomomyces raffaelei	Culicidae	Anopheles claviger	Coluzzi & Rioux (1962)
Coelomomyces raffaelei var. parvum	Culicidae	Anopheles sinensis	Laird et al. (1975)
Coelomomyces sculptosporus	Culicidae	Anopheles crucians	Couch & Dodge (1947)
Coelomomyces scultposporus	Culicidae	Anopheles punctipennis	Couch & Dodge (1947)
Coelomomyces sculptosporus	Culicidae	Anopheles walkeri	Laird (1961)
Coelomomyces solomonis	Culicidae	Anopheles punctulatus	Laird (1956a)

XXVIII.   COELOMOMYCES (continued)

Pathogen	Host group	Host	References
Coelomomyces stegomyiae	Culicidae	Aedes aegypti	Laird (1959a), Rajapaksa (1964)
Coelomomyces stegomyiae	Culicidae	Aedes albopictus	Keilin (1921), Rajapaksa (1964)
Coelomomyces stegomyiae	Culicidae	Aedes multiformis	Huang (1968)
Coelomomyces stegomyiae	Culicidae	Aedes polynesiensis	Laird (1966)
Coelomomyces stegomyiae	Culicidae	Aedes quadrispinatus	Briggs (1967)
Coelomomyces stegomyiae	Culicidae	Aedes scutellaris	Laird (1956a)
Coelomomyces stegomyiae var. rotumae	Culicidae	Aedes (stegomyia) sp.	Laird (1959b)
Coelomomyces stegomyiae	Culicidae	Aedes variabilis	Briggs (1967)
Coelomomyces stegomyiae	Culicidae	Armigeres obturbans	Laird (1959b)
Coelomomyces tasmaniensis	Culicidae	Aedes australis	Laird (1956b)
Coelomomyces uranotaeniae	Culicidae	Uranotaenia sapphirina	Couch (1945)
Coelomomyces walkeri (Walker's type 1)	Sierra Leone	Anopheles funestus	Walker (1938)
Coelomomyces walkeri	Culicidae	Anopheles gambiae	Walker (1938) /validated by van Thiel (1962)/
Coelomomyces walkeri	Culicidae	Anopheles tesselatus	van Thiel (1954)
Coelomomyces sp.	Culicidae	Aedes atropalpus epactius	Romney et al. (1971)
Coelomomyces sp.	Culicidae	Aedes cantans	Service (1974)
Coelomomyces sp.	Culicidae	Aedes caspius dorasalis	Khaliulin & Ivanov (1973)
Coelomomyces sp.	Culicidae	Aedes cyprius	Khaliulin & Ivanov (1973)
Coelomomyces sp.	Culicidae	Aedes excrucians	Briggs (1969)
Coelomomyces sp.	Culicidae	Aedes hebrideus	Genga & Maffi (1973)
Coelomomyces sp.	Culicidae	Aedes melanimon	Kellen et al. (1963)
Coelomomyces sp.	Culicidae	Aedes simpsoni	McCrae (1972)
Coelomomyces sp.	Culicidae	Aedes sollicitans	Chapman & Woodward (1966)
Coelomomyces sp.	Culicidae	Aedes togoi	Fedder et al. (1971)

XXVIII.   COELOMOMYCES (continued)

Pathogen	Host group	Host	References
Coelomomyces sp.	Culicidae	Anopheles farauti	Maffi & Genga (1970)
Coelomomyces sp.	Culicidae	Culex annulirostris	Maffi & Genga (1970)
Coelomomyces n. sp.	Culicidae	Culex gelidus	Rajapaksa (1964)
Coelomomyces sp.	Culicidae	Culex orientalis	Briggs (1969)
Coelomomyces sp.	Culicidae	Culex peccator	Briggs (1969)
Coelomomyces sp.	Culicidae	Culex pipiens fatigans	Rajapaksa (1964)
Coelomomyces sp.	Culicidae	Culex pipiens fatigans	Lacour & Rageau (1957)
Coelomomyces sp.	Culicidae	Culex portesi	Briggs (1969)
Coelomomyces sp.	Culicidae	Culex restuans	Chapman & Woodward (1966)
Coelomomyces sp.	Culicidae	Culex salinarius	Chapman & Woodward (1966)
Coelomomyces sp.	Culicidae	Uranotaenia barnesi	Maffi & Genga (1970)
Coelomomyces sp. (?)	Simuliidae	Simulium metallicum	Garnham & Lewis (1959)

XXIX.   NEMATODES

Pathogen	Host group	Host	References
Acarinocola hirsutus nematode	Acarina mites	Haemogamasus hirsutus	Warren (1941)
Agamomermis heleis	Ceratopogonidae	Culicoides pulicaris	Rubtsov (1967), Mirzaeva (1971)
Agamomermis sp.	Culicidae	Aedes canadensis	Jenkins (1964)
Agamomermis sp.	Culicidae	Aedes cantans	Shachov (1927)
Agamomermis sp.	Culicidae	Aedes cinereus	Frohne (1953), Jenkins (1964)
Agamomermis sp.	Culicidae	Aedes communis	Frohne (1953)
Agamomermis sp.	Culicidae	Aedes communis (Culex nemeralis)	Stiles (1903)
Agamomermis sp.	Culicidae	Aedes dorsalis	Shachov (1927)
Agamomermis sp.	Culicidae	Aedes excrucians	Frohne (1955a)
Agamomermis sp.	Culicidae	Aedes haworthi	Muspratt (1945)
Agamomermis sp.	Culicidae	Aedes heischi	Lumsden (1955)
Agamomermis sp.	Culicidae	Aedes impiger	Frohne (1953)
Agamomermis sp.	Culicidae	Aedes michaelikati	Lumsden (1955)
Agamomermis sp.	Culicidae	Aedes pionips	Frohne (1955b)
Agamomermis sp.	Culicidae	Aedes punctor	Frohne (1953), Jenkins (1964)
Agamomermis sp.	Culicidae	Aedes soleatus	Lumsden (1955)
Agamomermis sp.	Culicidae	Aedes stimulans	Anonymous (1970)
Agamomermis sp.	Culicidae	Aedes vexans	Stabler (1952)
Agamomermis sp. (Paramermis canadensis)	Culicidae	Aedes vexans	Trpis et al. (1968)
Agamomermis sp.	Culicidae	Anopheles annulipes	Laird (1956), Kalucy (1972)
Agamomermis sp.	Culicidae	Anopheles funestus	Hanney (1960)
Agamomermis sp.	Culicidae	Anopheles gambiae	Muspratt (1945), Hanney (1960), Service (1973)
Agamomermis sp.	Culicidae	Anopheles sp.	Johnson (1903)
Agamomermis sp.	Culicidae	Culex fatigans (pipiens quinquefasciatus)	Ross (1906)
Agamomermis sp.	Ceratopogonidae	Culicoides albicans	Callot (1959)
Larval mermithid (Agamomermis)	Glossina	Glossina palpalis	Foster (1963)

XXIX. NEMATODES (continued)

Pathogen	Host group	Host	References
Aproctonema chapmani	Ceratopo- gonidae	Culicoides arboricola	Nickle (1969), Chapman et al. (1969), Chapman (1973)
Aproctonema simuliophaga	Simuliidae	Boophthora erythrocephala	Rubtsov (1966c)
Atractonema sp.	Simuliidae	Boophthora erythrocephala	Rubtsov (1963)
Bathymermis sp.	Tabanidae	Chrysops furcata	Shamsuddin (1966)
Bathymermis sp.	Tabanidae	Chrysops mitis	Shamsuddin (1966)
Blatticola blattae	Blattidae	Blattella germanica	Cali & Mai (1965)
Blatticola tuapakae	Blattidae	Platyzosteria novaeseelandiae	Dale (1966)
Dipetalonema reconditum	Anoplura- Mallophaga	Heteroduxus spiniger	Nelson (1962)
Dirofilaria immitis	Siphonap- tera	Ctenocephalides canis	Jenkins (1964c)
Dirofilaria immitis	Siphonap- tera	Ctenocephalides felis	Jenkins (1964c)
Diximermis peterseni (Gastromermis sp.)	Culicidae	Anopheles barberi	Petersen & Chapman (1970)
Diximermis peterseni (Gastromermis sp.)	Culicidae	Anopheles bradleyi	Petersen & Chapman (1970)
Diximermis peterseni (Gastromermis sp.)	Culicidae	Anopheles crucians	Chapman et al. (1969), Petersen & Chapman (1970), Savage & Petersen (1971)
Diximermis peterseni (Gastromermis sp.)	Culicidae	Anopheles punctipennis	Chapman et al. (1969)
Diximermis peterseni (Gastromermis sp.)	Culicidae	Anopheles quadrimaculatus	Chapman et al. (1969), Petersen & Chapman (1970)
Filaria cypseli	Anoplura- Mallophaga	Dennyus hirundis	Dutton (1905)
Filaria stomoxeos (=? Habronema microstoma or ? Setaria cervi	Stomoxys	Stomoxys calcitrans	Von Linstow (1875)
Foleyella philistinae	Culicidae	Culex pipiens molestus	Schacher & Khalil (1968)
Gastromermis ambianensis	Simuliidae	Simuliidae	Rubtsov & Doby (1970)
Gastromermis boophthorae	Simuliidae	Eusimulium cryophilium	Welch & Rubtsov (1965)

XXIX.   NEMATODES (continued)

Pathogen	Host group	Host	References
Gastromermis boophthorae	Simuliidae	Boophthora erythrocephala	Rubtsov (1967a)
Gastromermis boophthorae	Simuliidae	Odagmia ornata	Welch & Rubtsov (1965)
Gastromermis boophthorae	Simuliidae	Simulium argyreatum	Welch & Rubtsov (1965)
Gastromermis boophthorae	Simuliidae	Simulium (Eusimulium) aurem	Welch & Rubtsov (1965)
Gastromermis boophthorae	Simuliidae	Simulium latipes	Welch & Rubtsov (1965)
Gastromermis boophthorae	Simuliidae	Simulium morsitans	Welch & Rubtsov (1965) Rubtsov (1967a)
Gastromermis boophthorae	Simuliidae	Simulium verecundum	Welch & Rubtsov (1965)
Gastromermis clingogaster	Simuliidae	Simulium aurem	Rubtsov (1967a)
Gastromermis crassicauda	Simuliidae	Simulium morsitans	Rubtsov (1967a)
Gastromermis crassifrons	Simuliidae	Boophthora erythrocephala	Rubtsov (1967a)
Gastromermis longispicula	Simuliidae	Simulium morsitans	Rubtsov (1967a)
Gastromermis odagmiae	Simuliidae	Odagmia ornata	Rubtsov (1967a)
Gastromermis rosalbus	Simuliidae	Eusimulium securiforme	Rubtsov (1967a)
Gastromermis rosalbus	Simuliidae	Simulium verecundum	Rubtsov (1967a)
Gastromermis virescens actipenis	Simuliidae	Boophthora erythrocephala	Rubtsov (1967a)
Gastromermis virescens virescens	Simuliidae	Boophthora erythrocephala	Rubtsov (1967a)
Gastromermis viridis	Simuliidae	Simulium corbis	Ezenwa (1974d)
Gastromermis viridis	Simuliidae	Simulium venustum	Ezenwa (1974d)
Gastromermis viridis	Simuliidae	Simulium vittatum	Anderson & DeFoliart (1962), Phelps & DeFoliart (1964), Welch (1962)
Gastromermis sp.	Reduviidae	Anopheles funestus	Coz (1966, 1973)
Gastromermis sp.	Simuliidae	Cnephia mutata	Anderson & DeFoliart (1962)
Gastromermis sp.	Simuliidae	Simulium corbis	Anderson & DeFoliart (1962)
Gastromermis sp.	Simuliidae	Simulium (nolleri) decorum	Anderson & DeFoliart (1962)
Gastromermis sp.	Simuliidae	Simulium jenningsi	Anderson & DeFoliart (1962)
Gastromermis sp.	Simuliidae	Simulium venustum	Anderson & DeFoliart (1962)
Gastromermis sp.	Simuliidae	Simulium vittatum	Anderson & DeFoliart (1962)
Habronema microstoma	Stomoxys	Stomoxys calcitrans	Hill (1919), Johnston & Bancroft (1920), Johnston (1920), Roubaud & Descazeaux (1922)

XXIX.   NEMATODES (continued)

Pathogen	Host group	Host	References
Habronema muscae	Stomoxys	Stomoxys calcitrans	Johnston (1913)
Habronema muscae (Carter), H. megastoma (Rudolphi) Seurat, H. microstoma (Schneider)	Stomoxys	Stomoxys calcitrans	Bull (1919)
Hammerschmidtiella diesingi	Blattidae	Blatta orientalis	Jarry & Jarry (1963), Groschaft (1965), Leong & Paran (1966)
Hammerschmidtiella diesingi	Blattidae	Periplaneta americana	Hominck & Davey (1972), Jarry & Jarry (1963), Kloss (1966), Groschaft (1965), Leong & Paran (1966)
Hammerschmidtiella diesingi	Blattidae	Periplaneta australasiae	Kloss (1966)
Heleidomermis vivipara	Ceratopogonidae	Culicoides nubeculosus	Rubtsov (1970, 1972)
Heleidomermis vivipara	Ceratopogonidae	Culicoides stigma	Rubtsov (1970, 1972)
Heterotylenchus autumnalis	Musca	Musca autumnalis	Chitwood & Stoffolano (1971), Jones & Perdue (1967), Stoffolano & Nickle (1966), Thomas et al. (1972), Treece & Miller (1968), Nappi (1973), Nappi & Stoffolano (1972), Nickle (1967), Stoffolano (1967, 1969, 1970a, 1970b, 1971, 1973), Stoffolano & Streams (1971)
Hexamermis sp.	Glossina	Glossina morsitans	Nickle (1973)
Hydromermis angusta	Simuliidae	Simuliidae	Rubtsov & Doby (1970)
Hydromermis churchillensis	Culicidae	Aedes communis	Welch (1960a)
Hydromermis churchillensis	Culicidae	Aedes impiger (nearcticus)	Welch (1960)
Hydromermis churchillensis	Culicidae	Aedes nigripes	Welch (1960)
Hydromermis sp.	Culicidae	Aedes communis	Jenkins & West (1954)
Hydromermis sp.	Culicidae	Aedes communis	Smith (1961)
Hydromermis sp.	Culicidae	Aedes impiger (nearcticus)	Jenkins & West (1954)
Hydromermis sp.	Culicidae	Aedes nigripes	Jenkins & West (1954)
Hydromermis sp.	Culicidae	Aedes pionips	Jenkins & West (1954)

XXIX.   NEMATODES (continued)

Pathogen	Host group	Host	References
Hydromermis sp.	Culicidae	Aedes pullatus	Smith (1961)
Hydromermis sp.	Simuliidae	Simuliidae	Anderson & Dicke (1960)
Hydromermis sp.	Simuliidae	Simulium damnosum	Briggs (1968)
Hydromermis sp.	Simuliidae	Simulium luggeri	Anderson & DeFoliart (1962)
Isomermis rossica	Simuliidae	Boophthora erythrocephala	Rubtsov (1967b)
Isomermis rossica	Simuliidae	Eusimulium cryophilium	Rubtsov (1968)
Isomermis rossica	Simuliidae	Eusimulium perteszi	Rubtsov (1968)
Isomermis rossica	Simuliidae	Simuliidae	Rubtsov & Doby (1970)
Isomermis rossica	Simuliidae	Simulium latipes	Rubtsov (1968)
Isomermis rossica	Simuliidae	Simulium morsitans	Rubtsov (1968)
Isomermis rossica	Simuliidae	Simulium verecundum	Rubtsov (1968)
Isomermis tansaniensis	Simuliidae	Simulium damnosum	Briggs (1968)
Isomermis wisconsinensis	Simuliidae	Prosimulium mixtum/ fuscum	Ezenwa & Carter (1975)
Isomermis wisconsinensis	Simuliidae	Simulium venustum	Ezenwa (1974d)
Isomermis wisconsinensis	Simuliidae	Simulium vittatum	Phelps & DeFoliart (1964), Anderson & DeFoliart (1962), Welch (1962)
Isomermis sp.	Simuliidae	Cnephia emergens	Anderson & DeFoliart (1962)
Isomermis sp.	Simuliidae	Cnephia mutata	Anderson & DeFoliart (1962)
Isomermis sp.	Simuliidae	Simulium decorum	Anderson & DeFoliart (1962)
Isomermis sp.	Simuliidae	Simulium (E.) latipes	Anderson & DeFoliart (1962)
Isomermis sp.	Simuliidae	Simulium luggeri	Anderson & Dicke (1960)
Isomermis sp.	Simuliidae	Simulium tuberosum	Anderson & DeFoliart (1962)
Isomermis sp.	Simuliidae	Simulium venustum	Anderson & DeFoliart (1962)
Isomermis sp.	Simuliidae	Simulium vittatum	Anderson & DeFoliart (1962)
Leidynema appendiculata	Blattidae	Gromphadorhina portentosa	Bhatnager & Edwards (1970)
Leidynema appendiculata	Blattidae	Periplaneta americana	Feldman (1972), Hominck & Davey (1972), Jarry & Jarry (1963), Kloss (1966), Pawlik (1966), Groschaft (1965), Leong & Paran (1966)
Leidynema appendiculata	Blattidae	Periplaneta australasiae	Kloss (1966)
Leidynema appendiculata	Blattidae	Polyphaga aegypticaca	Steinhaus & Marsh (1962)
Leidynema sp.	Blattidae	Leucophaea maderae	Thomas & Poinar (1973)

XXIX.   NEMATODES (continued)

Pathogen	Host group	Host	References
Leidynema appendiculata	Blattidae	Blatta orientalis	Jarry & Jarry (1963), Groschaft (1965), Leong & Paran (1966)
Leidynema appendiculata	Blattidae	Blattella fulginosa	Feldman (1972)
Limnomermis aculeata	Simuliidae	Simulium morsitans	Rubtsov (1967b)
Limnomermis aquatilis	Reduviidae	Anopheles spp.	Dujardin (1845)
Limnomermis cryophili	Simuliidae	Eusimulium cryophilium	Rubtsov (1967b)
Limnomermis cryophili	Simuliidae	Simulium latipes	Rubtsov (1967b)
Limnomermis lanceicapta	Simuliidae	Simulium morsitans	Rubtsov (1967b)
Limnomermis macronuclei	Simuliidae	Eusimulium cryophilium	Rubtsov (1967b)
Limnomermis macronuclei	Simuliidae	Simulium latipes	Jamnback (1970)
Limnomermis teniucauda	Simuliidae	Simulium morsitans	Rubtsov (1967b)
Mastophorus muris	Siphonaptera	Ceratophyllus laevicepts	Akopyan (1968)
Mastophorus muris	Siphonaptera	Ctenophthalmus avernus	Beaucournu & Chabaud (1963)
Mastophorus muris	Phlebotominae	Phlebotomus ariasi	Rioux & Golvan (1969)
Mastophorus muris (Gmelin)	Phlebotominae	Phlebotomus ariasi Tonn.	Golvan et al. (1963)
Mastophorus muris	Phlebotominae	Phlebotomus perniciosus	Rioux & Golvan (1969)
Mastophorus muris (Gmelin)	Phlebotominae	Phlebotomus perniciosus Newst.	Golvan et al. (1963)
Mastophorus muris	Phlebotominae	Sergentomyia minuta minuta	Rioux & Golvan (1969)
Mermis sp.	Culicidae	Aedes aegypti (Stegomyia fasciatus)	Gendre (1909)
Mermis sp.	Culicidae	Anopheles annularis (fulignosus)	Iyengar (1930)
Mermis sp.	Culicidae	Anopheles barbirostris	Iyengar (1930)
Mermis sp.	Culicidae	Anopheles hyrcanus	Iyengar (1930)
Mermis sp.	Culicidae	Anopheles leucosphyrus	Walandouw (1934)
Mermis sp.	Culicidae	Anopheles philippinensis	Iyengar (1930)
Mermis sp.	Culicidae	Anopheles ramsayi (pseudojamesi)	Iyengar (1930)
Mermis sp.	Culicidae	Anopheles sinensis	Iyengar (1930)
Mermis sp.	Culicidae	Anopheles tessellatus	Iyengar (1930)
Mermis sp.	Culicidae	Anopheles varuna	Iyengar (1930)

XXIX.   NEMATODES (continued)

Pathogen	Host group	Host	References
Mermis sp.	Ceratopogonidae	Atrichopogon sp.	Das Gupta (1964)
Mermis sp.	Ceratopogonidae	Culicoides alatus	Sen & Das Gupta (1958)
Mermis sp.	Glossina	Glossina morsitans	Thomson (1947)
Immature female Mermis	Glossina	Glossina palpalis	Leiper (1910)
Mermithonema acicularis	Simuliidae	Boophthora erythrocephala	Rubtsov (1966c)
Mermithonema brevis	Simuliidae	Boophthora erythrocephala	Rubtsov (1966c)
Mesomermis ethiopica	Simuliidae	Simulium damnosum	Briggs (1968)
Mesomermis simuliae	Simuliidae	Simuliidae	Rubtsov & Doby (1970)
Mesomermis sp.	Culicidae	Orthopodmyia signifera	Petersen & Willis (1969a)
Mermithid?	Culicidae	Aedes albopictus	Hovi & Ramachandran (1963)
Mermithid	Culicidae	Aedes communis	Artyukhovsky & Kolycheva (1965)
Mermithid?	Culicidae	Aedes flavescens	Hearle (1929)
Mermithids	Culicidae	Aedes impiger (nearcticus)	Gorham (1972)
Mermithid	Culicidae	Aedes laguna	Arnell & Nielsen (1972)
Mermithid	Culicidae	Aedes pullatus	Gorham (1972)
Mermithid	Culicidae	Aedes rusticus (maculatus)	Artyukhovsky & Kolycheva (1965)
Mermithid	Culicidae	Anopheles funestus	Nasr (1974)
Mermithid	Culicidae	Anopheles gambiae	Nasr (1974)
Mermithid?	Culicidae	Anopheles letifer	Hovi & Ramachandran (1963)
Mermithid	Simuliidae	Cnephia mutata	Anderson & DeFoliart (1962)
Mermithid	Siphonaphora	Ctenophthalmus avernus	Rothschild (1969)
Mermithid	Culicidae	Culex pipiens pipiens	Stabler (1952)
Mermithid	Culicidae	Culex salinarius	Stabler (1952)
Mermithid	Culicidae	Culex territans	Petersen et al. (1968)
Mermithid larvae	Glossina	Glossina brevipalpis	Molov (1972)
Larval mermithid	Glossina	Glossina palpalis	Nickle (1974)
Mermithid	Siphonaphora	Myoxopsylla laverani	Rothschild (1969)
Mermithid	Simuliidae	Prosimulium demarticulata	Anderson & DeFoliart (1962)
Mermithid	Simuliidae	Prosimulium alpestre	Bobrova (1971)

Pathogen	Host group	Host	References
Mermithid	Culicidae	Psorophora ciliata	Savage & Petersen (1971)
Mermithid	Simuliidae	Simuliidae	Rubtsov (1963), Briggs (1966)
Mermithid	Simuliidae	Simulium arcticum	Steinhaus & March (1962)
Mermithid	Simuliidae	Simulium aurem	Anderson & DeFoliart (1962)
Mermithid	Simuliidae	Simulium damnosum	Lewis (1965)
Mermithid?	Simuliidae	Simulium decorum	Anderson & DeFoliart (1962)
Mermithid	Simuliidae	Simulium latipes	Rubtsov (1968)
Mermithid	Simuliidae	Simulium rostratum	Bobrova (1971)
Mermithid	Simuliidae	Simulium tuberosum	Anderson & DeFoliart (1962)
Mermithid?	Simuliidae	Simulium venustum	Anderson & DeFoliart (1962)
Mermithid?	Simuliidae	Simulium vittatum	Anderson & DeFoliart (1962)
Mermithid?	Simuliidae	Simulium vulgare	Bobrova (1971)
Mermithid	Siphonaphora	Spilopsyllus cuniculi	Rothschild (1969)
Mermithidae	Ceratopogonidae	Culicoides buckleyi	Buckley (1938)
Mermithidae	Ceratopogonidae	Culicoides circumscriptus	Glukhova (1967)
Mermithidae	Ceratopogonidae	Culicoides crepuscularis	Smith & Perry (1967), Beck (1958)
Mermithidae	Ceratopogonidae	Culicoides grisecens	Glukhova (1967)
Mermithidae	Ceratopogonidae	Culicoides haematopotus	Smith & Perry (1967)
Mermithidae	Ceratopogonidae	Culicoides nanus	Chapman et al. (1968)
Mermithidae	Ceratopogonidae	Culicoides nubeculosus	Glukhova (1967)
Mermithidae	Ceratopogonidae	Culicoides obsoletus	Boorman & Goddard (1970), Service (1974)
Mermithidae	Ceratopogonidae	Culicoides orientalis	Buckley (1938)
Mermithidae	Ceratopogonidae	Culicoides oxystoma	Buckley (1938)
Mermithidae	Ceratopogonidae	Culicoides peregrinus	Buckley (1938)
Mermithidae	Ceratopogonidae	Culicoides pictipennis	Service (1974)
Mermithidae	Ceratopogonidae	Culicoides pulicaris	Glukhova (1967)

XXIX.   NEMATODES (continued)

Pathogen	Host group	Host	References
Mermithidae	Cerato-pogonidae	Culicoides puncticolis	Glukhova (1967)
Mermithidae	Cerato-pogonidae	Culicoides pungens	Buckley (1938)
Mermithidae	Cerato-pogonidae	Culicoides shortii	Buckley (1938)
Mermithidae	Cerato-pogonidae	Culicoides spp.	Mirzaeva (1971)
Mermithidae	Cerato-pogonidae	Culicoides stellifer	Smith (1966), Smith & Perry (1967)
Mermithidae	Cerato-pogonidae	Culicoides stigma	Glukhova (1967)
Mermithidae	Cerato-pogonidae	Dasyhelea obscura	Keilin (1921a)
Mermithidae	Cerato-pogonidae	Leptoconops kerteszi	Whitsel (1965)
Mermithidae	Cerato-pogonidae	Leptoconops sp.	Glukhova (1967)
Mermithidae	Cerato-pogonidae	Biting midges	Weiser (1963)
Microfilaria sanguinus equi africano	Stomoxys	Stomoxys calcitrans	Mitzmain (1914)
Neoaplectana carpocapsae (DD136)	Culicidae	Aedes aegypti	Welch & Bronskill (1962), Bronskill (1962)
Neoaplectana carpocapsae (DD136)	Culicidae	Aedes stimulans	Bronskill (1962)
Neoaplectana carpocapsae (DD136)	Culicidae	Aedes trichurus	Bronskill (1962)
Neoaplectana carpocapsae	Culicidae	Culex pipiens	Poinar & Leutenegger (1971)
Neoaplectana carpocapsae	Simuliidae	Simulium vittatum	Webster (1973)
Neomesomermis flumenalis	Simuliidae	Cnephia mutata	Ezenwa (1973)
Neomesomermis flumenalis	Simuliidae	Prosimulium mixtum/fuscum	Ezenwa & Carter (1975)
Neomesomermis flumenalis	Simuliidae	Simulium corbis	Ezenwa (1973), Ezenwa (1974d)
Neomesomermis flumenalis	Simuliidae	Simulium decorum	Ezenwa (1973)
Neomesomermis flumenalis	Simuliidae	Simulium latipes	Ezenwa (1974d)
Neomesomermis flumenalis	Simuliidae	Simulium venustum	Phelps & DeFoliart (1964), Welch (1962), Bailey et al. (1974), Ezenwa (1973), Ezenwa (1974d)

XXIX.   NEMATODES (continued)

Pathogen	Host group	Host	References
Octomyomermis troglodytis (mermithid)	Culicidae	Aedes sierrensis	Poinar & Sanders (1974)
Perutilimermis culicis (Agamomermis culicis)	Culicidae	Aedes aegypti	Petersen & Willis (1969a)
Perutilimermis culicis (Agamomermis culicis)	Culicidae	Aedes sollicitans	Smith (1904), Petersen et al. (1967)
Perutilimermis culicis?	Culicidae	Aedes sollicitans	Savage & Petersen (1971)
Paramermis canadensis	Culicidae	Aedes sticticus (aldrichii)	Hearle (1926)
Paramermis canadensis	Culicidae	Aedes vexans	Steiner (1924)
Perutilimermis culicis (Agamomermis sp.)	Culicidae	Aedes taeniorhynchus	Petersen & Willis (1969b)
Perutilimermis culicis (Agamomermis sp.)	Culicidae	Culex fatigans (pipiens quinquefasciatus)	Petersen & Willis (1969a)
Protrellina aurifluus	Blattidae	Periplaneta americana	Leong & Paran (1966)
Protrellina gurri	Blattidae	Platyzosteria novaeseelandiae	Dale (1966)
Psyllotylenchus (Heterotylenchus) pavlovskii	Siphonaptera	Ceratophyllus laevicepts	Kurochkin (1961)
Psyllotylenchus (Heterotylenchus) pavlovskii	Siphonaptera	Cotptopsylla lamellifera	Kurochkin (1961)
Psyllotylenchus viviparus	Siphonaptera	Catallagia sculleni rutherfordi	Poinar & Nelson (1973)
Psyllotylenchus viviparus	Siphonaptera	Catallagia sp.	Poinar & Nelson (1973)
Psyllotylenchus viviparus	Siphonaptera	Diamanus montanus	Poinar & Nelson (1973)
Psyllotylenchus viviparus	Siphonaptera	Monopsyllus ciliatus protinus	Poinar & Nelson (1973)
Psyllotylenchus viviparus	Siphonaptera	Monopsyllus wagneri	Poinar & Nelson (1973)
Reesimermis iyengari (Mermis sp.)	Culicidae	Anopheles subpictus	Iyengar (1930)
Reesimermis muspratti (Agamomermis sp.)	Culicidae	Aedes aegypti	Muspratt (1945), Reynolds (1972)
Reesimermis muspratti	Culicidae	Aedes aegypti	Obiamiwe & MacDonald (1973)
Reesimermis muspratti (Agamomermis sp.)	Culicidae	Aedes marshalli	Muspratt (1945)
Reesimermis muspratti (Agamomermis sp.)	Culicidae	Aedes calceatus	Muspratt (1945)
Reesimermis muspratti (Agamomermis sp.)	Culicidae	Aedes fulgens	Muspratt (1945)

XXIX.   NEMATODES (continued)

Pathogen	Host group	Host	References
Reesimermis muspratti (Agamomermis sp.)	Culicidae	Aedes metallicus	Muspratt (1945)
Reesimermis muspratti	Culicidae	Aedes polynesiensis	Obiamiwe & MacDonald (1973)
Reesimermis muspratti (Agamomermis sp.)	Culicidae	Aedes zethus	Muspratt (1945)
Reesimermis muspratti	Culicidae	Anopheles albimanus	Obiamiwe & MacDonald (1973)
Reesimermis muspratti	Culicidae	Anopheles rufipes	Muspratt (1945)
Reesimermis muspratti	Culicidae	Anopheles stephensi	Obiamiwe & MacDonald (1973)
Reesimermis muspratti (Agamomermis sp.)	Culicidae	Culex fatigans	Muspratt (1965) Reynolds (1972)
Reesimermis muspratti	Culicidae	Culex fatigans	Obiamiwe & MacDonald (1973)
Reesimermis muspratti (Agamomermis sp.)	Culicidae	Culex nebulosus	Muspratt (1945)
Reesimermis muspratti	Culicidae	Culex pipiens molestus	Obiamiwe & MacDonald (1973)
Reesimermis nielseni (Romanomermis sp.)	Culicidae	Aedes aegypti	Petersen et al. (1969)
Reesimermis nielseni	Culicidae	Aedes albopictus	Petersen (unpublished data)
Reesimermis nielseni (Romanomermis sp.)	Culicidae	Aedes atlanticus	Petersen et al. (1968)
Reesimermis nielseni	Culicidae	Aedes atlanticus	Petersen & Willis (1972b)
Reesimermis nielseni (Romanomermis sp.)	Culicidae	Aedes canadensis	Petersen et al. (1968)
Reesimermis nielseni	Culicidae	Aedes cinereus	Tsai et al. (1969), unpublished data
Reesimermis nielseni	Culicidae	Aedes communis	Petersen (unpublished data)
Reesimermis nielseni	Culicidae	Aedes dupreei	Petersen (unpublished data)
Reesimermis nielseni	Culicidae	Aedes fitchii	Tsai et al. (1969)
Reesimermis nielseni	Culicidae	Aedes fulvus pallens	Petersen (unpublished data)
Reesimermis nielseni	Culicidae	Aedes increpitus	Tsai et al. (1969)
Reesimermis nielseni	Culicidae	Aedes infirmatus	Petersen (unpublished data)
Reesimermis nielseni (Romanomermis sp.)	Culicidae	Aedes mitchellae	Petersen et al. (1968)
Reesimermis nielseni	Culicidae	Aedes nigromaculis	Hoy & Petersen (1973)
Reesimermis nielseni	Culicidae	Aedes polynesiensis	Petersen (unpublished data)
Reesimermis nielseni	Culicidae	Aedes pullatus	Tsai et al. (1969)
Reesimermis nielseni	Culicidae	Aedes scuttelaris	Petersen (unpublished data)
Reesimermis nielseni (Romanomermis sp.)	Culicidae	Aedes sierrensis	Petersen et al. (1969)

XXIX.   NEMATODES (continued)

Pathogen	Host group	Host	References
Reesimermis nielseni (Romanomermis sp.)	Culicidae	Aedes sollicitans	Petersen et al. (1969)
Reesimermis nielseni	Culicidae	Aedes sollicitans	Petersen & Willis (1971)
Reesimermis nielseni	Culicidae	Aedes sticticus	Petersen (unpublished data)
Reesimermis nielseni (Romanomermis sp.)	Culicidae	Aedes taeniorhynchus	Petersen et al. (1969)
Reesimermis nielseni (Romanomermis sp.)	Culicidae	Aedes thibaulti	Petersen et al. (1968, 1969)
Reesimermis nielseni	Culicidae	Aedes tormentor	Petersen & Chapman (1972)
Reesimermis nielseni (Romanomermis sp.)	Culicidae	Aedes tormentor	Petersen et al. (1969)
Reesimermis nielseni (Romanomermis sp.)	Culicidae	Aedes triseriatus	Petersen et al. (1969)
Reesimermis nielseni (Romanomermis sp.)	Culicidae	Aedes vexans	Petersen et al. (1968) Petersen & Willis (1972b)
Reesimermis nielseni	Culicidae	Anopheles albimanus	Unpublished data
Reesimermis nielseni (Romanomermis sp.)	Culicidae	Anopheles atropos	Chapman et al. (1970)
Reesimermis nielseni (Romanomermis sp.)	Culicidae	Anopheles barberi	Petersen et al. (1968)
Reesimermis nielseni (Romanomermis sp.)	Culicidae	Anopheles bradleyi	Petersen et al. (1968)
Reesimermis nielseni (Mermithid)	Culicidae	Anopheles crucians	Chapman et al. (1967a)
Reesimermis nielseni (Romanomermis sp.)	Culicidae	Anopheles crucians	Savage & Petersen (1971)
Reesimermis nielseni	Culicidae	Anopheles crucians	Petersen & Willis (1971, 1972b)
Reesimermis nielseni	Culicidae	Anopheles freeborni	Petersen et al. (1972)
Reesimermis nielseni	Culicidae	Anopheles p. pseudopunctipennis	Petersen (unpublished data)
Reesimermis nielseni (Mermithid)	Culicidae	Anopheles punctipennis	Chapman et al. (1967a)
Reesimermis nielseni (Mermithid)	Culicidae	Anopheles quadrimaculatus	Chapman et al. (1967a)
Reesimermis nielseni	Culicidae	Anopheles quadrimaculatus	Petersen et al. (1973), Petersen & Willis (1972)
Reesimermis nielseni	Culicidae	Anopheles stephensi	Petersen (unpublished data)
Reesimermis nielseni	Culicidae	Culex annulis	Mitchell et al. (1972)
Reesimermis nielseni (Mermithid)	Culicidae	Culex erraticus	Chapman et al. (1967a)

Pathogen	Host group	Host	References
Reesimermis nielseni	Culicidae	Culex erraticus	Petersen & Willis (1971, 1972b)
Reesimermis nielseni (Romanomermis sp.)	Culicidae	Culex fatigans	Petersen et al. (1969)
Reesimermis nielseni	Culicidae	Culex fatigans	Mitchell et al. (1972) Chapman et al. (1972)
Reesimermis nielseni	Culicidae	Culex fuscanus	Mitchell et al. (1972)
Reesimermis nielseni	Culicidae	Culex fuscocephalus	Mitchell et al. (1972)
Reesimermis nielseni (Romanomermis sp.)	Culicidae	Culex peccator	Petersen et al. (1968)
Reesimermis nielseni	Culicidae	Culex pipiens pipiens	Petersen (unpublished data)
Reesimermis nielseni (Romanomermis sp.)	Culicidae	Culex restuans	Petersen et al. (1968)
Reesimermis nielseni	Culicidae	Culex rubithoracis	Mitchell et al. (1972)
Reesimermis nielseni (Romanomermis sp.)	Culicidae	Culex salinarius	Petersen et al. (1969)
Reesimermis nielseni	Culicidae	Culex tarsalis	Petersen et al. (1972)
Reesimermis nielseni	Culicidae	Culex tritaeniorhynchus summorosus	Mitchell et al. (1972)
Reesimermis nielseni	Culicidae	Culiseta impatiens	Tsai et al. (1969)
Reesimermis nielseni (Romanomermis sp.)	Culicidae	Culiseta inornata	Chapman et al. (1969) Petersen et al. (1969)
Reesimermis nielseni (Romanomermis sp.)	Culicidae	Culiseta melanura	Petersen et al. (1968)
Reesimermis nielseni (Romanomermis sp.)	Culicidae	Orthopodomyia signifera	Petersen et al. (1968)
Reesimermis nielseni (Romanomermis sp.)	Culicidae	Psorophora ciliata	Petersen et al. (1968)
Reesimermis nielseni (Mermithid)	Culicidae	Psorophora confinnis	Chapman et al. (1967a)
Reesimermis nielseni (Romanomermis sp.)	Culicidae	Psorophora confinnis	Petersen et al. (1968, 1969)
Reesimermis nielseni	Culicidae	Psorophora confinnis	Petersen et al. (1973) Petersen & Willis (1972b)
Reesimermis nielseni (Romanomermis sp.)	Culicidae	Psorophora cyanescens	Petersen et al. (1968)
Reesimermis nielseni (Romanomermis sp.)	Culicidae	Psorophora discolor	Petersen et al. (1968)
Reesimermis nielseni (Romanomermis sp.)	Culicidae	Psorophora ferox	Petersen et al. (1969)
Reesimermis nielseni	Culicidae	Psorophora horrida	Petersen (unpublished data)

XXIX.   NEMATODES (continued)

Pathogen	Host group	Host	References
Reesimermis nielseni	Culicidae	Psorophora howardii	Petersen (unpublished data)
Reesimermis nielseni (Romanomermis sp.)	Culicidae	Psorophora varipes	Petersen et al. (1969)
Reesimermis nielseni (Mermithid)	Culicidae	Uranotaenia lowii	Chapman et al. (1967a)
Reesimermis nielseni (Mermithid)	Culicidae	Uranotaenia sapphirina	Chapman et al. (1967a)
Reesimermis nielseni (Romanomermis sp.)	Culicidae	Uranotaenia sapphirina	Petersen et al. (1968), Savage & Petersen (1971)
Reesimermis nielseni	Culicidae	Uranotaenia sapphirina	Petersen & Willis (1972b)
Reesimermis (Romanomermis sp.)	Ceratogoponidae	Culicoides nanus	Chapman et al. (1969)
Rhabditis axei (Cobbold)	Stomoxys	Stomoxys calcitrans	Hague (1963)
Rictularia proni Seurat	Phlebotominae	Phlebotomus ariasi	Rioux & Golvan (1969), Rioux et al. (1969)
Schwenkiella icemi	Blattidae	Blatta orientalis	Leong & Paran (1966)
Schwenkiella icemi	Blattidae	Periplaneta americana	Leong & Paran (1966)
Setaria cervi (Rudolphi) (=? Filaria labiato papillosa)	Stomoxys	Stomoxys calcitrans	Noè (1913)
Severianoia severianoi	Blattidae	Periplaneta americana	Leong & Paran (1966)
Spiroptera obtusa?	Siphonaptera	Nosopsyllus fasciatus	Jenkins (1964h)
Tetradomermis angusta	Simuliidae	Boophthora erythrocephala	Rubtsov (1966c)
Tetradomermis angusta	Simuliidae	Simulium morsitans	Rubtsov (1966c)
Tetradomermis decima	Simuliidae	Boophthora erythrocephala	Rubtsov (1966c)
Tetradomermis heterocella	Simuliidae	Boophthora erythrocephala	Rubtsov (1966c)
Tetradomermis heterocella	Simuliidae	Simulium morsitans	Rubtsov (1966c)
Tetradomermis isocella	Simuliidae	Boophthora erythrocephala	Rubtsov (1966c)
Tetradomermis longicorpis	Simuliidae	Simulium decorum	Rubtsov (1966c)
Tetradomermis isocella	Simuliidae	Simulium morsitans	Rubtsov (1966c)
Tetradomermis longistoma	Simuliidae	Boophthora erythrocephala	Rubtsov (1966c)
Tetradomermis polycella	Simuliidae	Boophthora erythrocephala	Rubtsov (1966c)
Tetradomermis varicella	Simuliidae	Boophthora erythrocephala	Rubtsov (1966c)

416

Pathogen	Host group	Host	References
Tetradonema sp.	Simuliidae	Boophthora erythrocephala	Rubtsov (1966c)
Thelastoma bulhoesi	Blattidae	Blatta orientalis	Groschaft (1965)
Thelastoma bulhoesi	Blattidae	Periplaneta americana	Groschaft (1965)
Thelastoma icemi	Blattidae	Blatta orientalis	Jarry & Jarry (1963)
Thelastoma icemi	Blattidae	Periplaneta americana	Jarry & Jarry (1963)
Thelastoma magalhaesi	Blattidae	Blaberus sp.	Kloss (1966)
Thelastoma singaporensis	Blattidae	Blatta orientalis	Leong & Paran (1966)
Thelastoma singaporensis	Blattidae	Periplaneta americana	Leong & Paran (1966)
Thelastoma sp.	Blattidae	Periplaneta americana	Jarry & Jarry (1963)
Thelazia sp.	Musca	Musca autumnalis	Chitwood & Stoffolano (1971)
Apparently close to genus Tylenchinema	Phlebotominae	Lutzomyia respertilionis	McConnell & Correa (1964)
Nematode larvae	Glossina	Glossina fuscipes	Carpenter (1912)
Nematode	Glossina	Glossina fuscipes	Carpenter (1913)
Nematode larvae (Mermis?)	Glossina	Glossina morsitans	Rodhain et al. (1913)
Immature nematode	Glossina	Glossina morsitans	Leiper (1912)
Nematode	Phlebotominae	Lutzomyia cayennensis braci Lewis	Lewis (1967b)
Nematode	Phlebotominae	Lutzomyia cruciata	Lewis (1965a)
Nematode	Phlebotominae	Lutzomyia panamensis	McConnell & Correa (1964)
Nematode	Phlebotominae	Lutzomyia sanguinaria	McConnell & Correa (1964)
Nematode	Phlebotominae	Lutzomyia shannoni (Dyar)	Rosabal & Miller (1970)
Nematode	Phlebotominae	Lutzomyia steatopyga (Ichld. & Hertig), as Brumptomyia beltrani	Lewis (1965a)
Nematode	Phlebotominae	Lutzomyia vespertilionis	McConnell & Correa (1964)
Nematodes?	Simuliidae	Odagmia ornata	Shipitsina (1963)

XXIX.   NEMATODES (continued)

Pathogen	Host group	Host	References
Nematode	Phleboto-minae	Phlebotomus orientalis Parrot	Ashford (1974)
Nematode	Phleboto-minae	Phlebotomus papatasii	Lewis (1967a), Adler & Theodor (1929), Mitra (1956)
Nematode	Phleboto-minae	Phlebotomus sergenti	Adler & Theodor (1929), Lewis (1967a)
Nematode	Phleboto-minae	Sergentomyia clydei (Sinton)	Lewis (1975b)
Nematode	Phleboto-minae	Sergentomyia clydei	Lewis & Minter (1960)
Nematode	Phleboto-minae	Sergentomyia clydei (Sinton)	Lewis (1967a)
Nematodes	Simuliidae	Simuliidae	Maitland & Penney (1967)
Nematode	Simuliidae	Simulium damnosum	Marr & Lewis (1964)
Nematode	Simuliidae	Simulium ornatum nitidfrons	Briggs (1967)
Nematode	Phleboto-minae	Sergentomyia graingeri Heisch, Guggisberg & Teesdale	Lewis & Minter (1960)
Nematode	Phleboto-minae	Sergentomyia schwetzi (Adler, Theodor & Parrot)	Lewis (1974c)
Nematode	Simuliidae	Simulium reptans	Shipitsina (1963)
Nematodes	Simuliidae	Tetanopteryx maculata	Shipitsina (1963)
Tylenchid	Phleboto-minae	Sergentomyia affinis (Theodor)	Ashford (1974)
Tylenchid	Phleboto-minae	Sergentomyia sp., probably S. schwetzi	Lewis & Minter (1960)

XXX.   OTHERS

Pathogen	Host group	Host	References
<u>Dipylidium caninum</u> Platyhelminth	Siphonaphera	<u>Stenocephalides felis</u>	Jenkins (1964d)
<u>Leucocytozoon simondi</u>	Simuliidae	<u>Simulium venustum</u>	Desser & Yang (1973)
<u>Litomosoides carinii</u> Cotton rat worm	Acarina mites	<u>Liponyssus (Ornitho-nyssus) bacoti</u> Tropical rat mite suspect vector of scrub and murine typhus	Bertram et al. (1946)
<u>Litomosoides carinii</u>	Acarina mites	<u>Liponyssus bacoti</u>	Williams & Brown (1945, 1946)
<u>Litomosoides carinii</u>	Acarina mites	<u>Liponyssus bacoti</u>	Bertram (1947), Hughes (1950), Quraishi et al. (1966)
<u>Litomosoides carinii</u>	Acarina mites	<u>Ornithonyssus bacoti</u>	Williams & Kershaw (1961)
<u>Moniliformis dubius</u>	Blattidae	<u>Periplaneta americana</u>	Mercer & Nicholas (1967)
<u>Wuchereria bancrofti</u>	Cimicidae	<u>Cimex hemipterus rotundatus</u>	Burton (1963), Gunawardena (1972)
Algal filaments	Glossina	<u>Glossina fuscipes</u>	Carpenter (1912)
Cestode, hymenolepid Cysticercoid	Phlebotominae	<u>Phlebotomus mascittii</u> Grassi	Quentin et al. (1971)
Cestode, hymenolepid Cysticercoid	Phlebotominae	<u>Phlebotomus orientalis</u>	Ashford (1974)
Cestode, hymenolepid Cysticercoid	Phlebotominae	<u>Phlebotomus perniciosus</u>	Quentin et al. (1971)
Crystal inclusions	Reduviidae	<u>Rhodnius prolixus</u>	Steinhaus & Marsh (1962)
Helminth	Phlebotominae	<u>Sergentomyia adleri</u>	Barnley (1968)
Helminth	Phlebotominae	<u>Sergentomyia adleri</u> (Theodor)	Barnley (1968)
Helminth	Phlebotominae	<u>Sergentomyia bedfordi</u> (Newst.)	Barnley (1968)
Helminth	Phlebotominae	<u>Sergentomyia schwetzi</u>	Barnley (1968)
Lecithodendrudae	Simuliidae	<u>Simulium exiguum</u>	Lewis & Wright (1962)
Unidentified parasites	Phlebotominae	<u>Lutzomyia cruciata</u>	Lewis (1965a)

XXX.   OTHERS (continued)

Pathogen	Host group	Host	References
Minute, unidentified Parasites	Phlebotominae	Lutzomyia ovallesi (Ortiz)	Lewis (1965a)
Unidentified abdominal parasite	Phlebotominae	Phlebotomus argentipes	Lewis & Killick-Kendrick (1973)
Unidentified parasite	Phlebotominae	Sergentomyia adleri (Theodor)	Kirk & Lewis (1947)
Plerocercoids	Phlebotominae	Phlebotomus argentipes	Subramaniam & Naidu (1944)
Symbiont	Glossina	Glossina morsitans	Hill et al. (1973)
Unidentified small bodies	Phlebotominae	Lutzomyia cruciata	Lewis (1965a)
Undetermined	Phlebotominae	Sergentomyia hirta (Parrot and Joliniere)	Theodor (1948), Kirk & Lewis (1951)